Mathematical Modeling

Problems, Methods, Applications

Mathematical Modeling

Problems, Methods, Applications

Edited by

Ludmila A. Uvarova

Moscow State University of Technology "STANKIN"
Moscow, Russia

and

Anatolii V. Latyshev

Moscow Pedagogical University
Moscow, Russia

Kluwer Academic / Plenum Publishers
New York, Boston, Dordrecht, London, Moscow

Proceedings of the Fourth International Mathematical Modeling Conference, held June 27 through July 1, 2000, in Moscow, Russia

ISBN 0-306-46664-3

233 Spring Street, New York, N.Y. 10013

http://www.wkap.nl/

10 9 8 7 6 5 4 3 2 1

A C.I.P. record for this book is available from the Library of Congress

Printed in the United States of America

PREFACE

This volume contains review articles and original results obtained in various fields of modern science using mathematical simulation methods. The basis of the articles are the plenary and some section reports that were made and discussed at the Fourth International Mathematical Simulation Conference, held in Moscow on June 27 through July 1, 2000. The conference was devoted to the following scientific areas:

- mathematical and computer discrete systems models;
- non-linear excitation in condensed media;
- complex systems evolution;
- mathematical models in economics;
- non-equilibrium processes kinematics;
- dynamics and structure of the molecular and biomolecular systems;
- mathematical transfer models in non-linear systems;
- numerical simulation and algorithms;
- turbulence and determined chaos;
- chemical physics of polymer.

This conference was supported by the Russian Ministry of Education, Russian foundation for Basic Research and Federal Program "Integration".

This volume contains the following sections:

1. models of non-linear phenomena in physics;
2. numerical methods and computer simulations;
3. mathematical computer models of discrete systems;
4. mathematical models in economics;
5. non-linear models in chemical physics and physical chemistry;
6. mathematical models of transport processes in complex systems.

In Sections One and Five a number of fundamental and sufficiently general problems, concerning real physical and physical-chemical systems simulation, is discussed.

Section Six is devoted to the problems of mathematical simulation of the transfer processes in the systems with non-linear characteristics. The phenomena of the transfer under the influence of the electromagnetic radiation with the change of phase in the group of spherical particles is considered.

The second section is devoted to the development of the modern computer experiment methods and their applications to study of the built mathematical models of the real systems.

A fundamentally new approach to the computer simulation of discrete systems is developed in Section Three and in Section Four the mathematical-economic models are represented.

The contents of the volume involve a rather wide range of the problems of modern science and their solutions with the help of the mathematical simulation methods—a kind of universal scientific approach. Both continuous and discrete mathematical models are considered and rapt attention is paid to the exploration and simulation of the non-linear effects.

We hope that this volume will be interesting to the specialists in mathematical simulation methods, mathematical physics methods, discrete mathematics, physics, biophysics, and to post-graduates and other students.

On the author's behalf we would like to thank Kluwer Academic/Plenum Publishers for giving us the opportunity to publish this volume. We would also like to tender thanks to the staff, post-graduates, and students of the Applied Mathematics Department of Moscow State University of Technology "STANKIN" for their help in artwork preparation.

We hope that the readers of this volume will find a lot of new and interesting information for their research.

Ludmila A. Uvarova
Anatolii V. Latyshev

CONTENTS

1. MODELS OF NONLINEAR PHENOMENA IN PHYSICS

2. NUMERICAL METHODS AND COMPUTER SIMULATIONS

6. MATHEMATICAL MODELS OF TRANSPORT PROCESSES IN COMPLEX SYSTEMS

1. MODELS OF NONLINEAR PHENOMENA IN PHYSICS

SPECTRAL CHANGES OF 1/F NOISE IN METALS AT CLUSTERIZATION OF LIGHT INTERSTITIALS

Alexei A. Berzin*

1. INTRODUCTION

Due to their quantum properties, light interstitial impurities remain mobile at low temperatures. A prominent example of such impurities is hydrogen isotopes in the matrix of a transition metal.

Although at high temperatures transitions of an impurity from one equivalent interstice to another occur primarily via activated over-barrier processes, tunneling between adjacent equivalent interstitial sites becomes increasingly dominant as the temperature decreases. Since the probability of coherent tunneling increases with decreasing temperature, one would expect the diffusion coefficient D to have temperature dependence like that shown in Fig. 1.

Experiments on hydrogen diffusion in metals do not, however, reveal anything of the kind.[1] The reason for this lies in the impurity clusterization phenomenon.

It is known that in an insulator the long-range part of the interaction between point defects is elastic, i.e., it is an indirect interaction via acoustic phonons. In a metal, one should add to this the indirect interaction via Friedel oscillations in electron density.

Since both these interactions have an alternating character, for any pair of defects in a metal matrix and a pair of neutral defects in an insulator a set of bound states develops, irrespective of the actual form of the short-range part of the interaction.[2,3] As the temperature is lowered, this leads inevitably either to capture of a mobile defect by a fixed one, or to clusterization of mobile defects. Our consideration below is limit to the latter case.

If cooling was performed in quasistatic conditions, clusterization would result in a large-scale separation of the system into phases, which would contain impurities in a high *(b)* and a low *(a)* concentration, with the equilibrium impurity concentration in the *a*

* Alexei A. Berzin, Moscow Institute of Electronic, Radioengineering and Automation, Moscow, Russia 117454.

Mathematical Modeling: Problems, Methods, Applications
Edited by Uvarova and Latyshev, Kluwer Academic/Plenum Publishers, 2001

phase tending to zero with decreasing temperature. However, the time required for the impurity subsystem to reach equilibrium at low temperatures, as a rule, is considerably longer than the duration of the experiment.

This is why in a crystal with a low concentration of interstitial impurities, $x \leq 10^{-2}$-10^{-1} (x is the dimensionless concentration per unit cell), small clusters, containing only a few impurity atoms each, appear in place of large-scale phase separation. These metastable states are long-lived, because clusterization reduces strongly the mobility of impurities. Metastable states differ from one another in the relative position and number of particles in a cluster, their location and concentration.

These metastable states are separated in phase space from one another and from the equilibrium state by high barriers, whose heights differ by many orders of magnitude. During an experiment, the impurity system undergoes averaging not over the whole phase space but only in the vicinity of the deep minimum, which the system reached under cooling. In other words, the behavior of the impurity system is not ergodic. In this sense its behavior is similar to that of spin glasses; only the potential barriers between metastable states in our system remain finite.

This work is aimed to studying the properties of the above metastable states by numerical simulation, because analytical treatment would involve considerable difficulties.

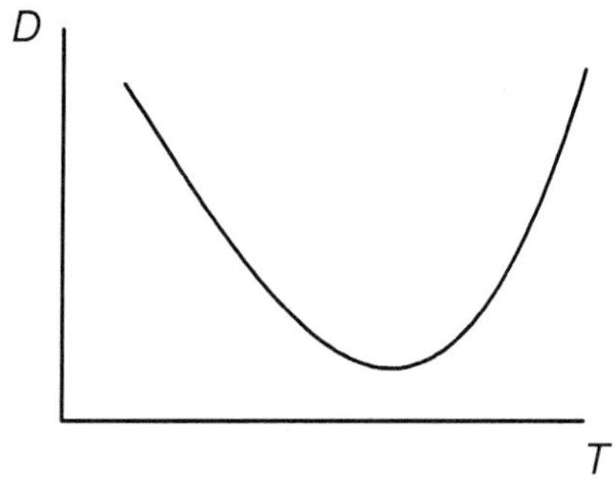

Figure1. Temperature dependence of the impurity diffusion coefficient in the absence of clusterization processes.

2. DESCRIPTION OF THE MODEL

2.1. Interaction Potential

Besides interaction with one another, impurities interact with the crystal matrix, and the potential of this interaction has sharp minima at interstices. We shall assume that the interaction of the impurities with the matrix is the strongest, and neglect the change in equilibrium positions of the impurities at interstitial sites caused by their interaction with one another. We take a cubic lattice of interstices with an edge a, which corresponds to tetrahedral pores in an fcc lattice. Interstitial impurities are distributed over the positions of the interstitial lattice. We shall assume the short-range part of the interaction among impurities to be repulsive and oppose the transition of an impurity to a site already occupied by another impurity.

The elastic interaction between impurities in a weakly anisotropic cubic crystal can be presented in the form[4]

$$W_{elas} = \alpha(a/r)^3\left[3/5-(X/r)^4-(Y/r)^4-(Z/r)^4\right], \tag{1}$$

where $\boldsymbol{r}=(X,Y,Z)$ is the distance between impurities in the coordinate frame whose axes coincide with the crystallographic axes of the cubic crystal, and the constant a has the same sign as the combination of the elastic constants $2c_{44}+c_{12}-c_{11}$. In the given interstice lattice, the dimensionless vector $\rho=r/a$ has integer coordinates, $\rho=(X,Y,Z)$.

The interaction via Friedel oscillations in electron density will be prescribed in a simple form corresponding to a spherical Fermi surface:

$$W_{el} = \beta(a/r)^3 \cos(2k_F r), \tag{2}$$

where k_F is the Fermi wave vector of the conduction electrons, and $\beta>0$. The constants α and β are of the same order of magnitude. For hydrogen in a metal, $\alpha\sim\beta\sim 10^{-2}$ eV.

After making all energy quantities dimensionless by dividing them by the constant $(\alpha+\beta)$, we finally arrive at the interaction potential acting on impurities at interstitial sites i and j:

$$\widetilde{W}_{ij} = \left\{b\left[3/5-(\widetilde{X}/\rho)^4-(\widetilde{Y}/\rho)^4-(\widetilde{Z}/\rho)^4\right]+(1-b)\cos(\gamma\rho)\right\}/\rho^3, \tag{3}$$

where $\gamma=2k_F a$, and $b=\alpha/(\alpha+\beta)$.

The value $b=1$ corresponds to the case of an insulator ($\beta=0$).

2.2. Transition Probability

The simulation was performed for a cube of 30×30×30 interstices, which was extended periodically to eliminate boundary effects. The number of impurity atoms was set equal to 30 and 100, which corresponds to concentrations $x=1.1\times10^{-3}$ and 3.6×10^{-3} per site (or to concentrations twice as large per matrix atom).

The behavior of the impurity system was studied using the Metropolis algorithm for the Monte Carlo method,[5] by which the impurity and the adjacent site j to which it could transfer from site i were generated randomly. Next the quantity ς_{ij}, the change in potential energy of the chosen impurity in the field of the other impurities, was calculated

$$\xi_{ij} = \sum_{m\neq i}(\widetilde{W}_{jm}-\widetilde{W}_{im}), \tag{4}$$

where the summation runs over all impurities with the exception of the one chosen in the beginning.

Let J_0 be the tunneling matrix element for impurity transition between adjacent equivalent interstices in the absence of disorder, i.e., for $\varsigma_{ij}=0$.

In the case $|\varsigma_{ij}| >> J_o$ transition of an impurity from one interstitial site to another is caused by either its interaction with conduction electrons (in a metal) or one-phonon processes. The transition probability determined by interaction with electrons is, to the order of magnitude,[6]

$$w_{el} \sim J_o^2 / \hbar\xi_{ij}[\exp(\xi_{ij}/T) - 1], \tag{5}$$

where T is the temperature.

The transition probability due to phonon emission or absorption can be written in the form[7]

$$w_{ph} \sim J_o^2 \widetilde{E}\xi_{ij} / \hbar\theta^3[\exp(\xi_{ij}/T) - 1], \tag{6}$$

where $\widetilde{E}$ is atomic-scale energy, and θ is the Debye temperature.

The total transition probability is the sum of w_{el} and w_{ph}. For $\varsigma_{ij}<0$, it depends weakly on ς_{ij}, while for $\varsigma_{ij}>0$ it falls off exponentially with increasing ς_{ij}. Therefore the total transition probability can be represented to a good approximation in the form

$$w_{ij} = \begin{cases} 1, \text{ if } \xi_{ij} \le 0, \\ \exp(-\xi_{ij}/T), \text{ if } \xi_{ij} > 0 \end{cases} \tag{7}$$

With this choice of the jump probability, the jump time is dimensionless (in the units of the jump duration in the absence of interaction between impurities, i.e., for $\varsigma_{ij}=0$). Its characteristic value is of the order of 10^{-10}-10^{-11} s. This does not affect the static characteristics of the system. When simulating the diffusion process, however, we obtain the ratio D/D_0, (D_0 is the diffusion coefficient for noninteracting impurities) in the place of the diffusion coefficient D.

2.3. Calculation of the Heat Capacity of the Impurity System

We simulated annealing from the high-temperature region with the initial impurity distribution chosen in a random way. The temperature of the system was varied linearly:

$$T = T_o - ct, \tag{8}$$

where T_o is the initial temperature of the system, c is the cooling rate, and t is the time (i.e., the number of steps).

The energy of the impurity system E was determined as the sum of their couple interaction energies. Different cooling rates were chosen, so as to generate from 100 to 5000 impurity jumps in each step in temperature ($\nu = 100 \div 5000$).

The $E(t)$ relation [or, after the corresponding transformation, $E(T)$] for a given initial realization of impurities was found to be an oscillating function because of the breakup and formation of clusters, which results from the boundness of the system under study. Therefore we performed averaging over a large number of realizations, thus smoothing the oscillations. The heat capacity of the impurity system at constant volume was found by differentiating with respect to temperature the relation thus obtained.

2.4. Diffusion coefficient

To determine the diffusion coefficient, we used the Einstein relation connecting the diffusion coefficient with the mobility μ :

$$D = T\mu, \tag{9}$$

where the quantity μ was defined as the coefficient of proportionality between the velocity of directed impurity motion and the applied force.

After initiating a weak potential gradient along one of the crystallographic axes, the initial impurity coordinates were generated. Next annealing was simulated from the high-temperature region to the final level T_f according to

$$T = T_f + (T_o - T_f)\exp(-ct). \tag{10}$$

This resulted in relaxation of the impurity subsystem to the steady state. Indeed, in the low-temperature domain the distribution of impurities is not random because of their interaction with one another (clusters appear). The time at which the steady state was reached was determined by monitoring the total energy of the impurity system.

After the steady state was reached, the impurity flux produced by the applied constant force F was determined. The ratio of the particle flux to F yielded the mobility. The force was chosen so as to meet the condition $Fa \ll T$. This is needed in order to be able to take into account the potential energy gradient only in the first nonvanishing approximation.

In this way the temperature dependence of the quantity D/D_0 was found.

2.5. 1/f noise spectrum

In metals the low-frequency resistance noise (with frequency dependence close to $1/f$) is caused by electron scattering from defects, changing their cross-section (so called defect-fluctuators (DF)) [8-10]. In elemental metals, the DF is a pair of neighboring point lattice defects, which changes its scattering cross section as one of the pair defects transfers to a new site.

At high temperatures, when interstitials are distributed randomly, a concentration of the pairs is of the order of x^2.

At low temperatures the majority of defects are clusterized and the concentration of DF takes the order of x. So in the case of low interstitial concentrations one would wait for the increase of the DF number due to clusterization. The occurrence of large clusters gives new DF frequencies.

The spectral noise density is equal to [8,9]

$$S_V(f) = 2\int_{-\infty}^{\infty} dt e^{i2\pi ft} < V(0)V(t) > = \frac{\alpha(f)}{fN_e}V^2, \tag{11}$$

where V is specimen voltage, t is the time, the averaging over initial moment is denoted by angular brackets, N_e is the charge carrier concentration, $\alpha(f)$ is the dimensionless Hooge parameter. The $1/f$ dependence corresponds to the case $\alpha(f)=const$.

We are interested in the frequency region $f < 10^6$ Hz, thus we consider the low temperatures when $\varsigma_{ij} >> 1$ and the characteristic DF frequency f can be represented to a good approximation by the Arrenius low

$$f = f_0 \exp(-E/T), \tag{12}$$

with the f_0=10^{10}-10^{11} Hz. So one can obtain the spectral density $F(E)$ of DF activation energies from the experimental Hooge parameter:

$$F(E) = \alpha(E)/TN_e, \tag{13}$$

where $\alpha(E)$ is the result of variable replacement in $\alpha(f)$ according to Eq. (12).

In our modeling the histograms of ς_{ij} values were obtained after the steady state was reached. We took into account only jumps, in which either initial or final interstitial position was nearer than r_0 distance from any other interstitial.

3. HEAT CAPACITY

Figures 2 and 3 display the temperature dependence of the heat capacity C for different cooling rates obtained for 30 and 100 impurity atoms. The fact that the impurity system did not reach equilibrium is evidenced by the hysteresis in the $E(T)$ relation observed under temperature cycling (Fig. 4). Note also that at equilibrium $C(T) \rightarrow 0$ for $T \rightarrow 0$.

For high cooling rates (small v) the heat capacity grows with $T \rightarrow 0$. The reason for this is that at such cooling rates particles do not have time enough to form clusters (Fig. 5a), although the nuclei of clusters are seen clearly. Some impurity atoms freeze out in the process.

As the cooling rate decreases (v increases), the heat capacity passes through its maximum at $T_{max} \neq 0$. It shifts with increasing v toward higher temperatures, finally freezing down at the true clusterization temperature $T_{max} \neq 0$. It can be estimated as[2,3]

$$T_{cl} = w_0 / |\ln x|, \tag{14}$$

where w_o is the specific binding energy of defects in a cluster.

Because the quantity w_o depends on the shape and number of particles in a cluster and increases as one goes over from small clusters to a homogeneous high-concentration b phase, clusterization will occur at different temperatures depending on the cooling rate.

For infinitely slow cooling, when there is enough time for averaging to extend over all of the phase space, there will be a first-order phase transition accompanied by large-scale phase separation. It is characterized by the maximum value of T_{cl} and a sharp peak in heat capacity at T=T_{cl}.

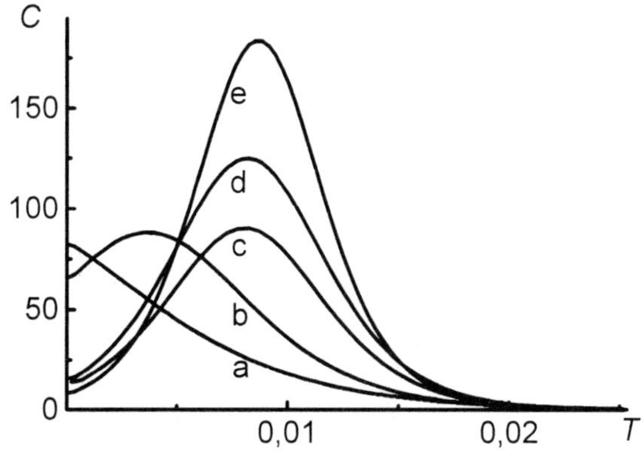

Figure 2. Heat capacity of the impurity subsystem for $x=3.6\times10^{-4}$, b=0.5, ν=: (a) 100, (b) 500, (c) 1000, (d) 2000 and (e) 5000.

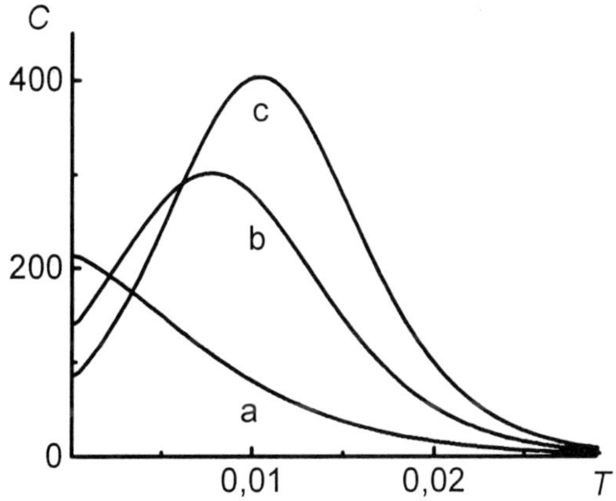

Figure 3. Heat capacity of the impurity subsystem for $x=1.1\times10^{-4}$, b=0.5, ν=: (a) 100, (b) 500, and (c) 1000.

For realistic cooling rates, averaging can encompass only a limited part of phase space, whose size increases as the cooling rate decreases. This is accompanied by an increase in T_{cl} and the height of the heat-capacity peak, while the peak width decreases. At ν=500 many small clusters arise (Fig. 5b), to coalesce at ν=2000 into one large cluster (Fig.5c).

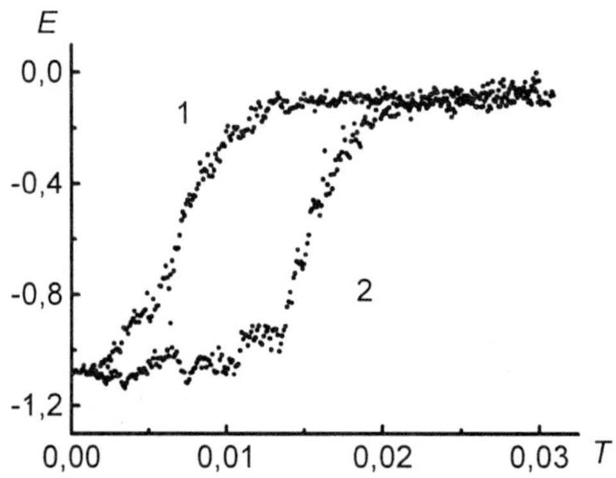

Figure 4. Energy of the impurity subsystem vs temperature under (1) cooling and (2) heating for x=3.6×10^{-4}, b=0.5, v=1000.

The shape of the resulting clusters and hence the physical characteristics of the impurity system depend substantially on the relative magnitude of the two contributions to the long-range interaction between them, i.e., on the constant b in Eq. (3). Substitution of b=0.8 insteed 0.5 causes the nearest-neighbor impurities in a cluster to occupy adjacent (Fig. 5d) rather than alternate (Fig. 5b) interstitial sites. Figure 6 presents for comparison temperature dependences of heat capacity relating to the same value of v but different b. A heat capacity peak similar to the one obtained by simulation was observed[11,12] in $ZrCr_2H_x(D_x)$ (0,27<x<0,45) for $T \leq 60$K. Note that in the phase diagram this concentration region (x<0,6) remains single-phase down to T=0,[13] in other words, no large-scale phase separation is observed here. One may thus conclude it is small hydrogen clusters that form ated in this concentration region.

4. DIFFUSION COEFFICIENT

Since the distribution of impurities over interstitial sites associated with cooling to $T<T_{cl}$ depends both on their initial distribution and on the cooling rate, simulation produces, generally speaking, not one D/D_0 curve but rather a set of them (Fig. 7). The leftmost curve corresponds to large-scale phase separation (T_{cl} is maximum), and the rightmost, to small-cluster formation and single impurities freezing out. In this case the diffusion coefficient will drop at a lower temperature.

The averaged D/D_0 ratio is displayed in Fig. 8 for different values of parameters x, b, and γ. The curves exhibit a common feature in the decrease of D/D_0 with decreasing temperature from one (high-temperature region) to practically zero ($T<T_{cl}$).

The theory yields the following estimate:[2,3]

$$D/D_0 = [1+\eta\, x \exp(w_o/T)]^{-1}, \qquad (15)$$

where $\eta \sim 1$.

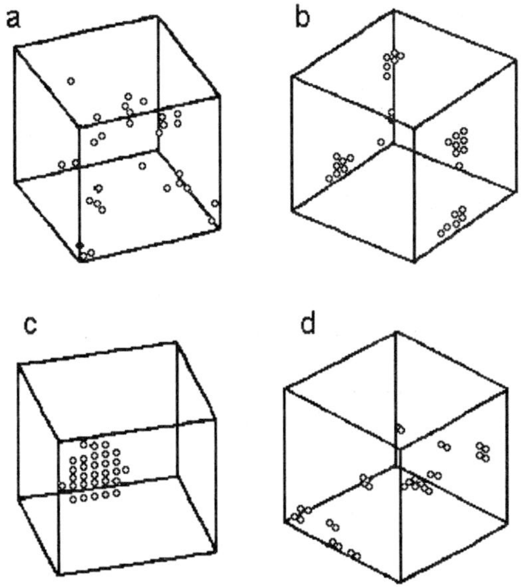

Figure 5. Impurity distribution produced by cooling for x=3.6×10^{-4}, b=0.5, v= (a) 100, (b) 500, (c) 2000 and (d) b=0.8, v= 1000.

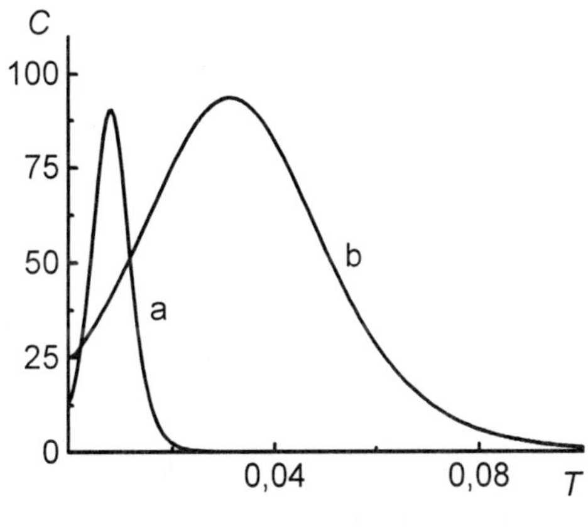

Figure 6. Heat capacity of the impurity subsystem for x=3.6×10^{-4}, b=: (a) 0.5, (b) 0.8, v=1000.

The error for the quantity D/D_0 in the high-temperature region, i.e., the scatter in the values obtained for different initial realizations, is 10%. As already mentioned, in the region where D/D_0 drops the error increases. For $T<T_{cl}$ the absolute error of D/D_0 determination is extremely small ($<10^{-4}$), but since different initial realizations give rise to different clusters and different values of w_0, the values of D/D_0 corresponding to different final realizations may differ by more than an order of magnitude.

Estimate (15) was obtained neglecting the cluster mobility. In actual fact, particle transport may occur in two ways:

1. The impurity breaks away from the cluster and diffuses a certain distance, where it is captured by another cluster or another impurity to contribute to formation of another cluster. We shall call this contribution to diffusion at the contribution of unclusterized impurities (i.e., of those external to the clusters). Since their fraction decreases exponentially with temperature, the same occurs with the quantity D/D_0.
2. Without breaking away from the cluster, the impurity moves along its boundary. It is followed by another impurity, and so on. This results in a shift of the cluster as a whole without evaporation of impurities out of it and condensation onto it.

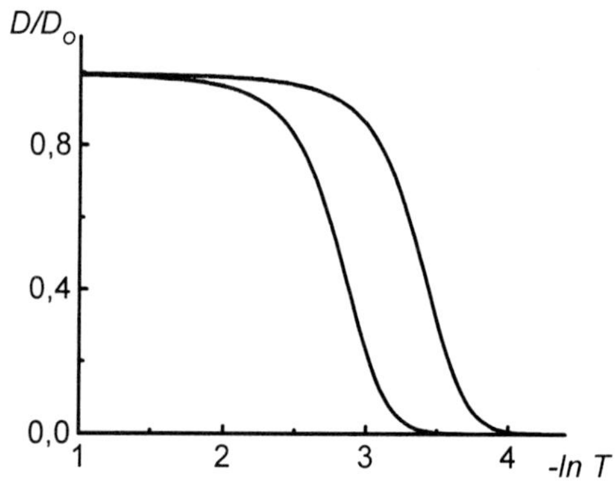

Figure 7. D/D_0 relations for different initial realizations and cooling rates.

Our simulation indicates that both in the clusterization region and for $T/T_{cl} > 0.3$ the main contribution to diffusion comes from the unclusterized impurities. For lower temperatures, processes of the second type should dominate, because their activation energy is lower than that of evaporation. In this region, however, diffusion is practically suppressed, and simulation requires a substantially longer time.

We are turning now to a comparison with experimental data available. Since in the case of quantum impurities $D_0(T)$ in the low-temperature domain grows according to a power law with decreasing temperature,[1] the exponential falloff of the diffusion coefficient is entirely associated here with clusterization processes. Experiments should exhibit a crossover from one activated dependence describing over-barrier transitions at high temperatures to another, with a lower activation energy, which corresponds to quantum diffusion at low temperatures. One may envisage different $D(T)$ relations depending on the relative temperature magnitude, at which tunneling begins to dominate over classical diffusion, and T_{cl}.[14]

The activation energy for protium in niobium and tantalum matrices was observed to decrease around 250 K.[15-17] For heavier hydrogen isotopes no such a decrease was found, which supports the quantum nature of this phenomenon.[16,18] We recall that J_0 falls of exponentially as the mass of the tunneling particle increases.

A similar effect was observed also in $ZrCr_2H_x$ ($0{,}25 \leq x \leq 0{,}5$) at 220 K.[19]

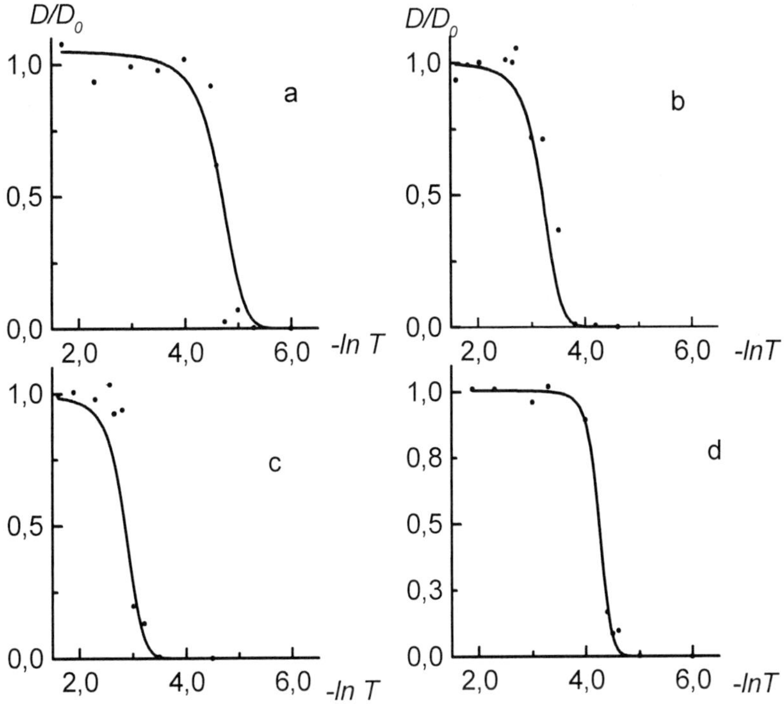

Figure 8. D/D_0 relations for (a) x=3.6×10^{-4}, b=0.5, γ=0.75; (b) x=3.6×10^{-4}, b=0.8, γ=0.75; (c) x=3.6×10^{-4}, b=0.8, γ=1.20; (d) x=1.1×10^{-4}, b=0.5, γ=0.75.

Because experiments on volume diffusion of hydrogen were carried out on macroscopic samples, they were not involved with the problem of dependence on initial realization, but the dependence of the final state on cooling rate remains an important aspect from the experimental standpoint.

5. 1/f NOISE SPECTRA

In the Figures 9-12 the interstitial distributions and corresponding activation energies spectra are present for the case r_0=2 and $\gamma = 0.75$.

In the high temperature phase (Fig. 9a, 10a) the interstitial distributions are random and 1/f noise intensity is low due to the low DF concentration. At the same time the DF activation energies are small.

After the clusterization the number of DF increases. In the case b=0.8 (fig. 9,11) the plane close paced clusters arise. From Fig. 9b,c and 11b,c one can deduce that a formation of larger clusters leads to the occurrence of DF with greater activation energies.

In the case b=0.5 the plane clusters arise in which interstitials form quadratic lattice with the period equal to 2 (Fig. 10). In this situation impurity can jump from one interstitial site to another in the cluster plane. The activation energies corresponding to the jumps form the second peak on the Fig. 12. The first one corresponds to the interstitial jumps on the cluster periphery and the jumps out of the cluster plane.

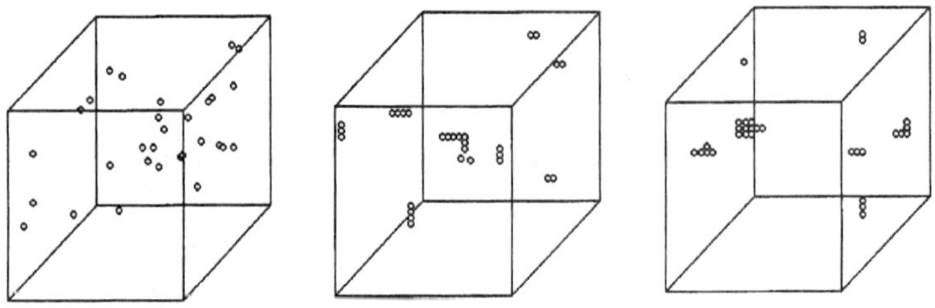

a b c

Figure 9. Interstitial distribution in the case $T>T_{cl}$ (a) and $T<T_{cl}$ (b,c) for b=0.8 and x=1.1×10^{-3}.

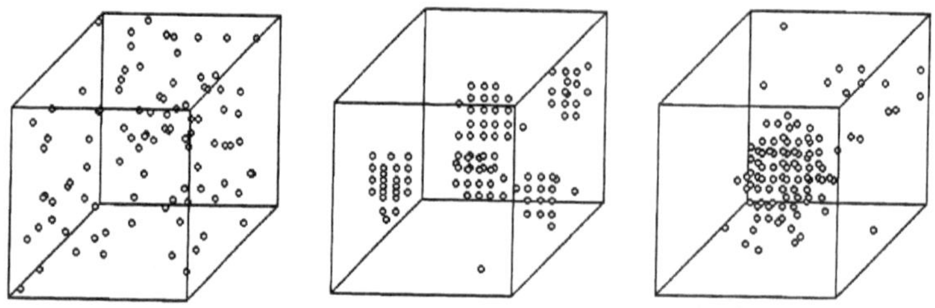

a b c

Figure 10. Interstitial distribution in the case $T>T_{cl}$ *(a)* and $T<T_{cl}$ *(b,c)* for b=0.5 and x=3.7×10^{-3}.

The concentration increase from 1,1×10^{-3} to 3.7×10^{-3} does not change the activation energy spectrum substantially.

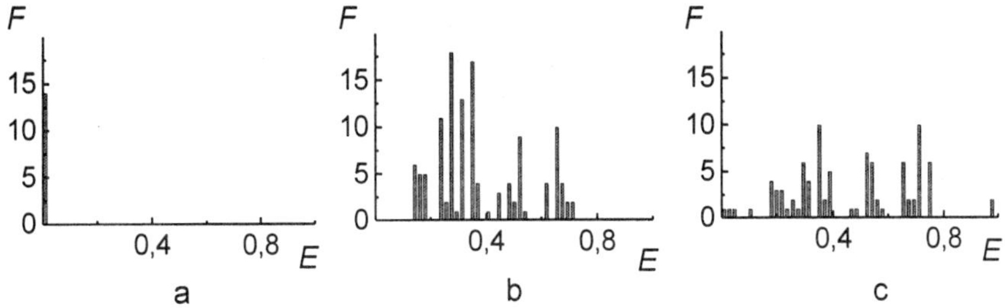

Figure 11. DF activation energy spectrum in the case $T>T_{cl}$ (a) and $T<T_{cl}$ (b,c) for b=0.8 and x=1.1×10^{-3}.

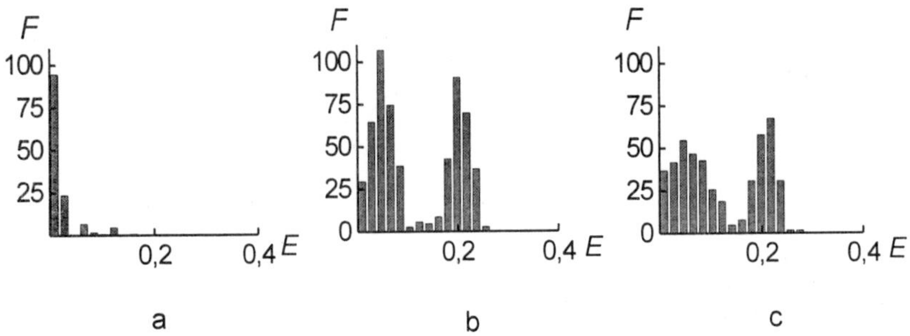

Figure 12. DF activation energy spectrum in the case $T>T_{cl}$ (a) and $T<T_{cl}$ (b,c) for b=0.5 and x=3.7×10^{-3}.

6. CONCLUSIONS

Clusterization of mobile interstitial impurities in a crystal matrix results, as a rule, not in a large-scale phase separation but rather in the onset of a metastable state characterized by a large number of small clusters. Their shape depends substantially on the cooling rate and parameters of the long-range interaction between impurities.

Clusterization manifests itself in a heat capacity peak, which was observed in a number of experiments.

Clusterization of impurities results in a strong depression of their diffusion coefficient. Therefore even in the case of quantum defects one observes with decreasing temperature not a rise in the diffusion coefficient but rather a replacement of one activated dependence of D by another, with lower activation energy.

As the temperature is decreased below the clusterization point, the diffusion begins to be dominated by cluster "creep", a process that should be accompanied by still another decrease of activation energy.

Clusterization leads to the abrupt increase of the DF concentration. The new DFs have large activation energies and produce the 1/f noise intensity increase in a low frequency region, where the experimental noise observation is possible.

The shape of activation energy spectrum depends upon the interstitial arrangement inside the clusters.

REFERENCES

1. Y. Fukai and H. Sugimoto, Diffusion of hydrogen in metals, *Adv. Phys.* **34**(2), 263-326 (1985).
2. A. I. Morosov and A. S. Sigov, Clusterization of quantum defects and quantum diffusion in metals, *Zh. Eksp. Teor. Fiz.* **95**(1), 170-177 (1989).
3. A. I. Morosov and A. S. Sigov, Kinetic phenomena in metals with quantum defects, *Phys. Usp.* **164**(3), 243-261 (1994).
4. R. A. Masumura and G. Sines, Elastic field of a point defect in a cubic medium and its interaction with defects, *J. Appl. Phys.* **41**(10), 3930-3940 (1970).
5. *Monte Carlo Methods in Statistical Physics,* edited by K. Binder (Springer, Berlin 1979; Mir, Moscow, 1982).
6. Yu. Kagan and N. V. Prokofev, Electronic polaron effect and quantum diffusion of heavy particle in metal, *Zh. Eksp. Teor. Fiz.* **90**(6), 2176-2195 (1986).
7. J. Jackle, L. Piche, W. Arnold and S. Hunklinger, Elastic effects of structural relaxation in glasses at low temperatures, *J. Non-Cryst. Solids* **20**(3), 365-391 (1976).
8. P. Dutta and P. M. Horn, Low - frequency fluctuations in solids: 1/f noise, *Rev. Mod. Phys.* **53**(3), 497-516 (1981).
9. Sh. M. Kogan, Low frequency noise with the spectrum of 1/f type in solid bodies, *Phys. Usp.* **145**(2), 285-328 (1985).
10. M. B. Weissman, 1/f noise and other slow nonexponential kinetics in condensed matter, *Rev. Mod. Phys.* **60**(2), 537-571 (1988).
11. A. V. Skripov and A. V. Mirmelstein, The heat capacity of C15-type $ZrCr_2H_x$ evidence for low-energy localized exatation, *J. Phys.: Condens. Matter* **5**(48), L619-624 (1993).
12. A. V. Skripov, A. E. Karkin, and A. V. Mirmelstein, Hydrogen-induced anomalies in the heat capacity of C15-type $ZrCr_2H_x(ZrCr_2D_x)$, *J. Phys.: Condens. Matter* **9**(6), 1191-1200 (1997).
13. V. A. Somenkov and A. V. Irodova, Lattice structure and phase transitions of hydrogen in intermetallic compounds, *J. Less-Common Met.* **101**, 481-492 (1984).
14. A. I. Morosov and A. S. Sigov, Quantum diffusion of hydrogen in transition metal hedrides, *Piz. Tverd. Tela* (Leningrad) **32**(2), 639-641 (1990).
15. G. Schaumann, J. Volkl, and G. Alefeld, The diffusion coefficients of hydrogen and deuterium in vanadium, niobium and tantalum by Gorsky - effect measurements, *Phys. Status Solidi* **42**(1), 401-413 (1970).
16. H. Wipf and G. Alefeld, Diffusion coefficient and heat of transport of H and D in niobium below room temperature, *Phys. Status Solidi* A **23**(1), 175-186 (1974).
17. D. Richter, G. Alefeld, A. Heidemann, and N. Wakabayashi, Investigation of the anomalous temperature dependence of the self-diffusion constant of hydrogen in niobium by quasielastic neutron scattering, *J. Phys.* F 7(7), 569-574 (1977).
18. Zh. Qi, J. Volkl, R. Lasser, and H. Wenzl, Tritium diffusion in V, Nb and Ta, *J. Phys.* F **13**(10), 2053-2063 (1983).
19. W. Renz, G. Majer, A. V. Skripov, and A. Seeger, A pulsed-field-gradient NMR study of hydrogen diffusion in the Laves-phase compounds, *J. Phys.: Condens. Matter* **6**(31), 6367-6374 (1994).

ANALYTIC SOLUTIONS OF BOUNDARY VALUE PROBLEMS FOR MODEL KINETIC EQUATIONS

Anatolii V. Latyshev and Alexander A. Yushkanov*

Our review will be dedicated to the analysis of the analytic solutions of model kinetic equations for one-component gas. Such solutions may be obtained only for linearized problems. So we will consider linearized form of kinetic equations. All equations will be represented in dimensionless form for simplicity. The most known model kinetic equations are: BKW-equation (Boltzmann, Krook, Welander)[1,2], ES-equation[3], and Shakhov equation[4]. These equations may be combined in the general equation

$$\frac{\partial h}{\partial t}+\vec{v}\frac{\partial h}{\partial \vec{r}}+h(\vec{r},\vec{v},t)=N(\vec{r},t)+2\vec{v}\vec{G}(\vec{r},t)+(v^2-\frac{3}{2})T(\vec{r},t)+$$

$$\omega\sum_{i,j=1}^{3} v_i\ v_j P_{i,j}(\vec{r},t)+(1-\frac{\Pr}{\beta})\sum v_i(v^2-\frac{5}{2})Q_i(\vec{r},t).$$

Here

$$N(\vec{r},t)=\pi^{-3/2}\int e^{-w^2}\ h(\vec{r},\vec{w},t)d^3w$$

– numerical gas density,

$$T(\vec{r},t)=\frac{2}{3}\pi^{-3/2}\int e^{-w^2}(w^2-\frac{3}{2})h(\vec{r},\vec{v},t)d^3w$$

* Moscow Pedagogical University, Moscow, Russia, 107005.

– gas temperature,

$$\vec{Q}(\vec{r},t)=\frac{4}{5}\pi^{-3/2}\int e^{-w^2}\,\vec{w}(w^2-\frac{5}{2})h(\vec{r},\vec{w},t)d^3w$$

– thermal flux vector,

$$P_{i,j}(\vec{r},t)=\pi^{-3/2}\int e^{-w^2}(w_iw_j-\frac{1}{3}\delta_{ij}w^2)h(\vec{r},\vec{w},t)d^3w$$

– the components of viscous tensor stress.

At the case of ω=0, β=Pr this equation transforms to BWK-equation, at the case β=Pr we obtains ES-equation, at the case – Shakhov equation.

C. Cercignani has obtained the analytic solution of the BWK-equation for Kramers problem[5]. This problem may be represented in the form

$$\begin{aligned}
&\mu\frac{\partial\varphi}{\partial x}+\varphi(x,\mu)=\frac{1}{\sqrt{\pi}}\int_{-\infty}^{\infty}e^{u^2}\varphi(x,\mu)du,\\
&\varphi(0,\mu)=0,\mu>0; \qquad\qquad (1)\\
&\varphi(x,\mu)=u_0+K(x-\mu)+o(1),\mu<0,x\to+\infty.
\end{aligned}$$

Here h=$c_y\,\psi(x,c_x)$, u_0 – an unknown velocity of viscous slip, K – mass velocity gradient far away from the surface.

The equation (1) has been used[6] for solving the thermal slip problem. In this case the surface conditions have the form

$$\varphi(0,\mu)=K_t(\mu^2-\frac{1}{2}),\mu>0;\quad \varphi(\infty,\mu)=2u_0\,,\mu<0,$$

where K_t – logarithmic temperature gradient, u_0 – thermal slip velocity.

In this problem

$h = c_y\psi(x, c_x)+c_y(c_z^2+c_y^2-2)\,\psi(x, c_x)\,\psi_1(x, c_x)$.

The temperature jump problem has some difficulties. This difficulties are concerned with the matrix Riemann-Hilbert problem. Methods for solving this problem have not been known in literature. Darrozes[7] obtained the solution of Riemann-Hilbert problem, but his solution has singular point at infinity. In[8] the half-analytic solution of temperature jump problem have been obtained.

The canonical solution and the fundamental solution of the Riemann-Hilbert problem has been constructed in[9]and [10]. But these solutions have not been applied to the temperature jump problem. The analytic solution of this problem has been obtained in[11] with the help of the fundamental matrix method.

In[11-15] the temperature jump problem and the weak evaporation problem have been analyzed at the same time. These two problems may be named as the generalized

Smoluchowski problem. In[15] the authors have taken into account the effect of energy accomodation.

The Kramers problem has been solved in[16] for Shakhov equation. The temperature jump problem has been solved in[17] for ES-model.

The strong evaporation problem has a long history[18-24]. We consider the strong evaporation as evaporation of material in its own vapor with velocities comparable with the sound velocity. C. Cercignani[18] proposed the "quasilinear " method for the problem solving. In this method distribution function is linearized with respect to the at infinit maxwellian . There is not linearization relatively evaporation velocity. This problem has been solved analytically for one-dimensional model in[25] and for three-dimensional model in[26].

The kinetic equation for this problem has the form

$$(U+c_x)\frac{\partial}{\partial x}h+h(x,\vec{c})=\pi^{-3/2}\int e^{-w^2}k_{BKW}(\vec{\mathrm{v}},\vec{w})h(x,\vec{w})d^3w,$$

where U – evaporation (condensation) velocity.

Generalized Smoluchowski problem has been solved in[27] for the case of molecular collision frequency being proportional to molecular velocity.

Boundary value problems for non-stationary BWK-equation have been considered in[28-30]. Laplace transformation converts the non-stationary equation into a stationary one. Half-space problem has been solved in[28-29]. Layer with a mirror surface conditions has been considered in[30]. This surface conditions have the form:

$$h(0,c_x,t)=h(0,-c_x,t)+2Uc_xc_ye^{i\varpi t},\ t>0,\ c_x>0;$$
$$h(d,c_x,t)=h(d,-c_x,t),\ t>0,\ c_x<0,$$

where d – the layer thickness, $Ue^{i\omega t}$ – motion law of the lower plate (x=0).

Kramers problem for the case, when molecular collision frequency is proportional to molecular velocity, has been solved in[31] The model kinetic equation for massless Bose-gas has been constructed in[32]. In this paper the analytic solution of the half-space boundary problem has been obtained. This problem can be reduced to the following one:

$$\mathrm{v}\frac{\partial}{\partial x}h(x,\mathrm{v},\omega)=\omega^{\alpha}\xi_0^{-1}\frac{1}{2}\int_0^{\infty}\int_{-1}^{1}u^{\alpha+4}E(u)\,du\int_{-1}^{1}h(x,y,u)\,dy,$$
$$h(0,\mathrm{v},\omega)=0,\ 0<\mathrm{v}<1,$$
$$h(x,\mathrm{v},\omega)=K_0+K(x-\mathrm{v}/\omega^{\alpha})+o(1),$$
$$x\to\infty,\quad -1<\mathrm{v}<0,$$
$$\xi_0=\int_0^{\infty}u^{\alpha+4}E(u)\,du.$$

Liu in[33] developed a new model kinetic equation. This equation has been considered in[34] in connection with the problems of thermal and viscous slip.

In[35] the Shakhov equation with molecular collision frequency proportional to molecular velocity has been solved analytically. As a result the exact expressions for viscous and thermal slip coefficients have been obtained. The Shakhov equation in these problems has the following form:

$$\vec{c}\nabla h(\vec{r},\vec{c}) = \nu c\{\frac{3}{4\pi}\int e^{-u^2} u\vec{c}\vec{c}'[1+\gamma_0(c^2-3)(u^2-3)]\times$$
$$\times h(\vec{r},\vec{c}')dud^3c' - h(\vec{r},\vec{c})\}.$$

ES-equation in the case when molecular collision frequency is proportional to molecular velocity for these problems has the form:

$$c_x\frac{\partial h}{\partial x} = \nu_0 c[2c_y u(x) + bc_x c_y P_{xy}(x) - h].$$

This equation may be solved exactly for the problem of thermal slip along a plane wall[36].

In[37] and[38] have been solved exactly Poiseuille and Couette problems. For the case of Couette problem the surface conditions have the following form:

$$h(-d,\vec{c}) = (1-q_1)h(-d,\vec{c}-2\vec{n}_1\vec{c}) - 2uc_z q_1, \; c_x > 0$$
$$h(d,\vec{c}) = (1-q_2)h(d,\vec{c}-2\vec{n}_2\vec{c}) + 2uc_z q_2, \quad c_x < 0.$$

Here ±u – velocity of lower/upper plane. The surface conditions for Poiseuille problem correspond the condition u=0. The problems gave been solved for the case, when accomodations coefficient q_1 and q_2 are small in wide range of Knudsen numbers.

The authors in[39] have obtained the exact solution of the BKW-equation for the problem of temperature and density jumps in the case of weak evaporation. In this work the effect of arbitrary evaporation coefficients has been taken into account. In this problem surface conditions have the form:

$$f^+(0,\vec{v}) = \alpha f_s(T_s,v) + (1-\alpha)f_s(T_0,v), \quad v_x > 0.$$

Here α - evaporation coefficients, f_s – maxwellian with surface parameters n_s, T_s, $f_0 = (n_0/n_s)f_s$, parameter n_0 corresponds the nonflow condition for the reflected molecules

$$(1-\alpha)\int v_x[f_0\theta_+(v_x) + f^-(0,\vec{v})\theta_+(-v_x)]d^3v = 0.$$

Here $\theta_+(x)$ – Heaviside function.

Model kinetic equations for molecular gas have been constructed in[40-42]. Authors of[41] have considered the case, when the collision frequency is proportional to molecular velocity. The case of a constant collision frequency has been analyzed in[42]. In this case kinetic equation has the form

$$c_x \frac{\partial h}{\partial x} + h(x,\vec{c},\vec{v}) = \int k(\vec{c},\vec{v};\vec{c}',\vec{v}')h(x,\vec{c}',\vec{v}')dm .$$

Here

$$k = 1 + 2c_x c_x' + \frac{1}{l+1/2}(c^2 + v^2 - l - \frac{1}{2})(c'^2 + v'^2 - l - \frac{1}{2}) .$$

Here l=2 for two-atomic gas and l=5/2 for N-atomic gas (N>2),

$$dm = 2\pi^{-3/2} \exp(-c'^2 - v'^2)v'dv'd^3c'$$

for l=2 and

$$dm = 2\pi^{-3} \exp(-c'^2 - v'^2)d^3v'd^3c'$$

for l=5/2.

Surface conditions in the case of full accomodation (α=1) have the form

$$h(0,\vec{c},\vec{v}) = 0, \quad c_x > 0 \text{ (by condition full accomodation).}$$

The distribution function in the distance from the wall behaves as follows

$$h(x,\vec{c},\vec{v}) = h_{as}(x,\vec{c},\vec{v}) + o(1), \quad x \to +\infty, \quad c_x < 0 .$$

Here

$$h_{as} = \varepsilon_n + 2uc_x + \varepsilon_t(c^2 + v^2 - l - \frac{1}{2}) + k_t(x - c_x)(c^2 + v^2 - l - \frac{3}{2}) .$$

For the case, when a collision frequency is proportional to molecular velocity, the kinetic equation may be transformed to the form[40]

$$\mu\frac{\partial h(x,\mu,c,v)}{\partial x}+h(x,\mu,c,v)=$$
$$=\int_{-1}^{1}\int_{0}^{\infty}\int_{0}^{\infty}k(\mu,c,v;\mu',c',v')h(x,\mu',c',v')dm,$$
$$\mu=c_x/c, k=1+\frac{3}{2}\mu c\mu'c'+$$
$$+\frac{1}{1+l}(c^2+v^2-l-1)(c'^2+v'^2-l-1), l=2;5/2,$$
$$dm=2\exp(-c^2-v^2)c^3vd\mu dcdv, \quad l=2,$$
$$dm=\frac{4}{\sqrt{\pi}}\exp(-c^2-v^2)c^3v^2d\mu dcdv, \quad l=5/2.$$

In the case of full accomodation the surface conditions have the form

$$h(0,\mu,c,v)=0, \quad 0<\mu<1,$$
$$h(x,\mu,c,v)=h_{as}(x,\mu,c,v)+o(1), \quad x\to+\infty, -1<\mu<0,$$
$$h_{as}(x,\mu,c,v)=\varepsilon_n+\left(2u+\frac{2k}{3\sqrt{\pi}}\right)\mu c+\varepsilon_t\left(c^2+v^2-l-\frac{1}{2}\right)+$$
$$+k(x-\mu)\left(c^2+v^2-l-\frac{3}{2}\right).$$

The results may be presented in the form

$$\varepsilon_t=C_tk_t+S_t(2u), \quad \varepsilon_n=C_nk_t+S_n(2u).$$

Values of coefficient C_t, C_n, S_t, S_n are represented in the table.

Table 1. Numerical calculations of temperature and concentration jumps

Problem	Coefficient	1-atomic gas	2-atomic gas	Polyatomic gas
Smoluchowski	c_t	0.79954	0.77187	0.76269
Problem	c_n	-0.39863	-0.37263	-0.37092
Weak Evaporation	s_t	-0.23687	-0.16330	-0.13888
problem	s_n	-0.82905	-0.89815	-0.90272

Authors of[43] have considered ncw model kinetic equation of BWK type. This equation leads to the true Prandtl number. In this work the exact solution of the slip problems have been obtained. New kinetic equation may be represented in the form

$$\mu\frac{\partial h}{\partial x}+h(x,\mu,c)=\frac{1}{2\pi}\int \exp(-c'^2)c'k(\vec{c},\vec{c}')h(x,\mu,\vec{c}')d^3c,$$

where

$$k(\vec{c},\vec{c}')=1+\frac{4a_2}{\sqrt{\pi}}\vec{c}\vec{c}'+\frac{1}{2}(c^2-2)(c'^2-2)+$$
$$+\frac{4a}{\sqrt{\pi}}\frac{\vec{c}\vec{c}'}{cc'}+\frac{4a_1}{\sqrt{\pi}}\frac{\vec{c}\vec{c}'}{c}+\frac{4a_1}{\sqrt{\pi}}\frac{\vec{c}\vec{c}'}{c'}.$$

Here

$$\mu=(\vec{c}\vec{n})/c,\ a_1=-2\alpha a,\ a_2=2\alpha(1+2\alpha a),\ \alpha=3\sqrt{\pi}/16.$$

Parameter a is related to the Prandtl number by the formula

$$\Pr=\frac{8\alpha(1+2a)-2a}{9\alpha-2a(1-9\alpha^2)}.$$

New equation transforms into the BKW-equation with the proportional velocity scattering frequency when a=0.

CONCLUSION AND ACKNOWLEDGMENTS

Discussed results may be applied to other types slip simulation. Analytical methods are useful for the determinate characteristics of molecular gas and gas mixtures. Numerical results may be used in aerosol science.

The work was supported by the Russian Foundation for Basic Researches (Grants 99–01–00336).

REFERENCES

1. P. L. Bhatnagar, E. P. Gross, M. Krook M., *Phys. Rev.* **94,** 511 – 525 (1954).
2. P. Welander, *Arkiv for Fysik*, Bd **44**, 7, 507 – 533 (1954).
3. L. H. Holway, Jr, Ph. D. Thesis, Harvard (1963).
4. E. M. Shakhov, *Izvestiya AN SSSR, ser. MZhG (in russian)* **5**, 142 – 145 (1969).
5. C. Cercignani, *Ann. Phys.* 20, 1, 219 – 233 (1962).

6. S. K. Loyalka, Phys. Fluids **14**, 1, 21 – 24 (1971).
7. J. S. Darrozes, *La Recherche Aerospatiale* **119** , 13 – 52 (1967).
8. J. T. Kriese, T. S. Chang, C. E. Siewert, *Intern. J. Eng. Sci.* **12** , 441 – 476 (1974).
9. C. Cercignani, *Transport Theory and Statistical Physics* **6** (1), 29 –56 (1977).
10. C. E. Siewert, C. T. Kelley, *J. Appl. Math. and Phys.* **31** , 344 – 351 (1980).
11. A. V. Latyshev, A. A. Yushkanov, *Math. modelling (in russian)* **1**, 6, 53 – 64 (1990).
12. A. V. Latyshev, A. A. Yushkanov, *Izvestiya AN SSSR, ser. MZhG (in russian)* **1**, 163 – 171 (1992).
13. A. V. Latyshev, A. A. Yushkanov, *Appl. math. and mech. (in russian)* **58,** 2, 70 – 76 (1994).
14. A. V. Latyshev, *Appl. math. and mech.* **54**, 6, 581 – 586 (1990).
15. A. V. Latyshev, A. A. Yushkanov, *Math. modelling (in russian)* **4**, 10, 61 – 66 (1992).
16. A. V. Latyshev, *In sbornik "Functions theory and appl." VINITI,* No. 2390 – V 91, 37 – 62 (1991).
17. A. V. Latyshev, *Izvestiya AN SSSR, ser. MZhG* **2**, 151 – 164 (1992).
18. M. D. Arthur, C. Cercignani, *J. Appl. Math. and Phys.* **31**, 5, 634 – 645 (1980).
19. C. E. Siewert, J. R. Thomas, Jr, *J. Appl. Math. and Phys.* **32**, 4, 421 – 433 (1981).
20. C. E. Siewert, J. R. Thomas, Jr, *J. Appl. Math. and Phys.* **33**, 2, 202 – 218 (1982).
21. W. Greenberg, C. van der Mee, V. Protopopescu, *Boundary value problems in abstract kinetic theory.* Basel: Birkhauser Verlag, 1987, 526 p.
22. C. Cercignani, A. Frezzotti, *Teor. and appl. mech.* **19**, 3, 19 – 23 (1988).
23. M. D. Arthur, *Transport Theory and Statistical Physics* **13** (1-2), 179 – 191 (1984).
24. C. E. Siewert, J. R. Thomas, Jr, *J. Appl. Math. and Phys.* **33**, 5, 626 – 639 (1982).
25. A. V. Latyshev, A. A. Yushkanov, *Appl. mech. and tech. phys. (in russian)* **1,** 102 – 108 (1993).
26. A. V. Latyshev, A. A. Yushkanov, *Fluid mech (in russian)* **6,** November – December, 861 – 871 (1992).
27. A. V. Latyshev, A. A. Yushkanov, *Fluid Dynamics* **31** (3), 454 – 466 (1996).
28. A. V. Latyshev, A. A. Yushkanov,. *Teor. & Math. Phys* **92**, 1, 782 – 790 (1992).
29 A. V. Latyshev, A. A. Yushkanov, *Teor. & Math. Phys* **116** , 2, 978 – 989 (1998).
30. A. V. Latyshev, A. A. Yushkanov, G. V. Slobodskoi, *Appl. mech. and tech. phys.* **38**, 6, 32 – 40 (1997).
31. A. V. Latyshev, A. A. Yushkanov, *Comput. Maths and Math. Phys.* **37** (4), 481 – 491 (1997).
32. A. V. Latyshev, A. A. Yushkanov, *Teor. & Math. Phys* **111** , 3, 762 – 770 (1997).
33. G. Liu, *Phys. Fluids A* **2,** 277 – 293 (1990).
34. A. V. Latyshev, A. A. Yushkanov, *Pisma v ZhTPh (in russian)* **23**, 14, 13 – 16 (1997).
35. A. V. Latyshev, A. A. Yushkanov, *Poverkhnost'* **1,** 92 – 99 (1997).
36. A. V. Latyshev, A. A. Yushkanov, *Ingeneer. Phys. J. (in russian)* **71**, 2, March – April, 353 – 359 (1998).
37. A. V. Latyshev, A. A. Yushkanov, *J. Tech. Phys. (in russian)* **68**, 11, 27 – 31 (1998).
38. A. V. Latyshev, A. A. Yushkanov, *Poverkhnost'* **10,** 35 – 41 (1999).
39. A. V. Latyshev, A. A. Yushkanov, *Ingeneer. Phys. J. (in russian)* **73**, 3, May – Yune, 542 – 549 (2000).
40. A. V. Latyshev, A. A. Yushkanov, Exact solutions of boundary value problems for molecular gases . Monograh (in russian). VINITI. No. 1725 – V 98. 1998. 186 p.
41. A. V. Latyshev, A. A. Yushkanov, *J. of experimental and theoretical physics* **87**, 3, 578 – 526 (1998).
42. A. V. Latyshev, A. A. Yushkanov, *in book "Mathematical Models of Non-Linear Excitations, Transfer, Dynamics, and Control in Condensed Systems and Other Media".* Kluwer Academic/Plenum Publishers. N.-Y. – Moscow. 1999, 3 – 16.
43. A. V. Latyshev, A. A. Yushkanov, *Pisma v ZhTPh (in russian)* **26**, 23, 16 – 23 (2000).

MATHEMATICAL MODELS IN NON-LINEAR SYSTEMS THERMODYNAMICS

Andrei V. Tatarintsev*

1. INTRODUCTION

As a well known, thermodynamic properties of various physical systems depend on atomic and molecular interactions in these systems and can be a good instrument of studying theirs internal properties. The departure of the potentials and spectra from the harmonic one and the need for taking additional intermolecular forces into account sometimes result in essential modifications of the thermodynamic equations of system and characteristics such as the heat capacity, chemical potential, and thermodynamic mean size of a, molecule (the bond length) etc. In addition, the number of excited degrees of freedom at different intervals of temperatures, a possibility for a quasiclassical description of the particle pair interaction in the system, and some other properties can be easily inferred from these characteristics [1, 2].

One of the most essential examples of non-linear interacting system is "common" water, which hydrogenous bounding led to the random three-dimension clastorized structure. The tunneling processes of proton upon hydrogenous bound exert influence on physical properties of this matter. The existence of intermolecular hydrogenous bounds in liquid phase of water led, as a result, to its anomalous properties and role in live. As a simple model of water properties describing commonly uses the model of interacting pairs with the potential

*Moscow State Institute of Radio Engeneering, Electronics and Automatics (Technical University), Vernadskogo st., Moscow, 1175454, Russia.
E-mail: avt@imamod.ru

$$U(x) = U_0 \left[\left(\frac{x}{d} \right)^2 - 1 \right]^2 . \tag{1}$$

This potential allows considering not only the nonlinearity of the interaction but also (when solving the problem on the quantum level) the proton transport processes [3, 4]. The quantum mechanical treatment of this problem reveals the fine structure splitting of the energy levels of the tunneling proton because of degeneracy of the spectrum for such a potential, which involves certain difficulties in constructing the thermodynamic theory of this effect.

Non-linear models with finite temperature are commonly used in quantum field theory (models with spontaneously broken symmetries, Nambu-Iona-Lasinio and Gross-Nevuie models[3]), in high-temperature superconductivity theory, Casemir effect, in order to get mass-spectrum of elementary particles and their properties in finite temperature [4].

This paper will be dedicated to describing properties of thermodynamical systems with non-harmonic (non-linear) type potentials of pair interaction. It allows us to get some interesting peculiarities of such systems in middle temperature region.

Here, we consider the thermodynamic properties of an "ideal gas" of pairwise-bonded atoms with an interaction of type (1) (the analogue of the hydrogen bond of water molecules). We disregard the interaction of pairs with each other, as well as the effects of dissociation and ionization of molecules, assuming that the considered temperature interval is below the typical dissociation temperatures. We also assume that the interaction of the constituents of a pair can be considered on the classical level using the standard Boltzmann distribution. We study only the contribution corresponding to the interaction of the constituents of a pair assuming that the rotational degrees of freedom and the translations of the pair as a whole can be taken into account in the standard way [1].

The necessity to take into account intermediate region of temperature led to use most exact approach in the consideration of non-linear pair interaction. To describe model properties, we need to get thermodynamical potential, such as free-energy density (for systems with fixed number of particles) $F(T,N,V) = -T \ln Z$. The expression for thermodynamical potential depend on statistical integral Z, which appearance (for quantum model):

$$Z \equiv Sp(e^{-\beta H}) = \sum_n e^{-\beta E_n} ,$$

where $\beta = 1/T$ - inverse temperature, $H(p,x)$ - is the Hamiltonian of system, E_n - its eigenvalues (energy spectrum), and $Sp(\ldots)$ can be calculate for any full system of functions. In classic models, statistical integral Z can been calculate like a phase-space integral:

$$Z = \frac{1}{N!} \int d\Gamma \exp[-\beta E(p,x)] ,$$

where $E(p,x)$ - full energy of system (dynamical invariant), and $d\Gamma$ - phase-space differential element.
For identical particle systems one can evaluate statistical integral to form in which summing up energy of individual particle must be happen. In particular, for trivial case of quantum harmonic oscillations of two atomic molecule we get statistical sum:

$$Z_{osc.} \equiv \sum_{n=0}^{\infty} e^{-\beta\omega(n+1/2)} = \frac{1}{\sinh(\beta\omega/2)} .$$

If calculate in quantum model, one can get free-energy density $F = -T \ln Z$ and heat capacity $C_V = -T\ \partial^2 F / \partial T^2$:

$$C_V^{(harm.)} = \left[\frac{\beta\omega/2}{\sinh(\beta\omega/2)} \right]^2 .$$

For classical approach, the heat capacity is to be equal $C_V^{(harm)} = 1$. Quantum value of heat capacity is less then classical one especially in law temperature region.

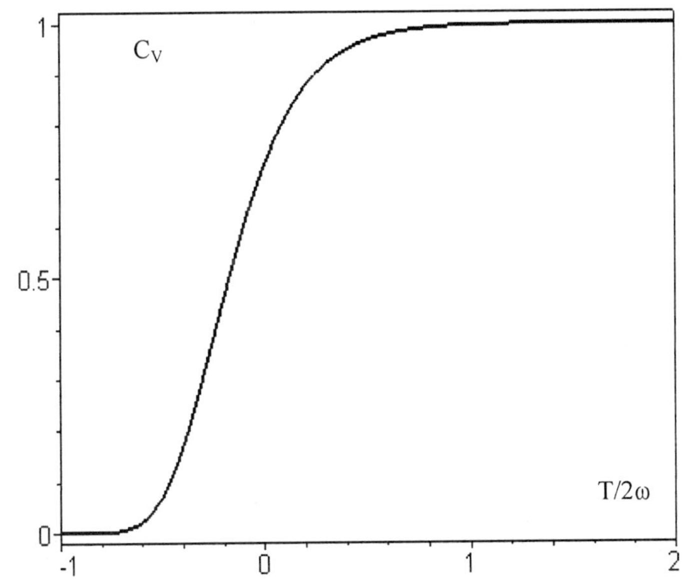

Figure 1. Heat capacity for quantum harmonic degree of freedom. Classical value of heat capacity is equal one.

2. CLASSICAL ANHARMONIC MODEL OF INTERACTION

For the first step, let us consider one-dimension classical anharmonic problem with two-atomic interaction potentials of two forms:

$$U_r^{(1)}(x) = \left(x^r - a^r\right)^2, \tag{2}$$

$$U_r^{(2)}(x) = x^{2r} + 2a^r x^r, \tag{3}$$

where $a > 0$, $r > 0$ - parameters, and in (3) r - even. The energy of particle is equal $E_r(p,x) = p^2 + U_r(x)$. Potentials (2,3) have equal minimum $U_{\min} = 0$. First of them get it in two points $\pm a$ (r-even) and in one point $+a$ (r-odd). Second potential has trivial minimum in point $x = 0$. Wide class of potentials in consideration allows us to illustrate general properties and peculiarities of such interactive nonlinear systems. Note, that if $r = 1$, we get standard classical harmonic potential with dislocate minimum $< x >= a$, and if $r = 2$ - potential (1).

The investment of particle pair interaction in statistical sum (or integral) has the form:

$$Z \equiv Sp\{\exp[-\beta H(p,x)]\} = \iint \frac{dx\,dp}{2\pi} \exp[-\beta E(p,x)]. \tag{4}$$

In particular, for potentials (2,3), one gets:

$$Z = C_r^{(0)}\, b^{-(r+1)/2r} \begin{Bmatrix} e^{-b} \\ 1 \end{Bmatrix} \Phi(b) \ . \tag{5}$$

Here and then, up value – for (2), down value – for (3). There are such meanings in (5):

$$\Phi(b) = C_r^{(1)}\, M[1/(2r);1/2;b] \pm 2\sqrt{b}\, C_r^{(2)}\, M[(r+1)/(2r);1/2;b],$$

$$C_r^{(0)} = \pi^{-1} a^{r+1}\, 2^{-1/r}\, \Gamma[(r+1)/r];$$

$$C_r^{(1)} = \cos(\pi/2r)\,\Gamma[(r-1)/(2r)];$$

$$C_r^{(2)} = \frac{1}{2}\left[1 + (-1)^r\right] \sin(\pi/2r)\,\Gamma[(2r-1)/(2r)].$$

Confluent hypergeometric function $M(a,b,z)$ (the Kummer function) is been used in the notations [5,6] and $b = \beta a^{2r}$ - variable, depending on inverse temperature. Note, that the coefficient $C_r^{(2)} = 0$ for odd r and potential (2). For $r = 1$ the expression for $C_r^{(1)}$ is not defined, but can be replaced by its corresponding limit value. For (3) potential r can de only even (in order to differ (2)), so we can modify formula (5) in this way. For $r = 2$, the partition function can be obtained using the partial value of the Kummer function and has the simpler form in terms of imagine argument Bessel functions. Using equality (5), one can get corresponding contribution to free-energy density and heart capacity for arbitrary r :

$$C_V(r) = \frac{r+1}{2r} + \frac{\Phi_1(b)}{4r\Phi(b)} - \left[\frac{\Phi_2(b)}{2r\Phi(b)}\right]^2 . \tag{6}$$

The functions $\Phi_{1,2}(b)$ in formula (6) are defined as:

$$\Phi_1(b) = C_r^{(1)} \left[M\left(\tfrac{1}{2r};\tfrac{1}{2};b\right) + (2b-1)\, M\left(\tfrac{2r+1}{2r};\tfrac{3}{2};b\right)\right] \pm$$

$$\pm\, 2\sqrt{b}\, C_r^{(2)} \left[2b\, M\left(\tfrac{r+1}{2r};\tfrac{3}{2};b\right) + r(2b-1) M\left(\tfrac{r+1}{2r};\tfrac{1}{2};b\right)\right],$$

$$\Phi_2(b) = C_r^{(1)} \left[M\left(\tfrac{2r+1}{2r};\tfrac{1}{2};b\right) - M\left(\tfrac{1}{2r};\tfrac{1}{2};b\right)\right] \pm 2\sqrt{b}\, C_r^{(2)}\, rM\left(\tfrac{r+1}{2r};\tfrac{1}{2};b\right) .$$

The dependence of the heat capacity on $\theta \equiv 1/b = T/a^{2r}$ (on logarithm scale) for $r = 2,4,6$ and potentials (2,3) is shown accordingly in Fig. 2, 3.

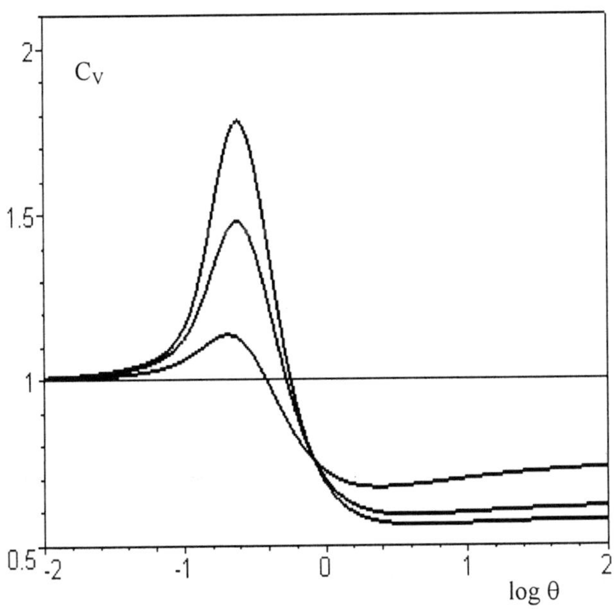

Figure 2. Heat capacity for classical anharmonic degree of freedom (2). Classical value is equal one.

Limit values of the heat capacity for potential (2), following from (6), does not depend on r in the low-temperature region (but sufficiently high to ignore quantum effects and the difference in statistics):

$$C_V = 1 + O(\theta), \ \theta << 1, \tag{7}$$

for large values of the parameter, $\theta >> 1$, the leading terms of the partition function and heat capacity are determined by the leading term of the polynomial potential, $U_r(x) \sim x^{2r}$, this term defines the convergence of integral (4):

$$C_V = \frac{r+1}{2r} + O(\theta^{-2}), \theta >> 1 . \tag{8}$$

The heat capacity (2) has a local maximum in the intermediate-temperature region, in the neighborhood of this maximum, C_V exceeds the standard value $C_V^{(harm)} = 1$ (by a quantity that grows with the nonlinear parameter r.

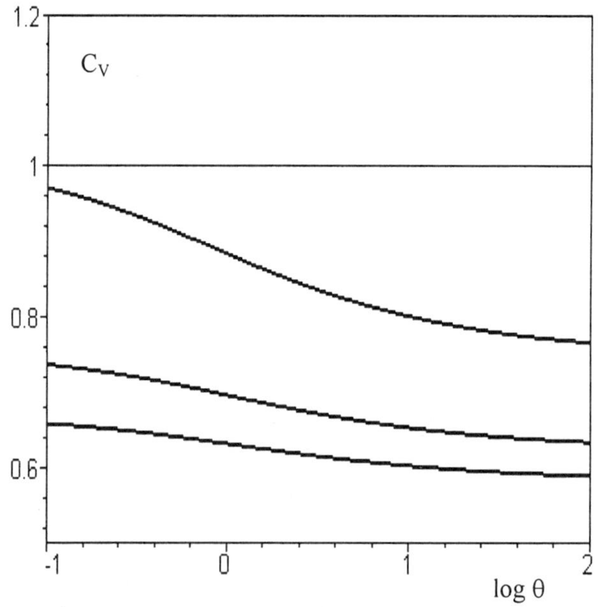

Figure 3. Heat capacity for classical anharmonic degree of freedom (3). Classical value is equal one.

Such a behavior of the heat capacity of the ideal gas with an anharmonic oscillatory molecular degree of freedom is rather interesting in connection with the existence of substances with an anomalous behavior of this thermodynamic characteristic. It is possible that the properties of these substances are just related to the essential non-linearity of the interaction.

We also consider some interesting features of the thermodynamical means for a gas with potentials of type (2,3). Assuming that the averaging is performed in the standard statistical way:

$$< x^n >_\beta \equiv \frac{1}{Z}\, Sp\left[x^n\, e^{-\beta H}\right]. \tag{9}$$

We introduce a dimensionless variable related to mean (9):

$$M_n(r,\beta) \equiv \frac{< x^n >_\beta}{a^n} \quad . \tag{10}$$

We call it the moment of order n. The moment M_1 describes the mean coordinate (in units of the scale a) and for symmetric potentials of type (2,3) with an even non-linear parameter is equal to zero. For odd r the potential (2) is asymmetric, and therefore the moment is not equal zero ($r = 1,3,\ldots$):

$$M_1(1,b) = 1\,, \qquad M_1(r,b) = 2^{2-1/r}\, \Gamma\begin{bmatrix} \frac{2}{r} & \frac{r-1}{r} \\ \frac{1}{r} & \frac{r-1}{2r} \end{bmatrix} \sin\!\left(\frac{\pi}{2r}\right) b^{\frac{r-1}{2r}}\, \frac{M\left(\frac{r+2}{2r};\frac{3}{2};b\right)}{M\left(\frac{1}{2r};\frac{1}{2};b\right)} \,.$$

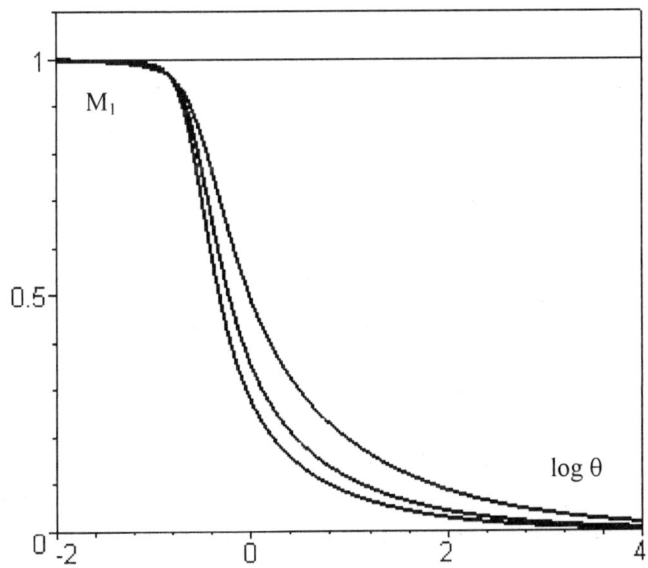

Figure 4. The moment M_1 for classical anharmonic degree of freedom with nonsymmetrical potential (3).

The dependence of the moment on temperature (on logarithm scale) for different $r = 1,3,5$ are presented in Fig.4. Here we can show, that moment M_1 (and other odd moments) vanishes with growing temperature θ and potential asymmetry can be neglected. The same behavior occurs in the theories with broken symmetry restoration, so we can call this effect – thermodynamic symmetry restoration.
The M_2 value determine x dispersion, or length of bond in pair for given temperature:

$$M_2(1,b) = 1 + \frac{1}{2b}, \qquad M_2(3,b) = \frac{2^{1/3}\pi}{\sqrt{3}\,\Gamma^2(1/3)}\, b^{-\frac{1}{3}} \frac{e^b}{M\left(\frac{1}{6};\frac{1}{2};b\right)}.$$

For the harmonic potential with $r = 1$, this moment is a monotonic function of temperature. For all other integer r the behavior of this moment is essentially different (nonmonotonic). The corresponding curves are presented in Fig. 5.

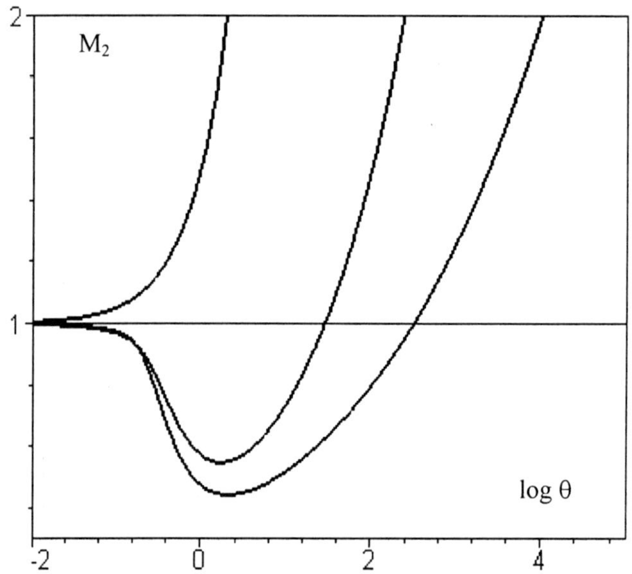

Figure 5. The moment M_2 for classical harmonic and anharmonic (nonmonotonous) potentials.

The analysis of these curves shows that, as was to be expected, the mean $\sqrt{< x^2 >}$ goes to the minimum points of the potentials (2) with decreasing temperature. Also note that in contrast to the harmonic potential, the function $M_2(r;b)$ has nonmonotonic behavior for arbitrary $r \geq 2$, and has a minimum at in the intermediate-temperature region in temperature point $\theta = \theta_{\min}(r)$. This minimum is an interesting peculiarity, which is unique to the potentials that have intervals of different convexity (the signs of $U_r''(x)$ are different). The value of the effective temperature

$\theta_{\min}$ increases with increasing r. For $\theta < \theta_{\min}(r)$ region the bond length of two interacting constituents of a pair decreases, which can lead to an anomalous behavior of some thermodynamic characteristics such as density. For $\theta > \theta_{\min}(r)$ the behavior of momentum is normal.

In general case for moments of order n one can get:

$$M_n(r,b) = C^{(0)}_{r,n}\, b^{-n/2r}\, \frac{C^{(1)}_{r,n}\, M\left(\frac{n+1}{2r};\frac{1}{2};b\right) \pm 2\sqrt{b}C^{(2)}_{r,n}\, M\left(\frac{r+n+1}{2r};\frac{3}{2};b\right)}{C^{(1)}_{r,0}\, M\left(\frac{1}{2r};\frac{1}{2};b\right) \pm 2\sqrt{b}C^{(2)}_{r,0}\, M\left(\frac{r+1}{2r};\frac{3}{2};b\right)}, \tag{11}$$

where the constant coefficients are not depending the temperature and have the forms:

$$C^{(0)}_{r,n} = 2^{-\frac{k}{r}}\, \Gamma\begin{bmatrix}\frac{n+1}{r}\\ \frac{1}{r}\end{bmatrix},$$

$$C^{(1)}_{r,n} = \frac{1+(-1)^n}{2}\cos\left(\frac{\pi(n+1)}{2r}\right)\Gamma\left(\frac{r-n-1}{2r}\right),$$

$$C^{(2)}_{r,n} = \frac{1+(-1)^{n+r}}{2}\sin\left(\frac{\pi(n+1)}{2r}\right)\Gamma\left(\frac{2r-n-1}{2r}\right).$$

Mention that when $n = r-1; 2r-1; 3r-1\ldots$ coefficients are not determined, but we can get them by limiting n to its respective value.

3. CLASSICAL ASYMMETRIC ANHARMONIC MODEL

Let us consider (in addition with potentials (2,3), studying before) properties of classical anharmonic model with small asymmetric potential:

$$U(x) \equiv U_2(x) + \Delta U_\varepsilon(x) = \left(x^2 - a^2\right)^2 + \varepsilon\, a^3 x\,. \tag{12}$$

We will be assume, that the value ε - sufficiently small, in order to keep the general form of potential (2), and number of its minimums. The statistical sum for this type of potential, can been evaluated to form of asymptotic series by $(\varepsilon\, b)$-parameter:

$$Z = \iint \frac{dxdp}{2\pi}\, e^{-\beta\left(p^2+U_2(x)\right)} \cosh\left(\varepsilon\beta a^3 x\right) = Z^{(0)} \sum_{k=0}^{\infty} \frac{(\varepsilon b)^{2k}}{2k!} M_{2k}(2,b).$$

Here we determine $Z^{(0)} = \frac{\sqrt{\pi}}{4} a^3 b^{-1/2} e^{-b/2} \left[I_{1/4}(b/2) + I_{-1/4}(b/2) \right]$ - statistical sum for even potential (2), with $r = 2$, and M_n - moments, defined in (12) expression. If temperature is not too small ($\theta >> \varepsilon$), the series are summing with out any problems, and one can take into account finite number of its addendum in order to get suitable asymptotic for statistical sum: $Z \sim Z^{(0)}[1 + O((\varepsilon b)^2)]$. In low temperature region ($\theta << \varepsilon$) the series can sum by using the asymptotic of moments M_n. Using Kummer function asymptotic $M(a;b;x) = \frac{\Gamma(a)}{\Gamma(b)} e^x x^{(a-b)} [1 + O(x^{-1})]$, and summing the series by (εb) parameter we get limiting behavior of statsum in form: $Z \sim Z^{(0)} \cosh(\varepsilon b)$. Uniting two cases, one can suppose, that total expression $Z = Z^{(0)} \cosh(\varepsilon b)$ gets not only correct asymptotic behavior in limits of high and low temperature, but may be suitable in the middle temperature θ region. Note, that the same form of statistical sum may been obtain by some other ways.

Additional coefficient in statistical sum will contribute some additional free energy and heat capacity:

$$\Delta C_V^{(\varepsilon)} = \left[\frac{\varepsilon b}{\cosh(\varepsilon b)} \right]^2 , \tag{13}$$

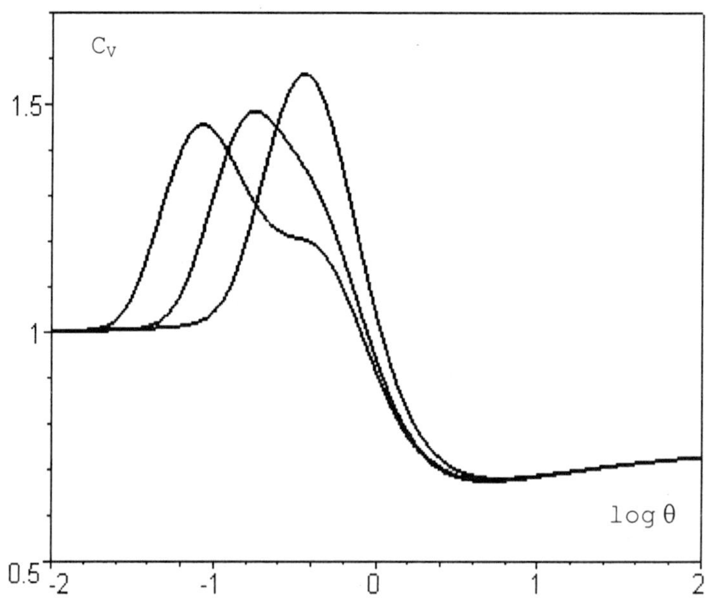

Figure 6. The heat capacity for classical asymmetrical anharmonic degrees of freedom.

which had been located in $\theta \sim \varepsilon$ temperature region. In Fig. 6 we can see total heat capacity (sum of two contributions – from symmetrical potential (2) and small asymmetrical addition in form (13)) for asymmetry parameter values: $\varepsilon = 0.05; 0.1; 0.2$: Essentially big contribution occurs when $\varepsilon \sim \theta_{\max} \approx 0.2$. There is about 50% increasing of maximum heat capacity in this point.

4. QUANTUM PROBLEM FOR ANHARMONIC POTENTIAL

Exact calculations in quantum problem with non-harmonic potentials meet some essential problems. One of them – to get potential's energy spectrum E_n for (2,3). For statistical sum:

$$Z = \sum_{n=0}^{\infty} e^{-\beta E_n} \quad ,$$

heat capacity depends on energy dispersion:

$$C_V = \frac{D(E)}{T^2} \;, \qquad D(E) = \langle E_n^2 \rangle - \langle E_n \rangle^2 \;,$$

where $\langle \ldots \rangle$ - is statistical mean value. For high energy levels spectrum can been calculate with quasiclassical approach [7,9]. Its asymptotic for polynomial potentials with even biggest power $U_r \sim x^{2r}$ has the form $E_n \sim (2n+1)^{\delta}$, where $\delta = \frac{2r}{r+1}$ and large n. This particular asymptotic form is important in the high-temperature limit. For statistical sum:

$$Z = \sum_{n=0}^{\infty} e^{-\beta(2n+1)^{\delta}} = \frac{1}{2}\Gamma\left(\frac{1}{\delta}\right)\beta^{-1/\delta} + \sum_{n=0}^{\infty} \frac{(-1)^n}{n!}\left(1 - 2^{-n\delta}\right)\zeta(-n\delta)\beta^n ,$$

one can get in high temperature limit of heat capacity: $C_V \to \frac{r+1}{2r}$ when $T = 1/\beta \to \infty$, that coincides with obtained before in the classical approach (8). We sum the series for the partition function using the standard Mellin transformation. In low temperature region, the low energy levels pay important investment in. For its values one can calculate using algebraic approach, see for example [7-8].

5. CONCLUSION

In present paper, we have discussed some thermodynamic properties of an ideal gas of molecules with an essentially nonlinear oscillatory degree of freedom. The behavior of the thermodynamic characteristics of the gas has some important features. Our analysis on the classical level (Boltzman statistic) is justified in the intermediate-temperature region because the molecular constituents are massive. As was shown for wide class of the one-dimensional potentials, the contribution C_V of the nonlinear atomic interaction to the heat capacity in the intermediate-temperature region can exceed the similar contribution of the harmonic interaction by a quantity that increases monotonically with the nonlinearity parameter r. In addition, the characteristic feature for the case of the anharmonic interaction with potentials (2), that have two minimums, is a nonmonotonic temperature dependence[10] of the mean squared coordinate $< x^2 >$, which, in the given problem setting, corresponds to the effective-temperature bond length for the molecular constituents. The presence of a minimum of this moment can yield an anomalous behavior of the substance density (especially in the liquid phase) with increasing temperature.

Also note, that the high-energy asymptotic form of the termodynamic potentials and other characteristics differs from the standard (such as Stefan-Boltzman law, etc.) This departure can be due to restrictions on the domain of applicability of the considered model. In the high-temperature region, the dissociation processes lead to breaking all intermolecular bonds, which means that the considered degrees of freedom entirely disappear. Therefore, strictly speaking, all obtained results are only valid in the region of the intermediate temperatures not exceeding the characteristic dissociation temperature.

REFERENCES

1. L. D. Landau and E. M. Lifshitz, Course of Theoretical Physics [in Russian], Vol. 5, Statistical Physics: Part 1 (4th ed.), Nauka, Moscow (1995); English transl. prev. ed., Pergamoii, Oxford (1980).
2. R. Balescu, Equilibrium and Nonequilihrium Statistical Mechanics, Wiley, New York (1975).
3. Ebert D., Klimenko K.G., Vdovichenko M.A., Vshivtsev A.S. Preprint CERN-TH/99-113; hep-ph/9905253.- 29 p.
4. Faustov R.N., Galkin V.O., Tatarintsev A.V., Vshivtsev A.S. Theor. Math. Phys. **113**, pp.1530-1542, (1997).
5. M. Abramowitz and I. Stegan, eds., Handbook of Mathematical Functions witli Formulas, Graplis, and Mathematical Tables, Wiley, New York (1972).
6. Erdelyi et al., eds., Higher 'Transcendental Functions (Based on notes left by H. Batcman), Vol. 2, McGraw-Hill, New York (1953).
7. A.S. Vshivzev, N. V. Norin, and V. N. Sorokin, Theor. Math. Phys., **109**, pp. 1329-1341, (1996).
8. Vshivtsev A.S., Sorokin V.N., Tatarintsev A.V. Physics of Atomic Nuclei, **61**, 1499-1506, (1998).
9. V. P. Barashev, V. V. Belov, A. S. Vshivtsev, and A. G. Kisun'ko, Theor. Math. Phys., **116**, pp. 1074-1082, (1998).
10. V.A.Vshivtsev, A.V.Prokopov, A.V.Tatarintsev, Theor. Math. Phys., **125,** pp. 1568-1577, (2000).

CRITICAL OPALESCENCE: MODELS—EXPERIMENT

Dmitri Yu. Ivanov*

1. INTRODUCTION

The critical opalescence investigations were always,[1, 2] and especially during last years, the subject of the top interest.[3, 4] The nature of critical opalescence as well as the nature of critical phenomena in general is determined by enormous growth of order parameter fluctuations with the approaching to the critical point.[5, 6] Light is a perfect instrument to study these fluctuations, while its influence on the medium is negligible. While the light wave length proves to be comparable with the typical size of fluctuations at temperatures close to the critical temperature (T_c) one can obtain the microscopic characteristics of the medium such as the correlation radius (ξ) of fluctuations and their relaxation time $\bar{\tau}$ ($\bar{\tau} \propto 1/\Gamma$, Γ being the scattering spectrum half-width) using light scattering.

The critical point being approached the scattering multiplicity grows reaching extremely high values in the case of "developed" opalescence. Scattered light evidently contains the information about static and dynamic properties of the scattering system in the latter case as well, but the challenge is to extract this information from the multiple-scattering spectra.[7] While studying the critical opalescence experimentalists usually try to decrease the influence of multiple scattering and to take it into consideration as small corrections. On the other hand, only double scattering could be described analytically (see e.g.[8]). However, studying the "developed" critical opalescence one has to deal with the scattering on the growing critical fluctuations – multiple-scattering, and therefore this phenomenon should be carefully examined. The series of our investigations were dedicated to this subject and this article contains a brief summary of the results obtained.

Due to the extreme complexity of the task the first step was to study the model systems of polystyrene latex. Submicron latex particles have almost an ideal spherical form, do not interact with one another and do not absorb in the visible part of the light spectrum.

---

* Saint-Petersburg State University of Refrigeration and Food Engineering, Lomonosov str., 9, Saint-Petersburg, 196135, Russia.
E-mail: Dmitri.IVANOV@pobox.spbu.ru

Mathematical Modeling: Problems, Methods, Applications
Edited by Uvarova and Latyshev, Kluwer Academic/Plenum Publishers, 2001

2.PHYSICAL MODELLING OF CRITICAL OPALESCENSE

Without going into details, the conceptions existed at the beginning of our investigations could be briefly summarized as follows: multiple scattering strongly influences on the scattered light intensity but its influence on the scattered light spectra (the critical opalescence spectra included) is quite small and is reduced to the appearance of only small distortion in the spectral line. Half-widths of single (Γ) and multiple (Γ_m) scattering spectra decrease similarly when temperature tends to the critical one and the angle dependence of the multiple scattering spectrum half-width is absent whatever the nature of scatterers is.[8] But our first experiments on the high-extinction model systems of Brownian particles demonstrated that under cylindrical geometry conditions the angle dependence of Rayleigh linewidth not only takes the place in this case, but strongly distinguishes from single scattering one.[9]

The quasi-elastic highly multiple scattering ($N >> 1$) spectra were investigated with the help of the single-bit photon 80 (72+8) channels correlator of our own made.[10] The geometry of the experiment and experimental setup did not differ from those usually adopted for angular measurements in the case of single scattering.[11] It has been found experimentally that the homodyne temporal autocorrelation function of the multiple-scattered light did not differ much from an exponential form typical to single-scattering. Therefore, the first cumulant of this function can be considered as the spectrum half-width too.

The experimental results obtained by the model systems are presented in Fig. 1-7.

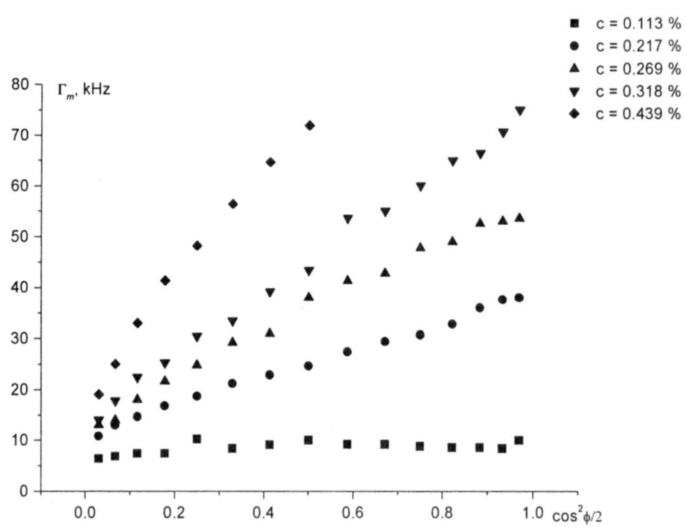

Figure 1. Γ_m vs $\cos^2(\varphi/2)$ for different concentrations of a latex. Radius of particles is 80 nm.

It was found[9, 12-15] that the multiple-scattering spectrum half-width of light scattered on the system of noninteracting Brownian particles mainly behaved as follows:

- it is proportional to the square of diffusing medium optical thickness $\gamma = L/l$. For a cylindrical cell $L = d\cos(\varphi/2)$ – the distance between the points where radiation enters and leaves the medium, d – the diameter of a cell, l – the mean free path of a photon in the medium, and φ – in contrast to single scattering is not scattering but an observation angle, (fig. 1-5);
- the temperature dependence of the single- and multiple-scattering spectra is the same and, consequently, $\Gamma_m \propto \Gamma(90°) \propto T/\overline{\eta}$, where $T, \overline{\eta}$ – temperature and viscosity of the medium, respectively, (fig. 6).
- In addition, a contribution (Γ_0) to the multiple-scattering spectrum half-width has been ascertained experimentally. This contribution did not depend on the optical thickness (γ), but it was directly related to the particle size (fig. 7).

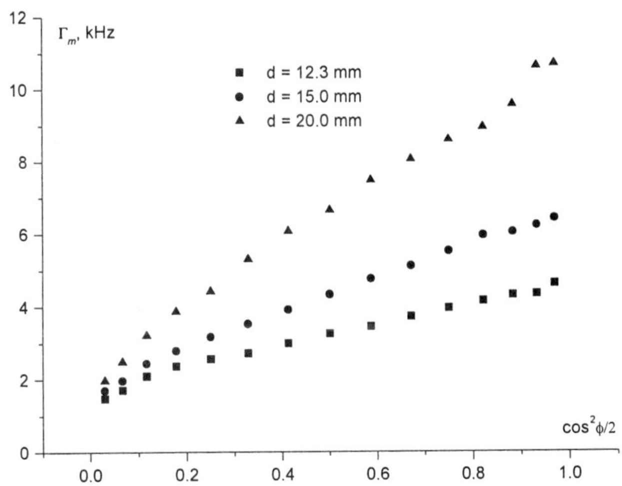

Figure 2. Γ_m vs $\cos^2(\varphi/2)$ for different diameters of the cylindrical cells. Radius of latex particles is 280 nm.

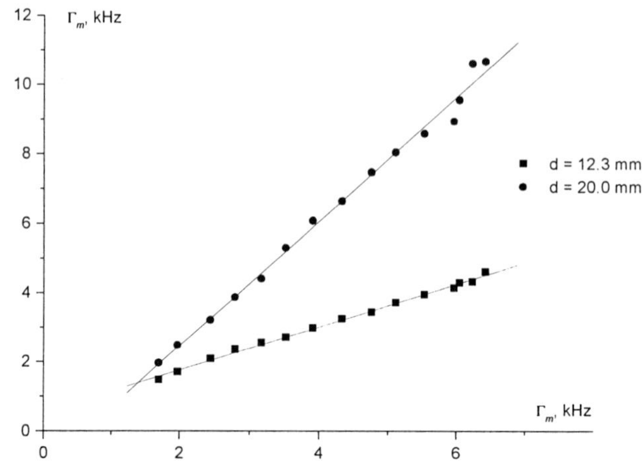

Figure 3. Another presentation of the same data as in fig. 2 (see the text and[13, 15])

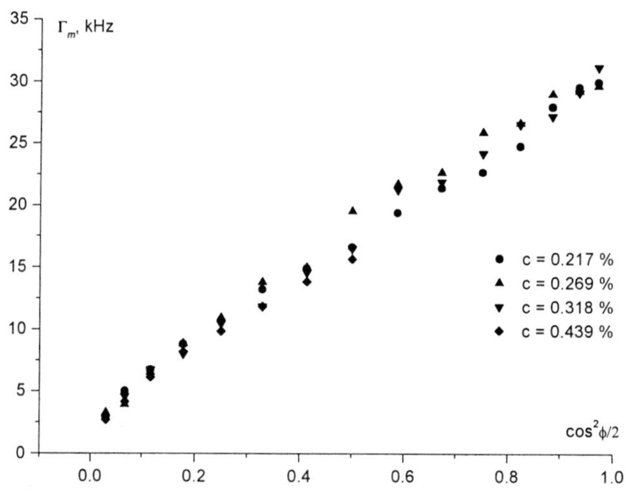

Figure 4. The same data as in fig. 1. The quadratic dependence of Γ_m on particle concentration and the presence of Γ_0 are taken into consideration.

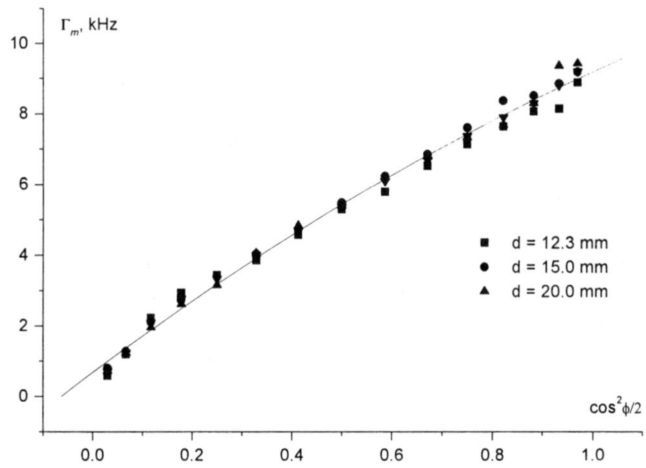

Figure 5. The same data as in fig. 2. The quadratic dependence of Γ_m on cell diameter and the presence of Γ_0 are taken into consideration.

3. THE SIMPLEST DIFFUSION MODEL OF MULTIPLE LIGHT SCATTERING

The interpretation of the physical modeling results has been performed in the framework of several mathematical models. It turned out that even a simple model of a random walk of a photon in a high-extinction dispersed non-absorbing medium presented in our first paper[9] was found adequate. Later a more profound approach – diffusion approximation of the radiation-transport theory[16] – has been used.[15] There is no doubt that this approach added some rigour, but it should be noted, however, that it has not changed principal conclusions of our first simple model.[9] Therefore, for the sake of simplicity solely, we shall examine here only such simplified diffusion model, more information can be found elsewhere.[15]

The intensity of light passing through scattering random media is known to decrease in accordance with Bouguer's law, the energy redistribution from incident wave to the scattering ones taking place (see e. g.[16]). There are two approaches to describe this process without the energy conservation law violation.

The first approach. The electromagnetic field propagating in medium is represented by the sum of fields of different multiplicity. The electromagnetic field between the successive scattering acts is commonly believed to correspond to the vacuum one. Of course, in the absence of absorption such assumption looks quite natural, however, for the reason of the waves of different multiplicity interference it is very difficult to be applied to turbid media. The extinction is known to be explained only by means of the interference of an incident wave with the scattering ones[17]. Obviously, in turbid milk-like media only taking into account the interference prevents the energy conservation law violation. Nevertheless, it is sometimes possible to meet an incorrect conception[18] that chaotic distribution of scatterers leads to vanishing of the fields of different multiplicity interference.

T h e s e c o n d a p p r o a c h . The difficulties mentioned above can be avoided if one takes extinction into consideration, having included it *ab initio* into the coherent field propagation law. Such an approach[19] has a significant advantage since all the effects of the interference will be taken into account automatically and, as a result, a statistical problem in the spirit of transport theory appears. Within the limit of high multiplicity of scattering this statistical problem can be reduced to the diffusion approximation.[20, 21]

Thus, the principal qualitative difference between strongly turbid media and transparent ones is that it is the extinction that determines the statistics of high multiplicity scattering radiation.[22] It is convenient, as Ishimaru[16] pointed out, to describe multiple scattering by means of correlation functions. So, a simplified diffusion model of multiple scattering can be based on the following assertions:[14, 22]

- a wave passing through the chains of the independent scatterers causes the multiplication of temporal autocorrelation functions, which define separate scattering acts;
- the autocorrelation function under study corresponds to the statistical averaging along the various chains of scatterers (trajectories of waves in turbid medium);
- between the successive scattering acts the attenuated waves propagate: light intensity $I \propto r^{-2}\exp(-hr)$, where h – the extinction coefficient.

Since such a statistical model corresponds exactly to the transport equation that Ishimaru[16] received for the temporal autocorrelation function, then the last statement can be formulated an another way:

- the statistics of chains is the same as with random walk of classical particles without interference. The interference under such conditions is taken into account by the attenuation.

The multiplication of autocorrelation functions is equivalent to the additivity of the spectral broadening:[9, 18, 23] multiple scattering spectrum half-width can be represented as

$$\Gamma_m = \sum_i \Gamma_i \,, \tag{1}$$

where Γ_i – the spectral broadening at the *i*-th scattering act, is determined by the scattering vector on the *i*-th scatterer. At the high multiplicity limit, when $hL >> 1$, Eq. (1) can be written as

$$\Gamma_m = N\bar{\Gamma} \,, \tag{2}$$

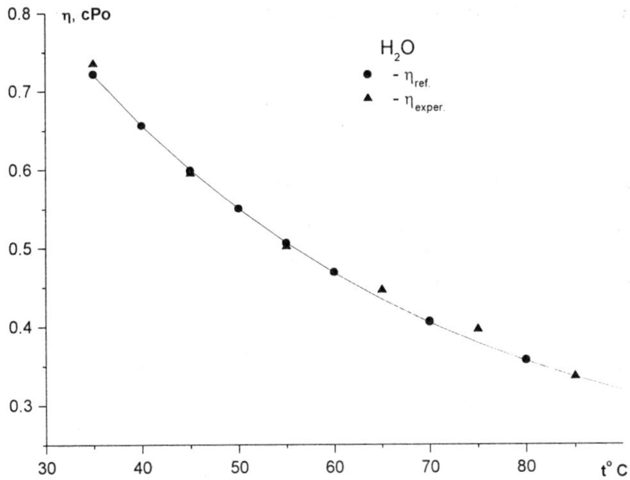

Figure 6. The temperature dependence of water viscosity measured by the multiple scattering correlation spectroscopy in milk-like latex system.[14] Radius of latex particles is 280 nm.

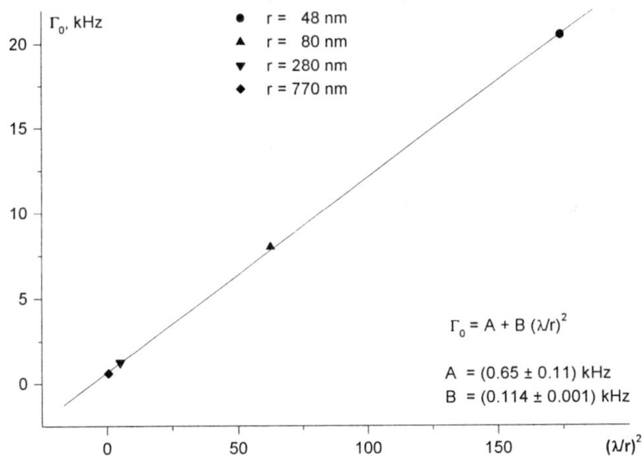

Figure 7. Dependence of Γ_0 on radius of scattering particles.

where N – is the mean scattering multiplicity at the point of observation and Γ – is the value of the spectral broadening in a single-scattering event averaged over a single scattering indicatrix.[9] The last Eq. (2), of course, is only the leading term of the asymptotical expansion for $N >> 1$, but even in such simplified form the diffusion model can answer many questions, if N is correctly determined.

To estimate N one can use the well-known Einstein relationship for the random-walk model

$$L^2 = 6D_p N \,, \tag{3}$$

In our case D_p – the mean square projection of the random step, has the following meaning: it is the product of the diffusion coefficient by the mean time of a photon free path, while L is the distance, already mentioned, between the points where the radiation enters and leaves the scattering medium.

To calculate D_p we make use of the fact that the trajectory of a random walk bears not only formal, but rather profound analogy with a polymer chain[24–26] – both have direct relation to the elementary stochastic diffusion problem.[27] Then the length of a photon free path corresponds to the length of a link in the chain, while the scattering angle θ_i corresponds to the coupling angle between links. The difference between these models lies in the fact that the coupling angle between the links of the polymer chain is fixed, while the scattering angle is a random quantity whose probability density is given by the form factor. Apart from that, the length of a link in the chain is fixed, while the distribution of the lengths of a photon free path is exponential. However, it can be shown strictly that the independence of the scattering angles for the different scattering events leads to identical solutions for both models if we replace $\mu \equiv \cos\theta$ appeared in the solution of the polymer chain problem, by its value $\bar{\mu}$ averaged over a single scattering indicatrix. As for taking into account photon free path length fluctuations, according to[28] it can be shown[12] that finally

$$L^2 = N l^2 \frac{2}{1-\bar{\mu}} + o(N) \tag{4}$$

Comparing Eq. 3 and Eq. 4, one can obtain a formula which is also well known from the theory of neutron transport[29]

$$D_p = \frac{l^2}{6} \cdot \frac{2}{1-\bar{\mu}} \,. \tag{5}$$

Equation (4) allows to receive an expression for the calculation of the mean scattering multiplicity

$$N = \frac{1-\bar{\mu}}{2} \left(\frac{L}{l} \right)^2 . \tag{6}$$

The last expression is the principal result of our simplified model of the photon diffusion in the infinite scattering medium. Comparing Eq. 2 and Eq. 6, we obtain

$$\Gamma_m = \frac{1-\bar{\mu}}{2} \left(\frac{L}{l} \right)^2 \bar{\Gamma} \,. \tag{7}$$

For the analysis of this formula first of all it is necessary to remind some facts. The mean free path of a photon in the scattering medium (l) is inversely proportional, by definition, to the extinction coefficient of the scattering medium (h), which, in turn, is proportional to the concentration of Brownian particles (c); it should also be noted that $\overline{\Gamma} \propto \Gamma = \frac{k_B T}{6\pi r \overline{\eta}} q^2$. Here k_B – the Boltzman constant, q – the scattering vector, r – radius of Brownian particles. After that let us turn to figures 1–7, where the results concerning experiments on the model systems of the monodispersed particles are shown. These results clearly confirm that the model (Eq. 7) fits the process in question for the most part adequately. Really, for the cylindrical geometry we immediately have $\Gamma_m \propto L^2 \propto \cos^2(\varphi/2)$ (see fig. 1, 2, 4, 5). Furthermore, we see that $\Gamma_m \propto L^2 \propto d^2$ (fig. 5) and $\Gamma_m \propto l^{-2} \propto c^2$ (fig. 4). As a result, the multiple-scattering spectrum half-width (Γ_m) is proportional to square of the scattering medium optical thickness $(L/l)^2$. It should be noted that such a peculiarity is an inherent characteristic of the multiple scattering spectrum only,[9] but not a single one. And finally, we notice that $\Gamma_m \propto \Gamma \propto T/\overline{\eta}$, i. e. the temperature and viscosity dependences of both Γ_m and Γ are the same, which is to say that multiple scattering spectra can be successfully used for viscosity measurements in strongly turbid milk-like systems (fig. 6).

Figure 3 needs an additional comment. Here the data as in fig. 2 are represented, but the values of spectrum half-width, Γ_m for $d = 15$ mm are taken as the abscissa and the other two (for $d = 12.3$ mm and $d = 20$ mm) – as the ordinates. Such a method of the data presentation permits to exclude an influence of unknown, but a fixed geometry factor, and to reveal the angular dependence of Γ_m in an explicit form. Moreover, such a plot allows to determine the value of Γ_0 more exactly as well.

As for Γ_0, its nature is not completely clear yet. It is found experimentally that Γ_0 doesn't depend on the medium optical thickness, but it is a function of the polarization of radiation and the size of the scatterers (fig. 7).

In conclusion of this paragraph, we would like to note that the attempts to interpret the multiple scattering spectra by means of a peculiar "successive approximation method" – single-, double-, triple scattering, and etc. because of insurmountable mathematical difficulties failed. However, as it may be seen from our works,[9, 12–15] considerable progress can be achieved if one starts from high values of the multiplicity from the very beginning.

4. THEORY OF THE CRITICAL OPALESCENCE SPECTRA

In contrast to dispersed turbid media where extinction coefficient is temperature independent and so the temperature dependence of both Γ_m and Γ is the same (fig. 6), in the vicinity of the critical point when $T \Rightarrow T_c$, rapid growth of the fluctuations is accompanied with the rise of the extinction and completely different temperature dependence of Γ_m should be expected. In accordance with Eq. (7), as the critical point is approached, the indicatrix is drawn out in the forward direction, monotonic decreasing of $\overline{\Gamma}$ and

$1-\bar{\mu}$ taking place. A simultaneous decreasing of the mean free path, l, complicates the analysis of Eq. (7), therefore, for further discussion of the limiting behavior of Γ_m it is more convenient to rewrite this formula as follows[7]

$$\Gamma_m = \frac{1}{2} L^2 \int (1-\mu)\, dh \int \Gamma dh \tag{8}$$

where dh is the differential extinction coefficient for scattering into solid-angle element $d\omega$, and the integration is over the total solid angle. For $T \Rightarrow T_c$ the growth of dh can be shown to exceed the decrease in Γ, and as a result both integrands increase monotonically, their limiting values being finite for any θ (multiplying dh by $(1-\mu)$ and Γ eliminates the divergence at $\theta = 0$).[7] Finally, we conclude that Γ_m increases monotonically with the approach to the critical point, reaching in the limit a value Γ_{max}. For comparison, we note that the width of the single-scattering spectrum (Γ), conversely, decreases on approach to the critical point.[11, 30, 31] It is worth adding that $\Gamma_m \propto L^2$, here as well, whereas Γ is completely independent on the dimensions of the critical system. For the sake of brevity to cite only final results for Γ_m dependences

$$\Gamma_{max} \propto \xi_0^{-4+2\eta+x_{\bar{\eta}}}, \quad \Gamma_{max} \propto \lambda^{-7-2\eta-x_{\bar{\eta}}}, \quad \Gamma_{max} \propto (k\xi_0)^{-1+2\eta+x_{\bar{\eta}}},$$

where ξ_0 – the amplitude of the correlation length, k – the wave vector, λ – the wave length, η and $x_{\bar{\eta}}$ – the critical exponents of the anomalous dimensionality and viscosity, respectively.

In such a manner the full temperature dependence of Γ_m in the vicinity of the critical point for the first time was estimated as well.[7] Figure 8 demonstrates the behavior of the calculated dependence of dimensionless quantity Γ_m^* (solid line) on usual variable $k\xi$, which has, in turn, the well-known temperature dependence (see e. g.[39]): $\xi = \xi_0 \tau^{-\nu}$, where ξ – the correlation length, $\tau = (T/T_c) - 1$, and ν – the critical exponent of the correlation length. The calculations were carried out for the strongly opalescent binary mixture aniline–cyclohexane (diameter of the cylindrical cell was 50 mm, the observation angle, $\varphi = 60^{\circ}$). It is seen that the rise of Γ_m becomes evidently slow after $k\xi \sim 10$. This, however, does not signify that the average scattering multiplicity N ceases to grow. Conversely, N continues to grow (nearly redoubling as $k\xi$ changes in the range of 10 and 100), but the growth is canceled by a simultaneous diminution in the $\bar{\Gamma}$.

Dotted line (fig. 8) shows the behavior under the same conditions of the theoretical curve calculated in accordance with well-known Kawasaki formula for the single-scattering regime.[30] It turned out that under these conditions for $k\xi \sim 10$ Γ_m is more than Γ by almost 20 times. Up to then, as $T \Rightarrow T_c$, the scattering spectrum half-width was believed to decrease only.[8, 11] This is, indeed, true for the scattering of any but the fixed multiplicity, provided that the multiplicity continuously growths the behavior of the Γ, as it was showed for the first time in our papers,[7] is essentially different (fig. 8, solid line).

It should be noted that one can not easily go over to the single-scattering regime by decreasing the thickness of the cell. As it was shown,[7] in the domain $10 << k\xi << 100$ the mean free path in the strongly opalescent binary mixture aniline–cyclohexane is $1\div 2$ mm. Alternatively, weakly opalescent systems don't permit to study light scattering in a wide neighborhood of the critical point successfully, which is necessary in order to take carefully into account the background contribution to the measured spectrum.[11]

5. EXPERIMENT

As the nature of the behavior of the Γ_m obtained was in a strong contradiction to the contemporary theories, an experimental test of it was of a particular interest. The above-mentioned experimental setup[10] has been modernized for the improvement of temperature stability up to ~ 0.3 mK, and for the elimination of the after-pulse problem by using the cross-correlation method. The strongly opalescent binary mixture aniline–cyclohexane having the considerable difference in the refractive indices ($\Delta n = 0.16$) and moderate one in the densities ($\Delta\rho = 0.24\,\mathrm{g\,cm^{-3}}$) of the components was proved to be an excellent object for the investigation.

The results of the study[32] are presented in figure 9. Solid circles are the experimental values of the scattering spectrum half-width ($\varphi = 60^{\circ}$) versus $\lg\tau$; dotted line is the same as in figure 8; solid line is the approximation of the experimental values. It is clearly seen that there are two regions of the change of the experimental values of the scattering spectrum half-width. In the domain of the temperature from $-3 > \lg\tau > -3.8$ the scattering spectrum half-width monotonically decreases, up to $\lg\tau = 3.2$ entirely corresponding to theoretical dependence for single-scattering regime by Kawasaki. As the critical point approaches, multiplicity grows, then the scattering spectrum half-width being now Γ_m, while passing the minimum, begins to increase in a qualitative agreement with our theory. In the temperature domain $\lg\tau < -4.3$ the experimental value of the "rate of the rise" of the Γ_m was found to be in a quite reasonable quantitative accord (particularly with regard to the difficulties of any experiment in the close vicinity of the critical point) with the theoretical one:

$$\left(\frac{\Delta\Gamma_m}{\Delta\lg\tau}\right)_{\mathrm{exp.}} = 6\cdot 10^3\ \mathrm{Hz}, \qquad \left(\frac{\Delta\Gamma_m}{\Delta\lg\tau}\right)_{\mathrm{calc.}} = 5\cdot 10^3\ \mathrm{Hz}.$$

6. CONCLUSION

We presented here the results of the extensive study on the critical opalescence spectra. This investigation was divided into some parts as follows: physical modeling of the critical opalescence by the monodispersed systems (latexes) with the controlled extinction coefficient, the creation on the basis of this the mathematical model of the multiple scattering spectra on the Brownian particles, the adaptation of this model to the critical opalescence spectra, and at last the special experiment. The correlation spectroscopy was chosen as the experimental investigation method of the broadening of the Rayleigh central line. Even simple model of a random walk of a photon in a high-extinction dispersed non-absorbing medium presented in our first paper[9] has been found adequate.

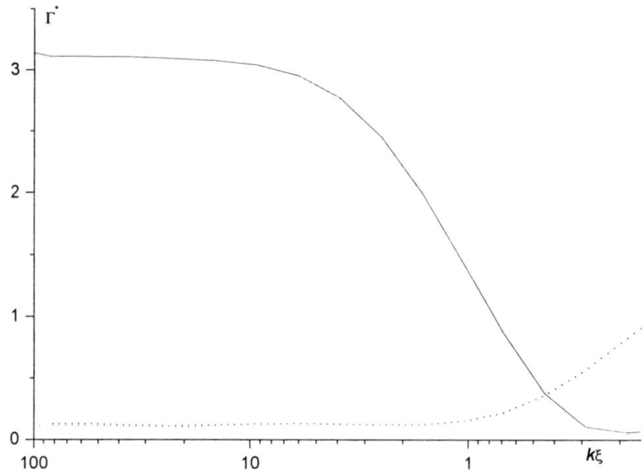

Figure 8. Theoretical dependences of the Γ for the single scattering (dotted line)[30] and the multiple one (solid line)[7] in the vicinity of the critical point (see the text).

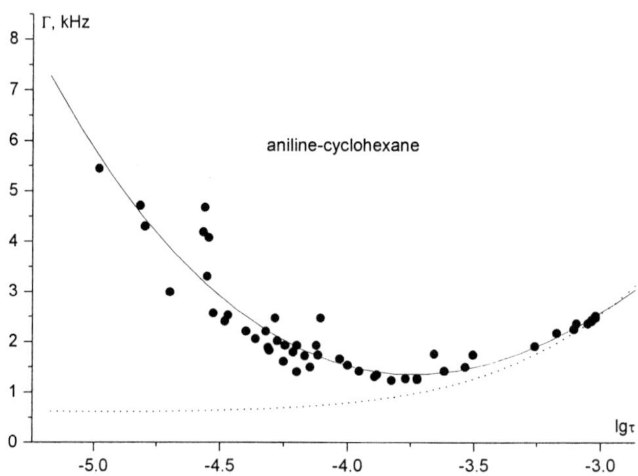

Figure 9. The experimental dependence of the Γ in the vicinity of the critical point of the binary aniline-cyclohexane mixture[32] (see the text).

As it was apparent afterwards, it has been just this research which has laid for the first time the foundation of the new scientific and applied method: the correlation spectroscopy of multiple scattering. In more recent time after the similar study in the slab geometry with the similar results this method has been named "*diffusing-wave spectroscopy*".[33] At the present time it continues to develop and to be used in both scientific and applied works.

As it was believed at the very beginning, the understanding of the nature of the multiple scattering spectra on the Brownian particles has given a key for a studying one of the most important nowadays problem – the problem of the phase transitions and critical phenomena.[34] We succeeded here in theoretical prediction of the temperature dependence shape of the multiple scattering spectrum half-width in the vicinity of the critical point[9, 10] and then in confirmation it under the experimental conditions that one generally tries to eliminate: strongly opalescent system, large size of the cell, close an approach to the critical point.[32]

Acknowledgements

The problem of the interpretation of the multiple scattering spectra on the Brownian particles and the critical fluctuations that was seemed practically insoluble at the beginning, nevertheless was successfully accomplished finally. A lot of the credit must go to my colleagues A. Kostko, V. Pavlov, S. Proshkin, and A. Soloviev. I profit by the occasion in order to express my sincere gratitude to all of them. Special thanks to A. Kostko, my former post-graduate, for correlator manufacturing.

REFERENCES

1. Critical Phenomena, M. S. Green & J. V. Sengers, ed. (NBS Misc. Publ. 273, Washington. 1966), pp. 1–242.
2. B.Chu, Critical opalescence, *Ber. Bunsenges. Phys. Chem.* 76 (3/4), 202–215 (1972).
3. M. A. Anisimov, Critical Phenomena in Liquids and Liquid Crystals, (Nauka, Moscow, 1987).
4. V. L. Kuz'min, V. P. Romanov, and L. A. Zubkov, Propagation and scattering of light in fluctuating media, *Physics Reports* 248(2–5), 71–368 (1994).
5. Z. Patashinskii and V. L. Pokrovskii, *Fluctuation Theory of the Phase Transitions* (Nauka, Moscow, 1975).
6. S. Ma , Modern *Theory of Critical Phenomena* (W. A. Benjamin, Inc., 1976).
7. D. Yu. Ivanov, A. F. Kostko, and V. A. Pavlov, Behavior of the width of the multiple-light-scattering spectrum in the immediate vicinity of the critical point, *DAN SSSR*, 282, 568–571 (1985) [*Sov. Phys. Dokl.*, 30, 397–399].
8. E. L. Lakoza and A. V. Chalyi, *Usp. Fiz. Nauk,* 140, 393 (1983) [Sov. Phys. Usp., 26, 573 (1983)].
9. D. Yu. Ivanov and A. F. Kostko, Spectrum of multiply quasi-elastically scattered light, *Opt. Spectrosk.*, 55, 950–953 (1983) [*Opt. Spectrosc.*, 55, 573–575].
10. D. Yu. Ivanov and A. F. Kostko, in: *Molecular Physics and Biophysics of water systems* (Leningrad State University, Leningrad, 1983), 5, 51–56.
11. *Photon Correlation and Light Beating Spectroscopy,* edited by H. Z. Cummins and E. R. Pike (Plenum Press, New York, 1974)
12. D. Yu. Ivanov, A. F. Kostko, and V. A. Pavlov, in: *XIV All-Union Conference on the Propagation of Radio Waves,* vol. 2, Moscow, Nauka, pp. 126–128 (1984)
13. D. Yu. Ivanov, A. F. Kostko, and V. A. Pavlov, in: *III All-Union Conference on the Propagation of Laser Radiation in Disperse Media,* part II, Obninsk (USSR), pp. 71–74 (1985).
14. D. Yu. Ivanov, A. F. Kostko, and V. A. Pavlov, in: *Molecular Physics and Biophysics of water systems* (Leningrad State University, Leningrad, 1986), 6, 145–152.
15. D. Yu. Ivanov, A. F. Kostko, and V. A. Pavlov, Multiple scattering spectra in strongly scattering media: Diffusion and non-diffusion contributions to spectrum halfwidth, *Phys. Lett. A* 138(6, 7), 339–342 (1989)

16. A. Ishimaru, *Wave Propagation and Scattering in Random Media* (Academic Press, New York, 1978)
17. M. Planck, Über elektrische Schwingungen welsche durch Resonanz erregt und durch Strahlung gedämpft werden, *S.–ber. Akad. Wiss* (Berlin) 10, 151–170 (1896).
18. C. M. Sorensen, R. C. Mockler, W. J. O'Sullivan, Multiple scattering from a system of Brownian particles, *Phys. Rev. A* 17(6), 2030–2035 (1978).
19. H. M. J. Boots, D. Bedeaux, P. Mazur On the theory of multiple scattering I, *Physica A* 79, 397–419 (1975); *ibid* On the theory of multiple scattering II. Critical scattering, 84, 217–255 (1976).
20. K. M. Case and P. F. Zweifel, *Linear Transport Theory* (Addison-Wesley, Reading, Mass, 1967; Moscow, 1972)
21. L. A. Apresyan and Yu. A. Kravtsov, *Radiation-Transport Theory: Statistical and Wave Aspects* (Moscow, 1983)
22. D. Yu. Ivanov, A. F. Kostko, and V. A. Pavlov, in: *Physics of Liquid State* (Kiev State University, Kiev, 1986), 14, 121–128.
23. F. Bøe and O. Lohne, Dynamical properties of multiply scattered light from independent Brownian particles *Phys. Rev. A* 17(6), 2023–2029 (1978).
24. M. V. Volkenstein, *Conformation Statistics of Polymer Chains*, (Academia Nauk SSSR, Moscow, 1959)
25. J. M. Ziman, *Models of Disorder,* (Cambridge University Press, Cambridge, 1979)
26. P.-G. de Gennes, *Scaling Concepts in Polymer Physics,* (Cornell University Press, Ithaca & London, 1979)
27. S. Chandrasechar, *Stochastic Problems in Physics and Astronomy*, (Moscow, 1947)
28. S. E. Bresler and B. L. Erusalimskii, *Physics and Chemistry of Macromolecules,* (Nauka, Moscow, 1963)
29. E. Fermi, *Nuclear Physics,* (University Chicago Press, Chicago, 1950)
30. K. Kawasaki, Kinetic equations and time correlation functions of critical fluctuations, *Ann. Phys N.Y.* 61, 1–56 (1970).
31. H. E. Stanley, *Introduction to Phase Transitions and Critical Phenomena,* (Clarendon Press, Oxford, 1971)
32. D. Yu. Ivanov, A. F. Kostko, and S. S. Proshkin, Critical opalescence investigation in the binary mixture aniline-cyclohexane by dynamic multiple light scattering, in: *13-th European Conference on Thermophysical Properties,* Lisboa, Portugal, pp. 377–378 (1993).
33. D. J. Pine , D. A. Weitz, P. M. Chaikin, and E. Herbolzhemer, Diffusing wave spectroscopy, *Phys. Rev. Lett.* 60(12), 1134–1137 (1988).
34. V. L. Ginzburg, *On the Perspectives of the Development of Physics and Astrophysics in the Late XX Century* in: *Physics of the XX Century Development and Perspectives*, (Nauka, Moscow, 1984)

2. NUMERICAL METHODS AND COMPUTER SIMULATIONS

METHANE COMBUSTION SIMULATION ON MULTIPROCESSOR COMPUTER SYSTEMS

B.N.Chetverushkin, M.V.Iakobovski, M.A.Kornilina, and S.A.Sukov*

1. PROBLEM FORMULATION

The problem of methane combustion in air is considered. Methane flows from the well and mixes with the air, then the gas gusher is ignited. More details about this problem are given earlier by Chetverushkin et al.[1]

The system of equations, governing gas-dynamic and chemical kinetic processes under combustion, expressed in operator form is the following:

$$\frac{\partial \mathbf{U}}{\partial t} + A\mathbf{U} = f, \qquad \mathbf{U} = \left(\rho, \rho y^{(i)}, \rho u, \rho v, E\right)^{\mathrm{T}}, \qquad f = (0, \omega_i, 0, 0, 0)^{\mathrm{T}},$$

Here A is a nonlinear operator, ρ - density, y_i - mass fraction of the i-th species, u, v - components of velocity along x and y respectively, p - pressure, E - total energy, ω_i - mass velocity of substance formation in all responses.

According to the method of summary approximation the system is split into two blocks of equations describing independently gas dynamic and chemical processes. GD block coincides with the quasi-gasdynamic system, where $\omega_i = 0$.

I. Gas dynamic block (GD):

$$\frac{\partial \mathbf{U}}{\partial t} + A\mathbf{U} = 0.$$

II. Block of chemical kinetics (CHEM):

* B.N.Chetverushkin, M.V.Iakobovski, M.A.Kornilina, and S.A.Sukov, Institute for Mathematical Modelling of RAS, Moscow, Miusskaya sq., 4-A, 125047.

Mathematical Modeling: Problems, Methods, Applications
Edited by Uvarova and Latyshev, Kluwer Academic/Plenum Publishers, 2001

$$\frac{d\,\mathbf{U}}{d\,t} = f, \text{ where } f = (0,\omega_i,0,0,0)^{\mathrm{T}}.$$

GD block is approximated via half-implicit finite-difference scheme:

$$\frac{\mathbf{U}^{j+1} - \mathbf{U}^{j}}{\Delta t} + \frac{1}{2}\left(A\mathbf{U}^{j+1} + A\mathbf{U}^{j}\right) = 0$$

The second accuracy order iteration process for this scheme is used:

$$\frac{\overset{(k+1)}{\mathbf{U}^{j+1}} - \mathbf{U}^{j}}{\Delta t} + \frac{1}{2}\left(A\overset{(k)}{\mathbf{U}^{j+1}} + A\mathbf{U}^{j}\right) = 0, \quad 0 \le k \le s-1; \quad \overset{(0)}{\mathbf{U}^{j+1}} = \mathbf{U}^{j}.$$

For integration of the stiff ordinary differential equations (ODE) different packages based on Gear C.W. and Adams G.C. methods are used (STIFF, DVODE[2]).

Chemical processes are represented by terms of production and reduction of mass. They are an example of local processes, which are independent from spatial gradients. New values of temperature and mass fractions determined in chemical reactions in each point of mesh depend only on their latter values. And thus they can be calculated regardless of adjacent points.

GD and CHEM blocks are calculated by turns. In doing so the values of y_i, p and T calculated in GD block are used to form initial values for CHEM block.

The main computational costs (over 90% of the processor time) are presented by solving of the stiff ordinary differential equations of chemical kinetics.

Parallel algorithm for GD block is based on domain-decomposition method. In case of proportional distribution of points in processors this guarantees obtaining of load balance in this block. CHEM block parallelization using domain decomposition is extremely inefficient for the following reasons:

1. The points with the most intensive chemical reactions ("hot" points) are located in relatively thin area - in the flame front and directly behind it (see Fig. 2). Generally, the points simulating the flame front, are located on relatively few processor nodes. As a result, processing time for CHEM block is considerably larger for those processors, which contain "hot" points in comparison with that one for other processors.
2. The flame front position moves in time, and "hot" points may settle in variable processors in different time steps.
3. As the reaction rates strictly dependent on temperature and mass fractions of species, calculating time for ODE greatly changes from point to point in space and in time. Calculating time for ODE in each "hot" point practically can't be estimated beforehand. Factors mentioned above result in inapplicability of static load balancing methods (including domain decomposition) for CHEM block parallelizing.

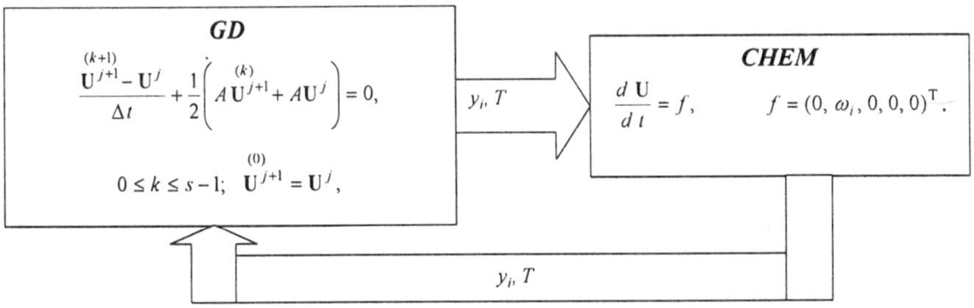

$$\frac{\overset{(k+1)}{\mathbf{U}^{j+1}} - \mathbf{U}^j}{\Delta t} + \frac{1}{2}\left(A\overset{(k)}{\mathbf{U}^{j+1}} + A\mathbf{U}^j\right) = 0,$$

$$0 \le k \le s-1; \quad \overset{(0)}{\mathbf{U}^{j+1}} = \mathbf{U}^j,$$

Figure 1. Algorithmic diagram.

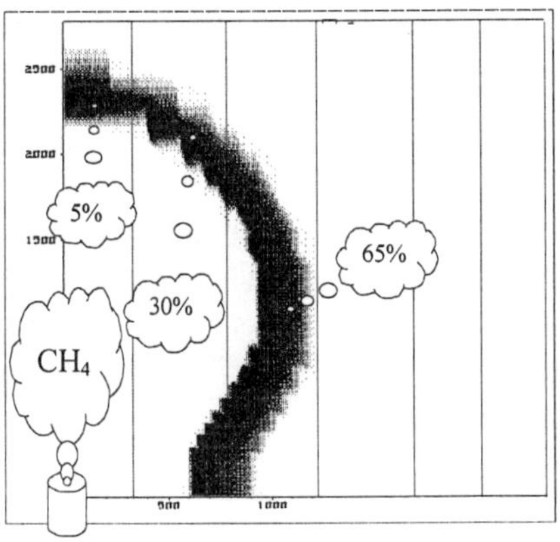

Figure 2. Domain decomposition in processors.

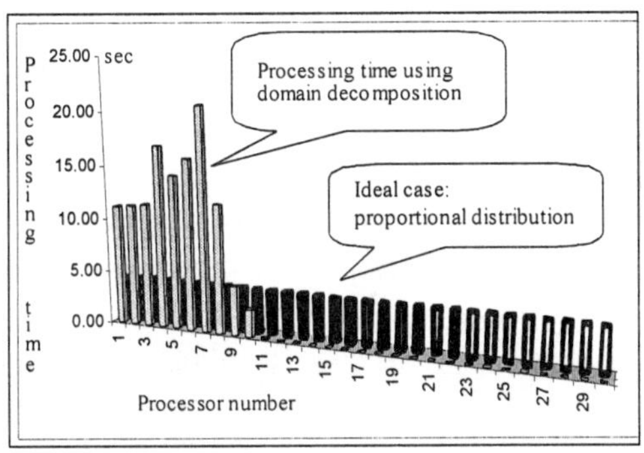

Figure 3. Loading distribution during "hot" points calculation.

For processor farm - well-know method of dynamic load balancing - the master node is supposed to contain the whole data necessary for calculating (temperature, pressure, et al.) in all mesh points. Efficiency of parallelizing is decreased unacceptably by collection of this data at each time step, its passing to worker nodes, receiving and passing back of calculated data for next GD step. The single master node is the bottle neck of the whole method.

Therefore an algorithm based on collective farm, but devoid of its shortcoming, is suggested. It is achieved due to great reduction of traffic as each processing node possesses equal controlling features.

2. PARALLEL PROGRAM STRUCTURE

Algorithm for load balancing is based on the following principles:

1. each processor primarily operates its local points (that are stored in its memory);
2. the processor can request points from the others provided that all local points are calculated or transferred for handling to other processors;

3. transmission of points for handling to other processors and handling of local points are fulfilled simultaneously
4. volume of passing data is considerably reduced in comparison with standard "processor farm";
5. control and redistribution cost is also reduced

Two processes are started on each processor (see Client on Figure 4). One process is control one and another is computational one, the last calculates ODE for each point passed to it for handling.

The following Server Parallelism algorithm, ensuring dynamic load balancing during CHEM block, is executed by each control processes:

1. If there are any unprocessed local points and the computational process is not busy, then one of those points is passed to the process for processing. When the processing is terminated the computational process sends a message to the corresponding control process.
2. In the case of absence of processed points the control process sends a request for unprocessed points to any other control process.
3. The control process proceeds to waiting for message either from local channel to its own computational process, or from any virtual channels to other control processes. The control process can receive either a message with local point processing result, or a message from a control process of another processor.
4. Received message is reading, handling and steps 1-4 are repeating.

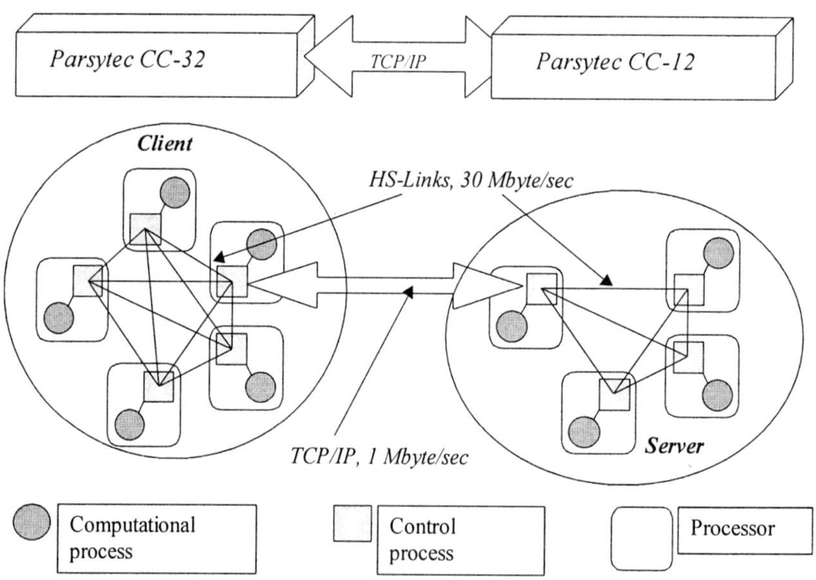

Figure 4. Load balancing program structure.

In whole the algorithm resembles search in distributed net of database servers, that led to the name "Server Parallelism" - each processor asks the others for work. Note that, if any processor ***A*** (Figure 5) has received several points for processing from processor ***B*** and still didn't process them until ***A*** received a request from processor ***C***, then processor ***A*** can send some points to ***C*** for processing. This operation insures the most balanced points distribution.

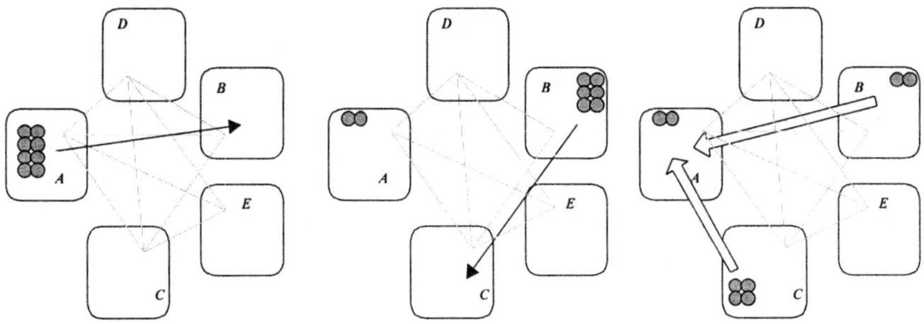

Figure 5. Redistribution of points in processors.

This algorithm is effective if operating system supports the following features:

- starting of two processes on one processor;
- the local link for connection between processes located at the same processor;
- asynchronous message passing from one process to another;
- possibility of alternative input from a set of channels.

For local processes (started at the same processor) the possibility of shared memory joint usage is desirable.

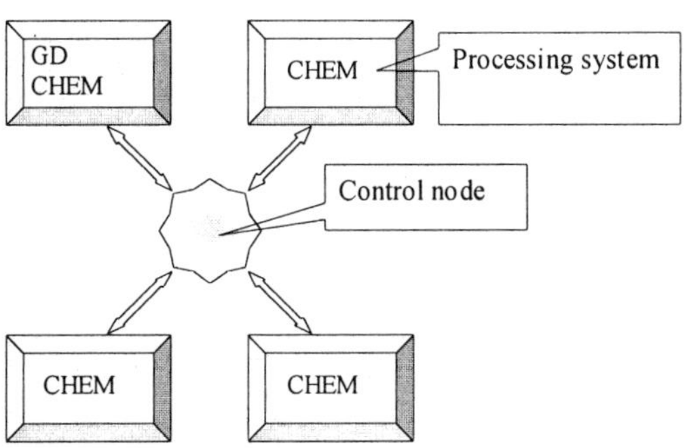

Figure 6. Program structure for joint using of several multiprocessor systems.

All stated above concerns to the case, when all processors belong to one computing system, and have a capability to send the messages directly to each other, using fast virtual channels. Yet, slight modification of the offered algorithm allows to use at CHEM stage of several multiprocessor systems connected by rather slow channels (see Figure 6). Use of a few multiprocessor systems for the solution of gas dynamic equations is ineffective in view of large size of data, passing through slow channels. Therefore these equations should be solved on the only one of computer systems. Just that system acts as a server, storing the points, which can be handled at CHEM stage either by processors of server-system, or by those ones of connected client-systems.

3. RESULTS

Two distributed memory multiprocessor systems Parsytec-CC (Germany) were used for calculating with 32 nodes PowerPC-604 (100MHz) and 12 nodes PowerPC-604 (130MHz) correspondingly.

The obtained efficiency of 4 processors PowerPC-604 (100MHz) is assumed to amount to 100 %, as calculations with use of less processors are ineffective for the limitation of random-access memory. This fact is plotted in Fig. 7-8 by a starting drop of efficiency.

As the GD block was fulfilled on a system Parsytec CC-12 only, one should not expect a raise of speedup at this stage of calculations, when the number of processors in another system increases. In this connection the results in parallelizing for CHEM block only, with the deduction of computing time for GD block, are shown in the Figure 8. Numerical results actually show practically linear growth of speedup at CHEM stage with the increase of total number of processors. Different curves in the figure are correspondent to different time in account.

The efficiency above 100% in Fig. 8 indicates that owing to Server Parallelism algorithm applied for dynamic load balancing proportional distribution of "hot" points in processors is achieved with use of rather large number of processors.

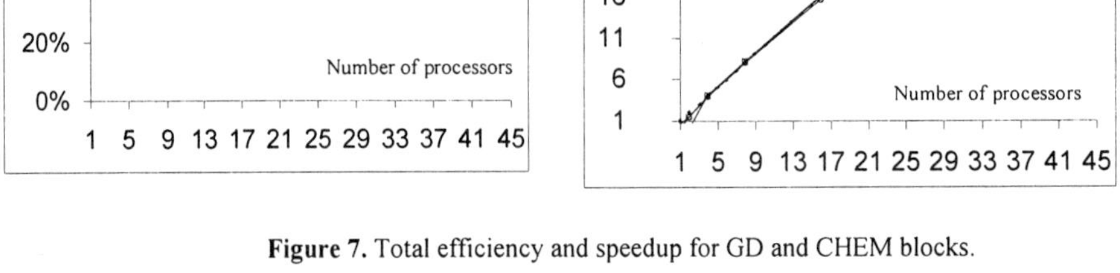

Figure 7. Total efficiency and speedup for GD and CHEM blocks.

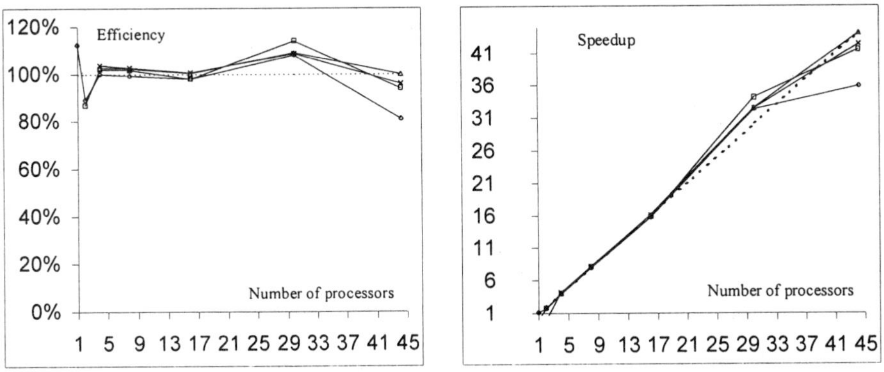

Figure 8. Efficiency and speedup for CHEM block only.

4. ACKNOWLEDGMENT

The Parsytec CC workstation used to provide all computations was delivered to the Institute for Mathematical Modelling in the framework of the equipment grant of European Economic Community (project No. ESPRIT 21042). The work is supported by Russian Foundation for Basic Research (grants No. 99-01-01215).

REFERENCES

1. B.N. Chetverushkin, M.V. Iakobovski, M.A. Kornilina, K.Yu. Malikov, N.Yu. Romanukha, Ecological after-effects numerical modelling under methane combustion, in: *Mathematical Models of Non-Linear Excitations, Transfer, Dynamics, and Control in Condensed Systems and Other Media,* Proc. of a Symp., June 29 – July 3 1998, Tver, Russia (Ed. by L.A. Uvarova, A.E. Arinstein, and A.V. Latyshev), Kluwer Academic, Plenum Publishers, New York, Boston, Dordrecht, London, Moscow. ISBN 0-306-46133-1, 1999, pp. 147-152.
2. P.N. Brown, G.D. Byrne, and A.C.Hindmarsh, 1989, VODE: a variable coefficient ODE solver, *SIAM J. Sci. Stat. Comput., 10, pp.1038-1051.*
3. M.A. Kornilina, E.I. Levanov, N.Yu. Romanukha, B.N. Chetverushkin, M.V. Iakobovski, Modelling of gas flow with chemical reactions on multiprocessor system, in: *Use of mathematical modelling for science and technology problems solution,* Trans. Int.. Conf. «Mathematical modelling for science and technology», Ed. M.Yu.Alies, Publishing house Inst. for Applied Math. Ural Dep. RA,. Izhevsk, 1999. pp. 34 – 48. (in Russian)

COMPUTER SIMULATION OF STRUCTURAL MODIFICATIONS IN THE METAL SAMPLES IRRADIATED BY PULSED BEAMS

Igor V. Puzynin and Valentin N. Samoilov*

1. INTRODUCTION

It is well known[1-3] that one of the effective methods of materials synthesis for modern technologies is the electron and ion surface treatment in a pulsed explosion mode. Energy deposition in a thin surface layer by the high energy electron and ion beams can lead to a completely new structure on the surface which can possess interesting physical and chemical properties[2, 4,]. For almost three decades the ion beams have been used for modification of materials in manufacturing integrated circuits[5-6]. One of the aspects of the electron and ion beam irradiation of materials consists in transforming the material parameters which are of particular interest for metallurgy. For example, the electron and ion beam irradiation of metals can change the metal hardening, fatigue, corrosion resistance and essentially increase their strength[2,7] We also should note using the beam modifications for hardening and improving the tribological properties of the surfaces, ion implantation, molecular epitaxy, etc.[8,9]. The ion beams used for these applications range from keV up to MeV of energy and penetrate the target material to the depths ranging from tens of nanometers to microns. The keV particle bombardment of solids is used for fabrication in semiconductor industry[10].

However, in spite of enormous experimental data and applications, it is not yet possible to say with confidence that there exists quite a detailed mathematical model of the phenomenon under study.

In this brief review we would like to describe three different approaches to the mathematical simulation of structural modifications in the metallic samples irradiated by electron and ion pulsed beams.

First is a fractal analysis of microphotographs of the surfaces of irradiated samples that confirms their structural modifications[11].

* Igor V. Puzynin, Joint Institute for Nuclear Research, Dubna, 141980 Dubna, Moscow Region, Russia.
Valentin N. Samoilov, Joint Institute for Nuclear Research, Dubna, 141980 Dubna, Moscow Region, Russia.

Then on the basis of a heat transfer equation with boundary conditions taking into account a finite width of the target and a contribution in cooling of the radiation heat exchange of the target bounds with vacuum, we consider results of the computer simulation and macroscopic peculiarities of the heating process of the material surface irradiated by a pulsed source. A temperature dependence of the kinetic coefficients characterizing the target's material was taken into account[12].

Finally, a model of the evolution of thermoelastic waves in the metal samples exposed to pulsed ion beams is investigated. We study a relation of the form of the thermoelastic waves and the form and the position of the pulsed source, an influence of temperature on the velocity of the thermoelastic waves and a condition of amplification and cancel of the waves[13].

In the future these effects can be used as an explanation of structural modifications in experimental samples.

2. FRACTAL ANALYSIS OF THE MICROPHOTOGRAPHS OF THE SURFACE OF IRRADIATED MATERIALS

Intensive investigations on the formation of condensed systems in non-equilibrium conditions when a condensed state of a substance is of a porous fractal structure rather than a continuous matter provide a fresh approach to the questions of the morphology of the sample's surface irradiated by high current pulsed electron and ion beams. Besides, the development of computational methods in physics has led to creation of the mathematical programs that allow simulation of fractal structures[14].

This section represents some results of numerical analysis on the fractal surfaces of the samples under irradiation.

We propose a new model for obtaining fractal structures with the help of pulsed high current electron and ion beams. Its idea is as follows. In the course of electron (or ion) pulsed beam exposure of the sample's surface we have a pulsed release of the beam energy in the sample at the depth determined by the kinetic energy of the particles. As a result, during the electron exposure of the sample the melting of the surface layer with a subsequent crystallization or amorphization takes place, and during the ion exposure - an ion stirring[15]. In both cases, intensive phase transformations of the substance take place which are characterized by formation of porous fractal aggregates.

In the experiments, the installation "ELIONA" intended for generating the electron and ion beams was used with the following parameters[16]: kinetic energy of ions and electrons – 100–500 keV; duration of the beam current pulse – 300 nsec; electron beam current density – 5-200 A/cm^2; species of ions C^+, Ti^+.

Metals, metallic films on substructures, and high temperature superconductors were selected for research. Microphotographs were taken from the irradiated samples with the help of the scanning electron microscope JSM870.

The conclusion that the images on the microphotographs taken during the experiments with the electron-ion source are fractals, is based on a comparison of the results of measuring the length of the curve "separating" the segments of various colour.

Having counted the quantity of squares of size δ required for covering the curve $N(\delta)$, we can determine the length as $L(\delta)=N(\delta)\delta$. For smooth curves, if δ decreases, the length $L(\delta)$ tends to a constant for $\delta \to 0$. This does not occur for the curves on the microphotographs. The graphics show a δ-dependence of L(δ) constructed in a logarithmic scale. From the figures one can see that for our curves the length is described by approximated formula $L(\delta)=\alpha\delta^{1-D}$. For the photographs under study, the index of exponent D is always more than 1. This confirms that the structures given in the photographs are of fractal character. To analyze the photographs, a computer program based on a box-counting algorithm has been developed[17].

Analysis has been performed on the number of the microphotographs taken from different irradiated samples. As an example, two of the photographs and some results of calculating the fractal dimensions are presented on Figs. 1, 2.

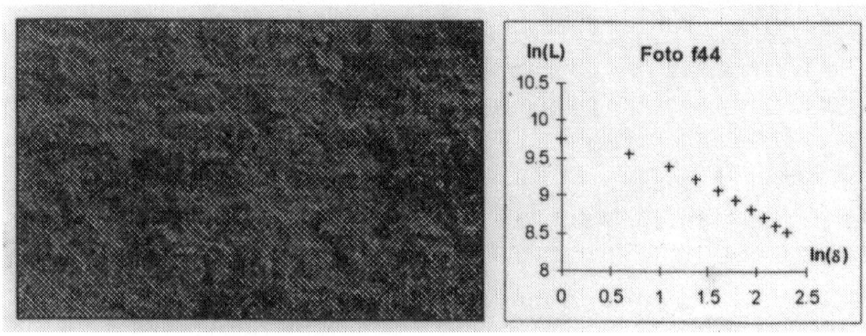

D=1.7

Figure 1. A microphotograph of the irradiated surface of tool steel.

The microphotograph of the surface of tool steel exposed to a 250 keV electron beam, current 1000 A and impulse time 300 nsec is shown on Fig.1. A dependence of length L of the isocolour line upon size δ of the cover mesh (in pixel) is presented on the left-hand graph. Calculated dimension of the isoline is D=1.7.

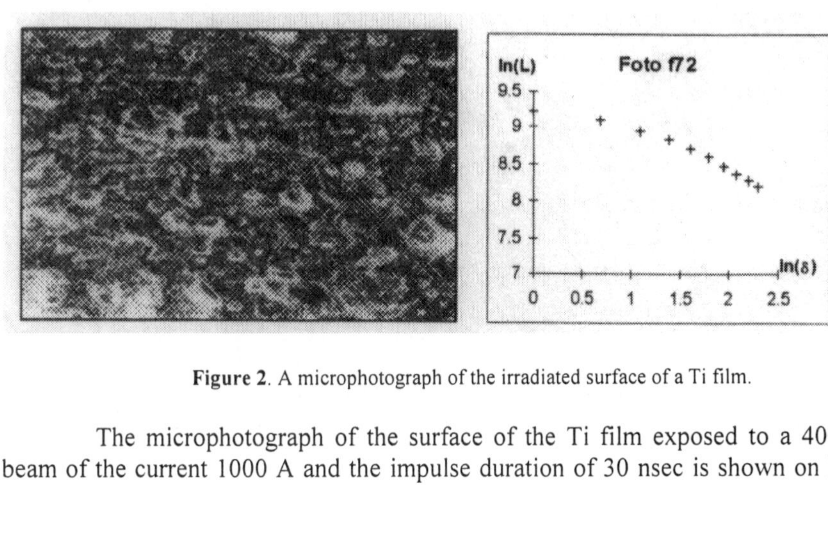

D=1.6

Figure 2. A microphotograph of the irradiated surface of a Ti film.

The microphotograph of the surface of the Ti film exposed to a 400 keV ion beam of the current 1000 A and the impulse duration of 30 nsec is shown on Fig.2. The results of the corresponding fractal analysis are given on the graph. Calculated dimension is D=1.6.

Let us note that we dealt not with the fractal dimensions of the irradiated surface itself but with its iso(colour) lines. For quite smooth in transversal direction fractal surfaces (for example, it occurs for the earthen landscapes), the dimension of the surfaces is exactly one unit more than the dimension of isolines[18]. Thus, one can estimate the dimension of the surface itself and compare the results with theoretical predictions. In case of the metals subjected to a surface modification, this relation is quite probable though but not vivid. Further investigations that need a precision electronic microscopy are of particular importance.

Therefore, in frames of the model considered it has been shown that as a result of irradiating the materials, the fractal structures of various fractal dimensions are generated on the samples' surface. There are some theoretical grounds in forming a connection between the structure generated on the materials' surface and the physical (chemical) process. However, this phenomenon needs a more detailed study.

3. NUMERICAL SIMULATION OF THE THERMAL TREATMENT OF METAL SURFACE BY HIGH CURRENT ION BEAM

A problem of modeling the behavior of the temperature profile on the shape and the power of the ion beam is considered in this section. The characteristics of the ion beams produced at the ELIONA installation were taken as the parameters specifying the source in the heat transfer equation.

We consider a one-dimensional heat transfer equation

$$\rho^*(T)\frac{dT}{dt} = \frac{\partial}{\partial x}\left(a(T)\frac{\partial T}{\partial x}\right) + Q(x,t), \tag{1}$$

with initial condition

$$T(x,0) = T_0 \tag{2}$$

and boundary conditions

$$a(T)\frac{\partial T(x,t)}{\partial x} - \alpha\sigma(T^4(x,t) - T_0^4)\big|_{x=0} = 0, \tag{3}$$

$$a(T)\frac{\partial T(x,t)}{\partial x} + \alpha\sigma(T^4(x,t) - T_0^4)\big|_{x=l} = 0, \tag{4}$$

The source had the form proposed in[12]

$$Q_i(x,t) = \begin{cases} E_0 j_i(t)/(ZeR), t \le \tau, x \le R, \\ \quad 0, \quad t > \tau, x > R, \end{cases}$$

where τ- time of operating the source; E_0, $j(t)$, ze, R - initial energy, current density, charge and average length of the ion run, respectively. The functions for the ion beam current density $j_i(t)$ have the form

$$j_1(t) = const = \frac{j_{\max}}{2},\ j_2(t) = \begin{cases} 2j_{\max}t/\tau, & t \le \tau/2, \\ 2j_{\max}(\tau - t)/\tau, & \tau/2 < t \le \tau \end{cases},\ j_3(t) = \frac{\pi j_{\max}}{4}\sin\left(\pi\frac{t}{\tau}\right)$$

j_{max} is varied in the range from 500 to 1000 A/cm^2, average ion energy $E_0 = 350$ kev. It should be noted that the total energy transferred by the target's ion beam at the momentum time τ is the same for all functions of the source $Q_j(t)$ and proportional to $j_{\max}\ \tau/2$. If $j(t)$=const, the source has the simplest form which corresponds to a greater degree to a uniform distribution of ions over the energy, i.e. it has a form of a small step in time and in coordinate x normal to the surface.

It should be underlined that in the model we took into account a temperature dependence of kinetic coefficients $\boldsymbol{a}$ and $\boldsymbol{c}$ so it has allowed us to take into account the finite rate of propagation of the temperature front that was necessary for a more real description of the thermal processes. The form of the functions $\boldsymbol{a}$(T) and $\boldsymbol{c}$(T) is built on the empirical data[19].

Varying the current density mainly at the expense of decreasing the beam radius, one can obtain the point when the maximal temperature of heating the surface of the irradiated sample reaches the melting temperature. To describe the evolution of the temperature profile taking into account the phase transformations on the sample surface, one needs to take into consideration the energy losses spent for melting and evaporating the target surface. These losses can be taken into account if to add to the heat capacity two terms describing the latent heat flows:

$$c^*(T) = c(T) + \frac{L_m}{T_0}\delta(\xi) + \frac{L_e}{T_0}\delta(\xi_1), \xi = \frac{T - T_e}{T_0}, \xi_1 = \frac{T - T_b}{T_0},$$

where L_m and L_e are specific heat of melting and evaporation, T_m and T_b are melting and boiling temperatures, respectively.

The boundary conditions (3), (4) take into account the heat exchange of the absorbent of the finite width $\boldsymbol{l}$ with vacuum due to the heat exchange on both sides of the target surface. Here σ is the Stefan-Boltsman constant, α is a coefficient of absorption in a gray body. At the initial moment of time it is assumed that the distribution of the temperature field in the target is homogeneous: $T(x,0)=T_0$. For a numerical solving of the initial-boundary problem (1)-(4) an absolutely stable implicit difference scheme in time was used. In order to approximate over a spatial variable x, difference schemes of the second order accuracy with a heterogeneous step and schemes of the finite element method with linear elements were used. The systems of nonlinear algebraic equations arising on each time layer were solved by a simple iteration method or by a continuous analog of the Newton's method. A sweep method was used for the solving of linear systems.

This model was used for the description of the evolution of the temperature field arising in the iron target exposed to carbon ions. The following thermo-physical properties of iron were taken into account for the computations: density - p=7870 kg/m^3, melting temperature - T_m=1799 K, boiling temperature - T_b=3149 K, specific melt heat - L_m=0.27 10^6 J/kg and specific evaporation heat is L_e=6.25 10^6 J/kg. Figs. 3a-3b give the results of calculating the spatial and time profile of $T(x,t)$ for various functions of the source $Q_i(t)$. One can see that the maximal heating is reached in the layer of a depth equal to the average length of the free run of ions in the substance. Two plateaus which can be seen in each figure correspond to boiling and melting the target. Upon switching off the source, at $t>\tau$, the surface gets cool quickly due to the thermal exchange of the target's bounds with vacuum and cooling deep into the sample. It is seen from the smoothing of the temperature profile for all source's functions. This is because the further cooling is described by Eq.(1) without a source for all three cases.

However, for each case the heat transfer equation is added by an initial condition in the form of the function $T(x, \tau)$ at the moment of switching on corresponding Q_i. Thus, the behavior of the temperature profiles for various functions of the source essentially differs only at the stage of heating up, the rate of which is specified by the form of function $j(t)$.

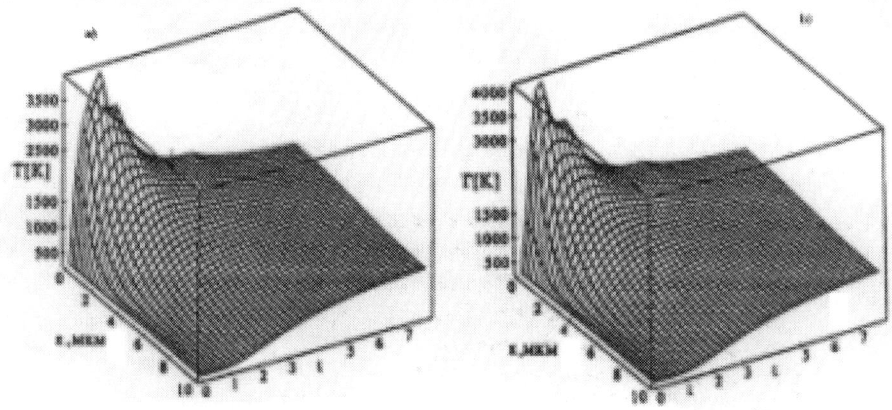

Figure 3. Spatial and time profiles of temperature in the iron target of the width l=10 mkm exposed to 350 KeV carbon ions, momentum duration - 300 ns, max. ion beam current density j_{max}=1000 A/cm^2 for $Q_1(t)$ (a), $Q_2(t)$ (b), time is in relative units t/τ.

Let us note that the interaction of the high current ion beams with substance is characterized by a quick heating up of the target surface with a subsequent quick cooling. This leads to damaging the crystal lattice and appearing defects on the surface and, thus, can essentially change the physical properties of the material. The results of the modelling are in agreement with the results of analogous computations and measurements presented in[20, 21].

4. NUMERICAL SIMULATION OF THERMOELASTIC EFFECTS IN METALS IRRADIATED BY PULSED ION BEAMS

In order to model the formation of the thermoelastic waves arising in the metals irradiated by intense ion beams, we analyze the thermoelastic processes on the basis of the following system of equations[22] ($\sigma=\sigma(x,t)$, $T=T(x,t)$):

$$\frac{\partial \sigma}{\partial t^2} = v_s^2 \frac{\partial^2 \sigma}{\partial x^2} - \alpha \frac{\partial^2 T}{\partial t^2}, \tag{5}$$

$$\left(1+g_0 T\right) = \frac{\partial T}{\partial t} = k_0 \frac{\partial^2 T}{\partial x^2} - \beta T \frac{\partial \sigma}{\partial t} + q, \tag{6}$$

for $0 < x < 1, t > 0$ with the initial and boundary conditions

$$T(x,0) = const;\ \sigma(x,0) = \frac{\partial\sigma(x,0)}{\partial t} = 0;\ 0 \le x \le 1; \tag{7}$$

$$\sigma(0,t) = \sigma(1,t) = 0;\ \frac{\partial T}{\partial x}\Big|_{x=0} = \frac{\partial T}{\partial x}\Big|_{x=1} = 0,\ t \ge 0, \tag{8}$$

which mean that till the moment of switching on the source in the sample it has no thermoelastic waves, and the boundaries of the sample are always free and thermal insulated. The system of equations (5)-(6) with the initial and boundary conditions (7)-(8) is written in the dimensionless values. Here $\sigma = \sigma(x, t) / \sigma_0$ ($\sigma = 2.2 \cdot 10^6$ Pa) - the strength, T = $T(x, t)/T_o$ (T_0 = 293K) - the temperature, $x = x/l_0$ ($l_0 = 10^{-5}$m - the thickness of the sample) - the distance from the surface of the sample, the time t = t/m (r = $3 \cdot 10^{-7}$s – the duration of the effect of the source; if t = 1, the source will be switched off), $x_l = Ro/I_0$ ($R_0 = 7 \cdot 10^{-7}$m - a free length of the run of the ions) - a depth of the ion penetration in the sample.

The dimensionless constants v_s^2, α, k_0, β, g_o are defined by the physical properties of the sample.

The function $q(x, t)$ describing the influence of the charged particle beams on the sample (the source function), is defined by the expression

$$q(x,t) = q_0\tilde{q}(x,t), \tag{9}$$

where

$$q_0 = \frac{E_0 j_{\max}}{2ZeR_0}\frac{\tau}{c_0\rho_0 T_0} \tag{10}$$

is a dimensionless constant that expresses the power of the thermal source, E_0, j_{max}, Ze are the ion energy, the current density and the ions charge, accordingly, ρ_0 is a specific density, c_0 – density. The function $\tilde{q}(x,t)$ describing the form and the way of the effect of the source, may be given from the physical viewpoint taking into account the characteristics of the beam and the properties of the sample.

To solve numerically the system of equations (5)-(6) with the initial and boundary conditions (7)-(8), we use a finite difference method. Let us introduce a uniform rectangular net $\{x_i = ih_x\ (i = 0, 1, \text{---}, m),\ t_j = jh_t\ (j = 0, 1, \text{---}, n)\}$; where h_x and h_t are steps for the variables x and t, accordingly. Then we use the following explicit scheme of the accuracy order $O(h_x^2 + h_t)$:

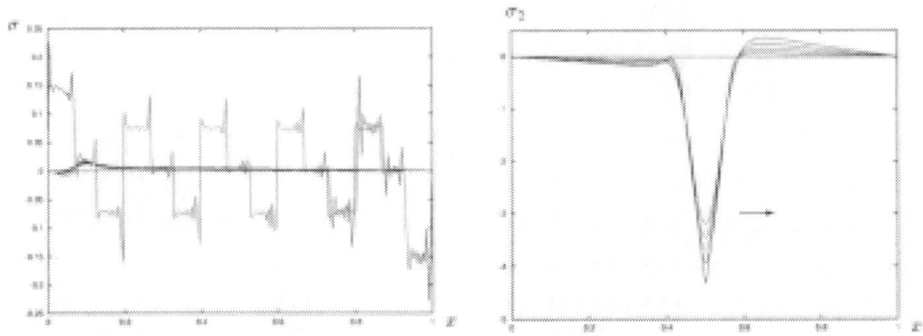

a) b)

Figure 4. Variation of amplitudes of thermoelastic waves as a result of their interaction: a)damping, b)amplification.

In frames of the mathematical model we study the thermoelastic waves arising in the iron sample irradiated by carbon ions. The calculations imply the characteristics of the ion beam going from the ELIONA set-up and the thermal physics parameters of iron.

The following effects have been found by numerical simulation:

- thermoelastic wave arises immediately after switching on and switching off the source, and their interaction takes place in a counter-direction;
- the amplitude of the wave decreases in time, and in the leading part of the wave in the direction of its motion a wave front of the opposite sign appears;
- the form of the wave, when switching on and switching off the source, depends upon the form and the location of the source;
- by successful choice of the moment of switching on and switching off the source, i.e. regulating the pulsed duration of the source, one can reach putting out or extension of switching on and switching off the waves (see Fig.4);
- the wave velocity is more where the temperature is higher.

5. CONCLUSION

This paper discusses the sequential development of the mathematical model for simulation of the macroprocesses of thermoelastic effects in the metal samples irradiated by pulsed beams. The future development of the model is related to the introduction of the factors modeling the plasticity effects of materials.

6. ACKNOWLEDGEMENTS

The work was supported by the Russian Foundation for Basic Research (grant 00-01-00617). The authors gratefully acknowledge Dr. T.A. Strizh and Mrs. M.V. Aristarkhova for the assistance.

REFERENCES

1. R.W. Stiimett et al., Thermal surface treatment using intense, pulsed ion beams, in: *Proc. Materials Research Society Symp.: Materials Synthesis arid Processing Using Ion Beams*, v.316, pp. 521-532 (Boston, 1994).

2. A.N. Didenko, A.E. Ligachev, and I.B. Kurakiri, The interaction of charged beams with the metalls and alloys surfaces (Moscow, Energoatomizdat, 1978), (in Russian).
3. S.A. Korenev, Pulse explosion ion vacuum condensation, *Preprint of JINR,* 12-89-615 (Dubna, 1989), (in Russian).
4. Yu.A. Bikovskiy, V.N. Nevolin, and V.Yu/ Fominskiy, Ion and laser implantation of metallic materials (Moscow, Energoatomizdat, 1991), (in Russian).
5. I. Yamada, Structures and dynamics of clusters, in: *Procs. of Yamada conference XLIII on structures and dynamics of clusters* (Universal Academy Press, Inc., Tokyo, Japan,1995).
6. D. F. Downey, M. Parley, K. S. Jones, and G. Ryding in: *Proc. Int. Conf. Ion Impl. Technol.-92* (Gainsville, North Holland, Amsterdam 1993).
7. A. .Zangwell, *Physics at Surfaces* (Cambridge University Press, New York, 1988).
8. S.A. Korenev, A.E. Ligachev, I.N. Meshkov, and V.I. Perevodchikov, in: *1st Int. Symp. "Beam technologies (BT'95)(* Dubna, 1995).
9. H. Hsieh, R. Averbach, H. Sellers, C.P. Flynn, Molecular-dynamics simulation of collisions between energetic clusters of atoms and metal substrates, *Phys. Rev.* B , vol. 45, No.8, 1992, pp. 4417-4431.
10. R. Ge, P.C. Clapp, J.A. Rifkin, Molecular dynamics of a molten Cu droplet spreading on a cold Cu substrate, *Surface Science* 426 (1999), L413-L419.
11. M.V. Altaisky et al., Fractal Structure Formation on on the Surface of Solids Subjected to High Intensity Electron and Ion Treatment. *JINR Rapid Communications*, No.2 [82] – 97, (Dubna, 1997), pp.37-46 (in Russian).
12. E.A. Airyan et al., Numerical Simulation of Thermal Treatment of Metal Surface by Means of High Current Ion Beam, *JINR Rapid Communications*, No.6{86] –97, (Dubna, 1997), pp.103-110 (in Russian).
13. I.V. Amirkhanov et al., Numerical Simulation of the Thermoelastic Effects in Metals Irradiated by Pulsed Ion Beams, *Comm. of JINR*, P11-2000-263, (Dubna, 2000), (in Russian).
14. Fractals in Physics, in: *VI Intern. Symp. on Fractals in Physics* (Eds. L. Pietronero and E. Fosatti, Russian trans. by J.G. Sinai and I.M.Khalatuikov), (Triest, Italy), 1985.
15. S.A. Korenev, A.J. Perry, *Vaccum*, 1996, p. 1089.
16. S.A. Korenev, *Preprint of JINR*, P13-94-192, (Dubna, 1994).
17. D. Oliver, *Fractal Vision: Put Fractals to Work for You*, (SAMS Publishing, Cornel, USA, 1992).
18. E.Feder, *Fractals* (Russian trans.), (Mir, Moscow, 1991).
19. The Tables of Physics Values, Ed. I.K. Kikoin, (Nauka, Moscow, 1976) (in Russian).
20. R. Rastov, Y. Maron, J. Mayer, *Phys. Rev.* B, 1985, vol..31, p..893.
21. R.G. Stinnett et al., in: *Proc. Materials Research Society Sypm.: Materials Synthesis and Processing Using Ion Beams*, (Boston, 1994), vol.316, p.521.
22. I.V. Amirkhanov et al., *Comm. of JINR*, P2-98-201, (Dubna, 1998).

VISUALISATION OF GRAND CHALLENGE DATA ON DISTRIBUTED SYSTEMS

M.V.Iakobovski, D.E.Karasev, P.S.Krinov, S.V.Polyakov*

1. INTRODUCTION

The problem of pictorial presentation of simulation data arises out of computational modeling of a wide range of scientific and applied problems. The problem of visualisation rises especially sharp while large-scale numerical experiments in three-dimensional simulation of fluid dynamics, combustion, microelectronics and some other problems on multiprocessor computing systems. At present practically there are no reasonable standard universal tools for visualisation of large size grid data.

Rapid growth of supercomputer centres, giving their resources to remote users through relatively slow networks, demands development of interactive network tools for visualisation of large size grid data. As a result of limitation of carrying capacity of local and global nets, passing of full simulation data through them is not reasonable and does not allow their interactive analysis. It should be noted that even when working through rather fast local network, big size of simulation data does not allow to use for visualisation directly personal computers of users. Thus, by now, absence of convenient interactive tools for evident representation of three-dimensional large size data is one of the major factors limiting use of high-performance multiprocessor systems for analysing of complex non-linear problems.

In what follows the system for three-dimensional stereoscopic visualisation of large size scalar data is described. The system is oriented at service of local and remote users of the super-computer centres. Visualisation program is divided into Server and Client parts (see Figure 1). This allows to carry out main data processing on computers of super-computer centres, so that the least data, required immediately for image preparation, is passed to the user's workplace. Such approach assumes that the image is finally formed at a user's workplace. In this case modern multimedia hardware (helmets, lr-glasses, three-

* M.V.Iakobovski, D.E.Karasev, P.S.Krinov, S.V.Polyakov, Institute for Mathematical Modelling of RAS, Moscow, Miusskaya sq., 4-A, 125047.

dimensional manipulators and etc.) can be used in order to increase the clearness of data presentation. Described visualisation system is not oriented to any concrete applied problem, and thus can be successfully used for analysis of simulation data for wide range of three-dimensional problems.

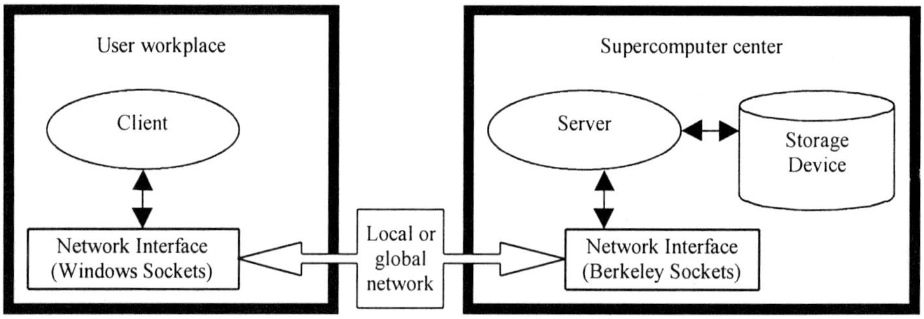

Figure 1. Structure of system of distributed visualization.

In present system three-dimensional scalar data visualisation is produced by means of iso-surfaces - surfaces, on which the function under study (temperature, density, cmass fractions etc.) has a fixed value. The problem of construction of each iso-surface is divided into following stages:

1. determination of iso-surface points;
2. search of a topology connecting these points;
3. compression of obtained surface;
4. construction of compressed iso-surface image[1] and its screen output at the user's workplace.

A considerable amount of operating memory is necessary for implementation of the first three stages, therefore the appropriate programs constitute a server part of the visualisation system, located on the multiprocessor computer system where the problem of lack of memory can be solved by addition of processors. The server part of system operates as TCP/IP server.

The experience of visualisation system has showed that number of points, describing iso-surface constructed at the third stage is comparable to number of points of original three-dimensional grid. And more over the ribs between iso-surface points are required. Taking this into account, total data size required for describing iso-surface, can exceed both size of original grid, and size of random-access memory of user's computer. That proves the necessity of the fourth stage – compression of constructed iso-surface.

The Client part of visualisation system operates on personal computer with Windows.

2. INPUT-OUTPUT OF LARGE SIZE GRID-FUNCTIONS

The large-scale numerical experiment and analysis of simulation data can't be carried out without tools for writing and quick readout of large size grid functions.

As a rule, hard disk space is insufficiency for storing of uncompressed data. The modern multiprocessor systems with performance over 1 Tflops, in principle allow numerical experiments with use of the billion nodes grids. However, the file size for only one function using single accuracy (4 bytes per number) is about 4Gbyte, that is unacceptable. Reasonable way out is the compression of grid functions before data storing. It is better to compress the data directly on each processor. For data intended for visualization lossy compression may be used, as the effect of losses on the visualization quality is invisible.

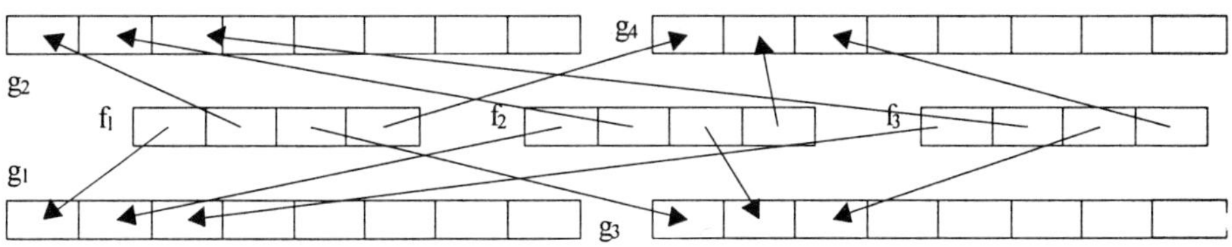

Figure 2. Reorder of float point array f_i bytes to bytes arays $g_{1\text{-}4}$.

Compression of real arrays using well known methods is ineffective. The considerable gain can be provided by reordering of binary representation of real numbers (see Figure 2). For smooth in space grid functions good compression of most of 4 formed arrays G can be obtained by means of libraries ZLIB[2] ((C) 1995-1998 Jean-loup Gailly and Mark Adler)). But an array with lower numbers in mantissas can hardly be compressed. Fortunately, it nearly has now effect on the quality of visualization and can be ignored.

Compression ratio for smooth functions (as in the problem of optical bistability concerned below) is within the limits from 5 to 7 for lossless compression with and is more than 20 for lossy compression, with relative error about 0.2% in the latter case.

3. COMPRESSION OF ISO-SURFACES FOR VISUALIZATION

The above compression algorithm ensures concise storing and quick readout within one computer complex, with fast channels. Obtained iso-surface constitute an unstructured grid. The major of date need for its describing are the numbers of neighbors of each points. This component of data cannot be lossy compressed, but known lossless compression methods[5] for this data are ineffective.

In this connection another approach is used in the visualizations system. The basic idea is to transfer instead of the iso-surface F, approximating surface G. Consider for the sake of simplicity a case, when iso-surface F is simply connected and single-valued function of two coordinates $F(x,y)$ (see Figure 3a). Then, all points, forming iso-surface can be divided into internal and boundary ones. Having removed all internal points and a part of boundary points (see Figure 3bc) we shall obtain basic reference points of our approximating surface G (see Figure 3c). We shall construct now a triangulation in the obtained coarsened area, using only coordinates (x,y) of basic reference points, leaving out of account values of functions $F(x,y)$. Further, with the help interpolation of function F we shall find values of function G in nodes of a new grid (see Figure 3d).

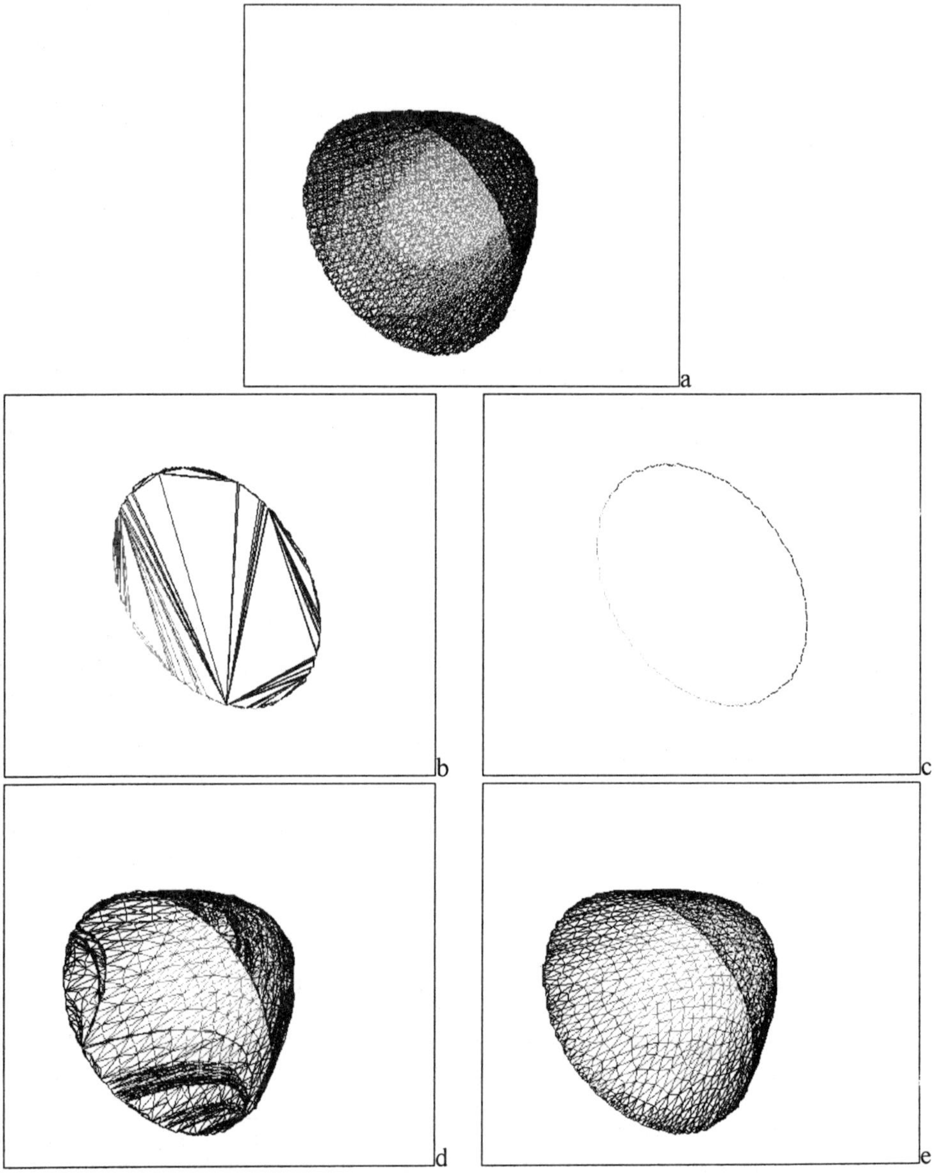

Figure 3. Steps of iso-surface compression.

Such approach gives double gain. Firstly, in surface *G* we can use much less points, than in iso-surface *F*. Secondly, there is no necessity to transfer both coordinates of all points of *G* triangulation, and appropriate topology. As it is possible to repeat construction of *G* triangulation, starting from reference points and connections between them, it is enough to transfer on Client basic reference points of grid *F* and the values of function *G(x,y)*, obtained on Server part of system.

It is easy to see, that quality of approximation of the initial surface essentially depends on the choice of triangulation algorithm. In the elementary case while construction of new triangulation the data on values of function *F* is not used at all (see the result of using simple fast method on Figure 3d or more accurate algorithm[6] on Figure 3e). In order to get more close approximation the data on grid refinement near added points should be transferred to Client. The refinement of grid *G* should be carried out in the area of big gradients of function *F(x,y)*. As *G* triangulation is constructed with the use of only the data on basic reference points of *F* and the data on grid refinement for already determined points of triangulation *G*, the process of grid refinement can be executed both on Server and on Client of system of visualization equally.

4. USER INTERFACE

In the visualisation system there is a facility of stereo scenes construction supported by appropriate stereo devices with help of OpenGL[2,3]. This allows to increase clearness of iso-surfaces presentation.

The visualisation system gives a possibility of any rotation of constructed iso-surface in the screen. Rotation can be executed with the help of standard computer mouse. On the one hand it is convenient, as mouse is a wide-spread device. On the other hand mouse is a two-dimensional device, so it can immediately change the orientation of observed object only in two directions. In order to simplify rotation control the device with three-degrees of freedom - tracker[4] - is used. This tools is capable to keep track change of the orientation in orthogonal axes. The tracker allows to establish the position of an object in screen. As a result, is possible to study the object image by rotating tracker.

5. APPLICATION OF PROPOSED VIDEO SYSTEM TO 3D SIMULATION OF OPTICAL BISTABILITY EFFECT IN SEMICONDUCTOR

The constructed program complex were probed successfully at analysis of 3D data obtained in computations of optical radiation propagation in nonlinear absorptive semiconductor crystal with bistability properties (see Figure 4). The selection of input radiation and semiconductor crystal parameters that initiate the formation of light beam contrast structures on the output crystal side is the main motivation for simulations. The contrast structure is the result of formation of high absorption domain into crystal. This domain is characterized by high concentration of free charge carriers. Under these conditions the maximum of absorption is reached in this domain (see Figure 5).

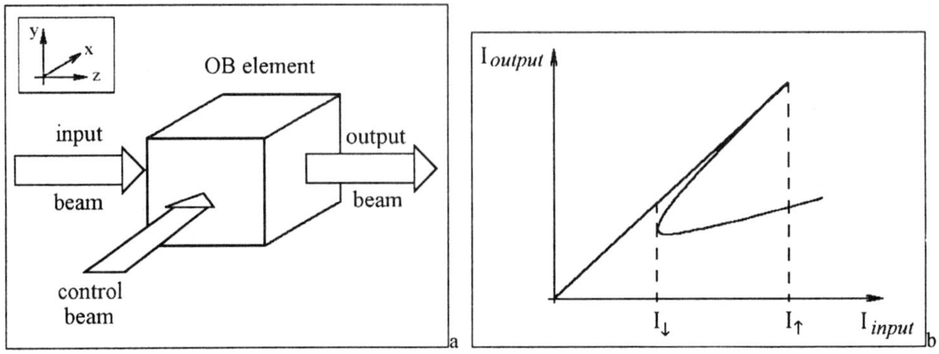

Figure 4. The scheme of optical memory element with two active light beams (a) and the optical bistability effect (b).

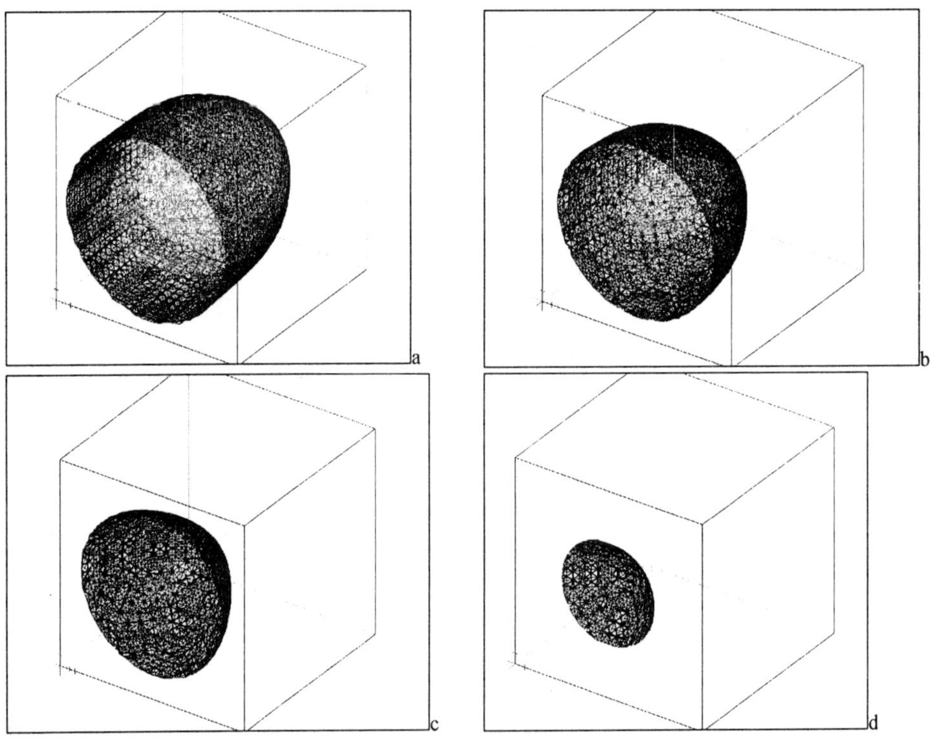

Figure 5. The iso-surfaces of charge carriers concentration n(x,y,z) = C in the high absorption domain for the case of low mobility of charge carriers and for different values of constant C = 0.15, 0.3, 0.45, 0.6 (a,b,c,d).

As the result, the light intensity passed through high absorption domain is lower essentially than output light intensity in transparent crystal zones. In practice this effect allows to realize the two stable states of crystal: logical zero (no high absorption domain into the crystal) and logical one (the domain exists). Any external action that transfers the crystal from state "0" to state "1" and back is named as "switching". The high absorption domain is localized usually in space of the crystal and it have very complex form (see Figures 5, 6). Perhaps, this form can be predicted easily (by experimentally or numerically), and we can to construct any space profiles of output light beams and to use its for the store of binary information. This physical principle is basis for the modern all-optical memory elements. The details of realization of all-optical memory and numerical analysis of the problems were discussed in Karamzin[7]. It is pointed out that in the case of full 3D numerical simulation of the optical bistability effect parallel computations are needed because in general case we wish to know both the stationary structure of output light beam and the switching dynamics.

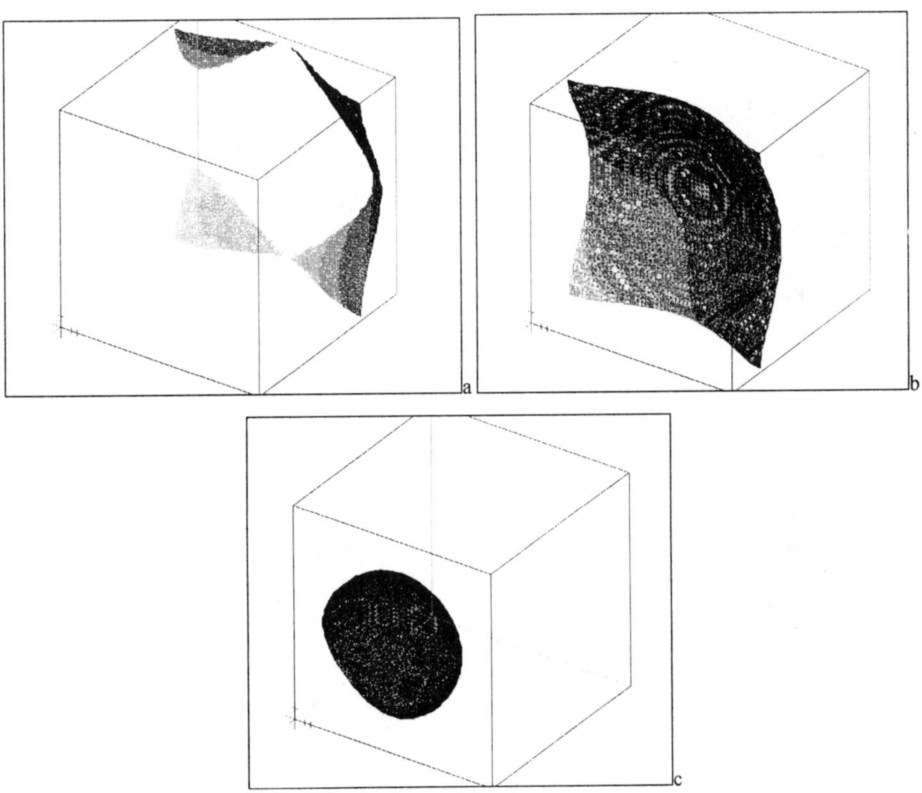

Figure 6. Iso-surfaces of charge carriers concentration n(x,y,z) = C in the high absorption domain for the case of high mobility of charge carriers and for different values of constant C = 0.15, 0.25, 0.35 (a,b,c).

We have tested proposed program complex on the data computed for described problem. In these tests we have used the property of the complex that is video-analysis of numerical data after many computations on the multiprocessor computer system. At the testing we had the possibility to view the distributed numerical data fields and select some interest results for detail analysis. It is needed to say that space grid in computations consists of about 10^6 nodes and more. Typical information size processed by complex was equal to $8*n$ Mbytes (where $n \approx 1000$ is the 3D image number) for each computation. The using of this high performance visualization system it was allow accelerate the finding of some optimal switching regimes that have low energy characteristics.

6. ACKNOWLEDGEMENTS

The work is supported by Russian Foundation for Basic Research (grants No. 99-01-01036) and purpose program "Integration" (project No. A0106).

REFERENCES

1. Tihomirov Y.V., *Programming of three-dimensional graphics*, BHV, Saint Petersburg, 1998, 256p. (in Russian)
2. http://www.info-zip.org/pub/infozip/zlib/
3. I.Tarasov. *Bases of OpenGL*, http://www.opengl.org, http://nehe.gamedev.net/ (in Russian)
4. http://www.isense.com/products/pro/index.htm
5. *Formats of graphic files* – K.:NIPF «DiaSoft Ltd.», 1995.-480p. (in Russian)
6. Thir A.V. Displaced front method for 2D unstructed grids, in: *Numerical methods and applications* / (Ed. by U.A. Kuznetcov). Institute of computational mathematics of RAS, 1995 (in Russian).
7. Yu.N. Karamzin, T.A. Kudryashova, S.V. Polyakov and I.G. Zakharova. *Simulation of 3D absorption optical bistability problems on multiprocessor computer systems*. In "Fundamental physics and mathematics problems and simulation of technique and technology systems", N. 2, (Ed. by L.A. Uvarova), pp. 117-124. Published by MGTU "STANKIN", Moscow, 1999 (in Russian).

SIMULATION OF ELECTRON TRANSPORT IN SEMICONDUCTOR MICROSTRUCTURES: FIELD EMISSION FROM NANOTIP

V.A. Fedirko[1], Yu.N. Karamzin, and S.V. Polyakov[2]

1. INTRODUCTION

Semiconductor submicron structures and nanostructures are now the basis of modern solid state electronics. Numerical simulation of their static and dynamic characteristic is an actual practical problem and adequate mathematical modeling of electron transport in these structures based on clear understanding of its fundamental physical peculiarities is vital for successful development of submicron devices.

The size of an active space in such micro-devices is comparable to a charge carrier characteristic free path and the electric field in the space due to its small size is very high for actual bias. For that reason electron transport in active domain is usually quite non-equilibrium one and widely used drift-diffusion model, DDM (see, e. g.[1]), fails to be adequate. In this paper we report on development of quasi-hydrodynamic approach to strongly non-equilibrium electron transport in semiconductor in application to modeling of field emission from semiconductor microcathode.

Quasi-hydrodynamic model is physically based on inequalities (see, e. g.[2]):

$$\tau_p < \tau_{ee} < \tau_\varepsilon , \tag{1.1}$$

between the characteristic times of free charge carrier system in the active space of semiconductor structure. Here τ_p, τ_ε are momentum and energy relaxation times respectively, and τ_{ee}^{-1} is electron-electron scattering rate. The inequalities (1.1) results in almost maxwellian form (or Fermi-type) of the symmetrical part of charge carriers distribution

[1] V.A.Fedirko, Moscow State University of Technology "STANKIN", Moscow, Vadkovskii per., 3-A, 101472.

[2] Yu.N.Karamzin, S.V.Polyakov, Institute for Mathematical Modelling of RAS, Moscow, Miusskaya sq., 4-A, 125047.

function with non-equilibrium temperature T_e. That enables in case of an elastic momentum scattering to describe the anti-symmetrical part of the distribution function in terms of two variables: local charge carrier density and their local temperature - e. g. electron concentration $n(\boldsymbol{r},t)$ and electron temperature $T_e(\boldsymbol{r},t)$ for unipolar electron semiconductor. The latter obey the continuity equation and the energy balance equation closed with the material equations for the current and energy flux with concentration and temperature included into the corresponding kinetic coefficients.

In the conditions of strong electron heating the process of impact ionization becomes very important. In this case hole conductivity is strongly different from electron conductivity, and the semiconductor exhibits bipolar properties. In this situation the continuity equation for hole concentration $p(\mathbf{r},t)$ is added to the model.

The Poisson equation for the self-consistent electric field $\mathbf{E}(\mathbf{r},t)$ and corresponding boundary and initial conditions close the model.

The resulting 3D evolutionary differential problem is strongly nonlinear one. The non-linearity is taken place both in equations and boundary conditions. The numerical analysis of the problem in quasi-classical approximation and possible applications to silicon field emitter simulation are considered in the work.

2. DIFFERENTIAL PROBLEM AND NUMERICAL SCHEMES

In general case the discussed differential problem can be written by following dimensionless quasi-classical deterministic equations:

$$\frac{\partial n}{\partial t} = div\,\boldsymbol{j}_e + G(n,p,T_e) - R(n,p,T_e), \tag{2.1}$$

$$\frac{\partial p}{\partial t} = -div\,\boldsymbol{j}_h + G(n,p,T_e) - R(n,p,T_e), \tag{2.2}$$

$$\frac{\partial w_e}{\partial t} = -div\,\boldsymbol{Q}_e + \widetilde{G}_e(n,p,T_e,\boldsymbol{E}) - \widetilde{R}_e(n,p,T_e), \tag{2.3}$$

$$-div(\kappa \boldsymbol{E}) = 4\pi(n - p - N_D), \tag{2.4}$$

where

$$\mathbf{j}_e = n\mu_e(T_e)\mathbf{E} + \nabla(D_e(T_e)n), \quad \mathbf{j}_h = -p\mu_h(T_e)\mathbf{E} - \nabla(D_h(T_e)p),$$

$$\mathbf{Q}_e = -n\widetilde{\mu}_e(T_e)\mathbf{E} - \nabla(\widetilde{D}_e(T_e)n), \quad w_e = \frac{3}{2}nT_e, \quad \mathbf{E} = -\nabla\varphi.$$

Here w_e is the local electron energy density, $\mathbf{j}_e$, $\mathbf{j}_h$ are electron and hole current densities, G and R are generation and recombination terms, $\mathbf{Q}_e$ is the electron energy flux, $\widetilde{G}_e$ and $\widetilde{R}_e$ are generation and relaxation energy rates, φ is the electric field potential, N_D is the effective donor concentration. The μ_j, D_j, and $\widetilde{\mu}_j = \chi\mu_j$, $\widetilde{D}_j = \chi D_j$ $(j = e, h)$ are the kinetic coefficients for the corresponding transport equations: μ_e, μ_h are respectively the electron and hole mobility, D_e, D_h are the diffusion coefficients, χ is the Peltier coefficient. The constant $\varkappa$ is static dielectric permittivity of the material. The div, ∇ are standard differential operators determined in the domain Ω, time $t > 0$.

The boundary and initial conditions can be written in form

$$(\mathbf{j}_e \cdot \mathbf{v})\big|_{\partial\Omega} = j_{es}(n, T_e, \mathbf{E}), \quad (\mathbf{j}_h \cdot \mathbf{v})\big|_{\partial\Omega} = j_{hs}(p, T_e, \mathbf{E}), \tag{2.5}$$

$$(\mathbf{Q}_e \cdot \mathbf{v})\big|_{\partial\Omega} = Q_{es}(n, T_e, \mathbf{E}), \quad (\mathbf{E} \cdot \mathbf{v})\big|_{\partial\Omega} = E_s, \tag{2.6}$$

$$n\big|_{t=0} = n_0, \quad p\big|_{t=0} = p_0, \quad T_e\big|_{t=0} = T_0, \quad \mathbf{E}\big|_{t=0} = 0, \tag{2.7}$$

where $\partial\Omega$ is the domain bound, $\mathbf{v}$ is the normal to $\partial\Omega$, n_0 and p_0 are equilibrium distribution of electron and hole concentrations, T_0 is the environment temperature.

The numerical realization of proposed difference problem is very complex computational problem. We consider the simple geometry case when Ω is rectangular domain. In the case to solve the problem (2.1)-(2.7) we can use the original additive difference method proposed in[3].

To solve transport equations (2.1)-(2.3) we use the finite difference schemes that include the splitting and smoothing procedures and guarantees the conservativity and weak monotonisity of the solution. For grid analogue of the Poisson equation with discontinuous dielectric permittivity coefficient $\varkappa$ the standard approximation is used. For the solution of constructed non-linear finite difference schemes we employ the uniform external iterative process. The finite difference flow equations and Poisson equations are solved with the help of independent internal iterative procedures based on the Lantzosh and Cholessky factorizations and gradient method.

The parallel algorithm is realized through Schwartz method when calculation domain Ω_h is divided on the rectangular sub-domains $\Omega_h^{(k)}$ corresponding to the k-th processor. The boundary conditions of desired functions in $\Omega_h^{(k)}$ for each external iteration are taken from preceded iteration obtained on the adjacent processor. This results in the independent processor work. The resulting algorithm efficiency is thus determined by load-balancing and data exchange between the processors. We have shown that for the small number of processors n_p ($n_p << N_x N_y N_z$ that is typical situation for real computaputations) and about 5-7 nodes sub-domains overlapping the 80-90 % efficiency can be achieved.

3. SIMULATION OF SILICON FIELD MICROEMITTER

In recent years vacuum microelectronics (VM) based on field emitter array[4] (FEA) concept has experienced tremendous growth. VM devices benefit consists in faster modulation and higher electron energies. These properties are achieved by using solid-state structures. In addition VM devices can operate in a wide temperature range, 4K<T<1000K, and in high-radiation environment. Applications include FEA flat-panel displays, microwave amplifiers, digital IC, electron and ion guns, sensors, high energy accelerators, electron beam lithography, free electron lasers, and electron microscopes and microprobes. Innovative approaches to FEA's were made in latest years with respect to FEA materials, structure, and applications. A number of VM devices have by now moved beyond the research laboratories to actual prototypes and commercial products.

Now silicon is considered as one of the most suitable material for FEA fabricating in batch technology (see, e.g.[5]). Silicon, though has orders of magnitude fewer conduction band electrons than metals, has emitting characteristics comparable to metals, and highly developed silicon batch technology can be applied to make various VM devices, including transistor-like structures.

There are many specific aspects and special requirements for VM, and comprehension of physics of VM devices functioning play a key role in its successful development. One of the most important problem is the creation of FEA with sufficiently high, controlled, and stable emission ability. Field emission from semiconductors has some peculiar features vital for successful development of VM devices.

High electric field penetrates deep enough into the semiconductor and results in intense electron heating near the emitting surface. Since the tunneling coefficient exponentially depends on energy, this drastically affects the emission characteristics and heat dissipation. Thus the electron transport in semiconductor field emitter is in fact highly non-equilibrium hot electron process. It was well demonstrated in[6,7] for one-dimensional model. Electrons with the energy higher than the semiconductor band gap contribute essentially to the emission current. So impact ionization can play an important role and should be taken into account. Besides, the quasi-electrostatic problem in a microcell space outside the cathode must be solved self-consistently to obtain current-voltage characteristics.

For the above reasons two-dimensional (2D) approach is principal for real cathodes microcell modeling[8-10]. It is an actual physical and practical problem that is rather complicated and can be solved accurately enough by means of numerical methods only.

3.1. MODEL

We use quasi-hydrodynamic model (2.1)-(2.7) to describe hot electron transport inside the semiconductor cathode body. In the model the some terms have the following concrete form:

$$G(n,p,T_e) = G(n,T_e), \quad R(n,p,T_e) = 0, \tag{3.1}$$

$$\widetilde{G}(n,p,T_e,\mathbf{E}) = \mathbf{j}_e \cdot \mathbf{E}, \quad \widetilde{R}(n,p,T_e) = \gamma \cdot G(n,T_e) + \frac{3}{2} n \frac{T_e - T_0}{\tau_\varepsilon}. \tag{3.2}$$

The electron temperature T_e is supposed to be much higher than the lattice temperature T_0 and we put T_0 be room temperature in our calculations. Relaxation time approximation for lattice relaxation term in electron energy balance equation is admitted[6,7] with the characteristic relaxation time τ_ε. We add the relevant impact ionization terms[11] - $G(n,T_e)$ into the electron continuity equation and $\gamma \cdot G(n,T_e)$ into the electron energy balance equation. The Einstein relation between $\mu_{e,h}$ and $D_{e,h}$ are supposed to be valid and the electron temperature dependence of electron mobility is approximated by the empirical relation[6,7]

$$\mu_j(T_e) = \mu_{0j}\left[1 + \beta(T_e - T_0)\right]^{-1} \tag{3.3}$$

where μ_{0j} are the low-field mobility coefficients. We do not take the hole heating into account because holes do not contribute to the tunneling current. However there is no principal problem except increasing the computational time to include hot-hole effects in the model using the same quasi-hydrodynamic approach.

An example of typical real FEA microcell cross-section geometry is shown schematically in Fig. 1. 2D approach is quite reasonable for the wedge-type cathode as its size in z-direction is usually much greater than in an its (x-y)-cross-section.

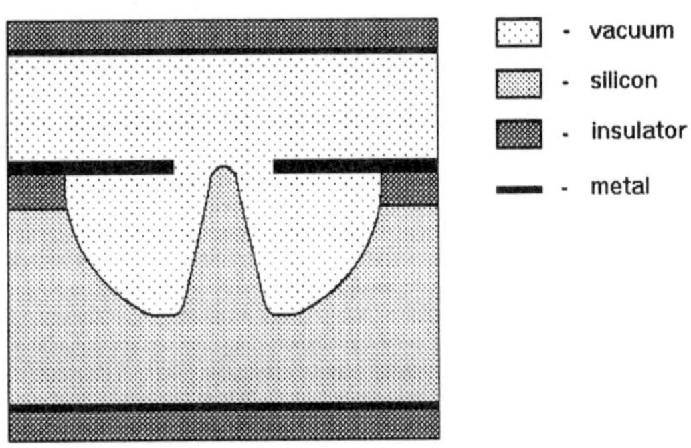

Figure 1. The typical real FEA microcell cross-section.

For typical FEA operation condition the electron space charge can be neglected outside of the cathode. So in that domain only Laplas equation is to be solved self-consistently with the equations (2.1)-(2.3).

The computation domain is shown in Fig. 2. It is a half of a cell for symmetry reason. For that domain we put zero normal components of $\mathbf{j}_e$, $\mathbf{j}_h$, $\mathbf{Q}_e$, $\mathbf{E}$ on lateral surfaces, $x = 0$ and $x = L_x$, and electrostatic potential $\varphi = 0$ at the back surface ($y = 0$), $\varphi = V_g$ at the grid plates, and $\varphi = V$ at the anode plate. The standard boundary condition are used for the electric field at the cathode surface, so the normal component of an electric field at the surface inside semiconductor is found as E_s / κ. At the emitting surface of a cathode tunneling boundary conditions[7,9] are admitted for normal components of $\mathbf{j}_e$ and $\mathbf{Q}_e$. As in [9,10] we used well-known model of quasi-classical dependence for energy dependent electron tunneling coefficient, and put the hole tunneling current to zero.

On the backside surface ($x = L_x$) Ohm's relation between j_{es} and E_s and corresponding relation between Q_{es} and E_s are supposed. The normal component of the hole current j_{hs} is set to be controlled by surface recombination with high recombination velocity *s*:

$$j_s = s(p - p_0). \tag{3.4}$$

The equilibrium values n_0, p_0, T_0 are taken as initial conditions for n, p and T_e.

3.2. RESULT AND DISCUSSION

Our main aim in this work part was to study numerically physical aspects of hot electron transport inside semiconductor. The realization of the above model for the calculation domain with non-rectangular geometry of Fig. 2 is rather time-consuming computational problem. It is however physically obvious that due to the wedge form of a microcathode surface electric field distribution on the emitting surface is highly inhomogeneous with its peak at the apex of the cathode tip. So we simplified the problem by introducing model dependence of an electric field $E_y(x)$ on the external side of the emitting surface for the rectangular domain ABCD (Fig. 3) with an alternating rectangular mesh.

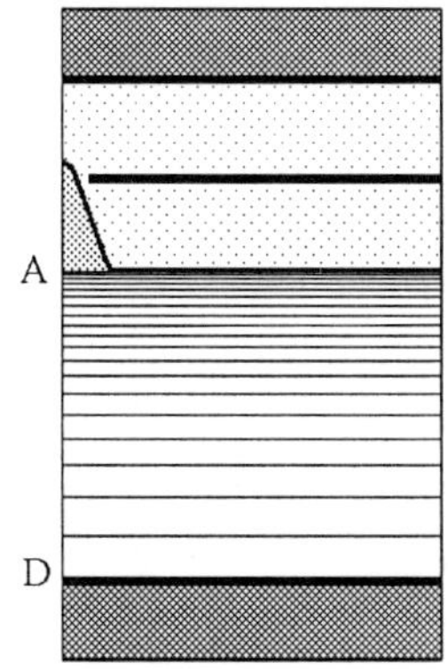

Figure 2. The calculation domain ABCD with rectangular alternating grid for model surface field.

We take $E_y(x)$ in the form alike that at a conducting wedge surface:

$$E_y(x) = E_m f(x), \quad f(x) = \begin{cases} 1, & x \leq \rho, \\ \sqrt{\dfrac{\rho}{x}}, & x > \rho, \end{cases} \tag{3.5}$$

where ρ is the tip radius. Parallel processing with relevant parallel algorithm has been applied.

We calculated 2D steady-state distributions of the electric field, charge carriers density and electron temperature in the semiconductor cathode using the following typical values of silicon parameters: $N_D = 10^{18}$ cm^{-3}, $\tau_\varepsilon = 0.3$ ps, $\beta = 9 \times 10^{-4}$ K^{-1}, $\rho = 10$ nm, $s = 10^4$ cm/s, $\gamma = 7E_g/6$, E_g being the energy gap.

High non-homogeneous surface electric field on the emitting surface results in high spatial non-uniform distributions of the electric field inside semiconductor cathode, which causes strong electron heating near the emitting surface and charge carriers redistribution near the emitting surface.

The local electron temperature at the cathode face increases monotonously with the increase of a surface electric field (Fig. 3a) while the electron concentration changes non-monotonously (Fig. 3b). That can be explained by de-localization of hot electrons from the potential well near the surface when their temperature increases. Near the backside of the cathode ($y = 0$) the electric field and current density are practically homogeneous and obey Ohm's relation.

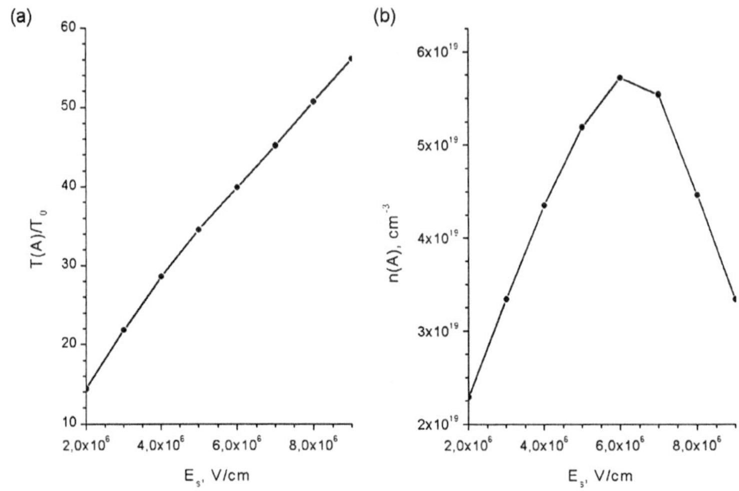

Figure 3. The dependencies of electron temperature (a) and electron concentration (b) at the cathode face from applied electric field.

Electron heating manifests itself pronouncedly in emission current energy distribution. Our theoretical estimate shows that the peak of emitted electrons energy distribution shifts with an increase of surface electric field to as high energy as compared with electron affinity in semiconductor (about 4 eV for silicon) within very narrow range of electric field values (near $\approx 10^6$ V/cm for silicon). This is well seen in fig. 5 and is in agreement with the experimental results of Gray and Shaw[12].

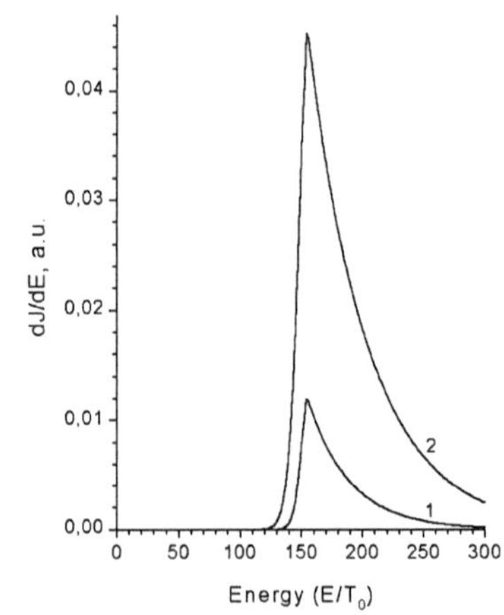

Figure 4. The energy distribution of emitted electron current.
1 - $E_m = 5\text{x}10^6$ V/cm, 2 - $E_m = 8\text{x}10^6$ V/cm.

So, for actual applications the electron field emission is dominated by hot electrons, and impact ionization plays important role in electron transport in semiconductor field emitter. Our numerical calculation shows that impact ionization alongside with the electron phonon scattering control electron heating - and thus electron emission current.

In Fig. 5 calculated electrical potential, electron temperature and charge carrier spatial distributions density are shown.

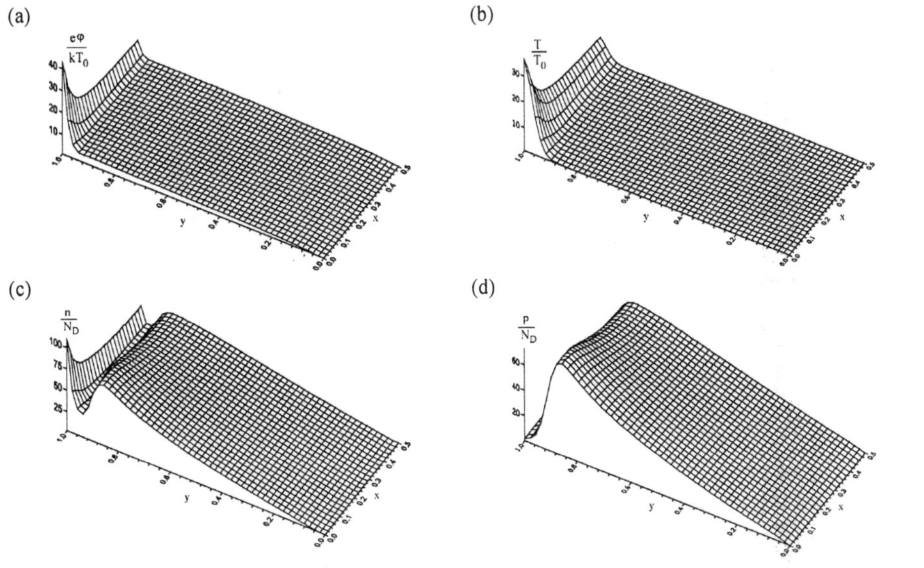

Figure 5. The electric potential (a), electron temperature (b), electron (c) and hole (d) densities distributions near the emitting surface inside the semiconductor cathode.

In Fig. 6 steady-state distributions of the electric field, electron density and electron temperature in the cathode calculated without impact ionization terms are shown for comparison.

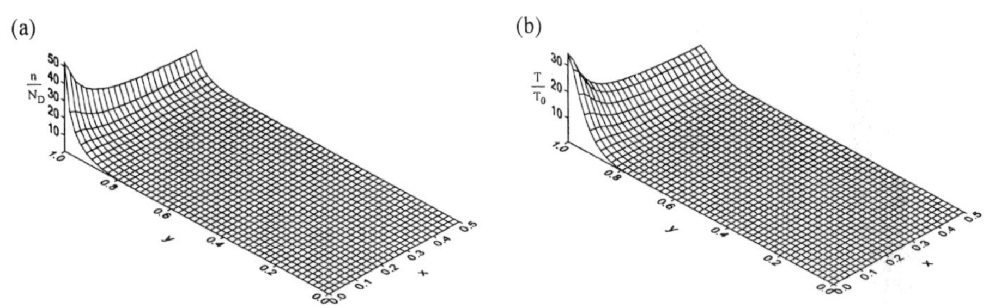

Figure 6. Electron density (a) and electron temperature (b) distributions near the emitting surface inside the semiconductor cathode calculated without impact ionization.

One can easily see that for electric field higher than 10^6 V/cm impact ionization contribute markedly to the electron transport in a silicon field emitter. It limits the infinite electron heating while the electron concentration gets higher.

Strong electric field is concentrated in submicron domain near the apex of the cathode tip where electron concentration and electron temperature are much higher then their equilibrium values. The intense electron heating and impact ionization near the emitting surface results in a high emission current density due to the exponential increase of the electron tunneling coefficient with electron energy. Emission current density distribution over the emitting surface of the cathode is shown in Fig. 7 together with surface electric field distribution. The calculated emission current dependence on external electric field shown in Fig. 5 (a, b).

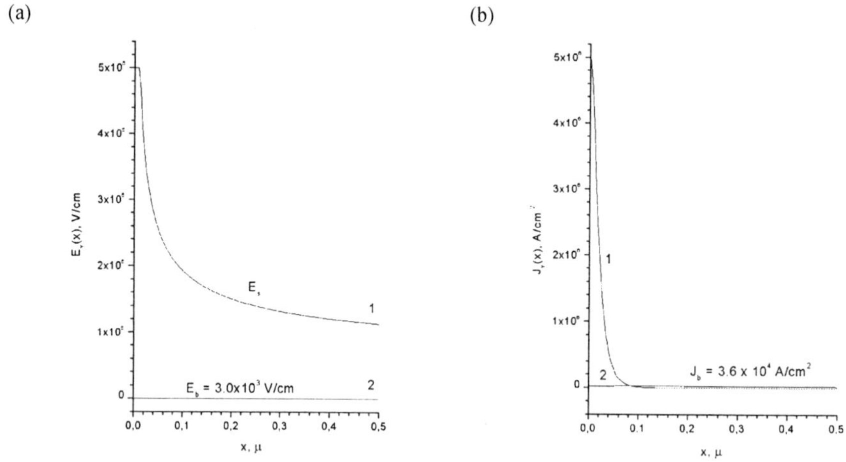

Figure 7. Electric field (a) and emission current density distribution (b) over the emitting surface (1) and over the back surface (2).

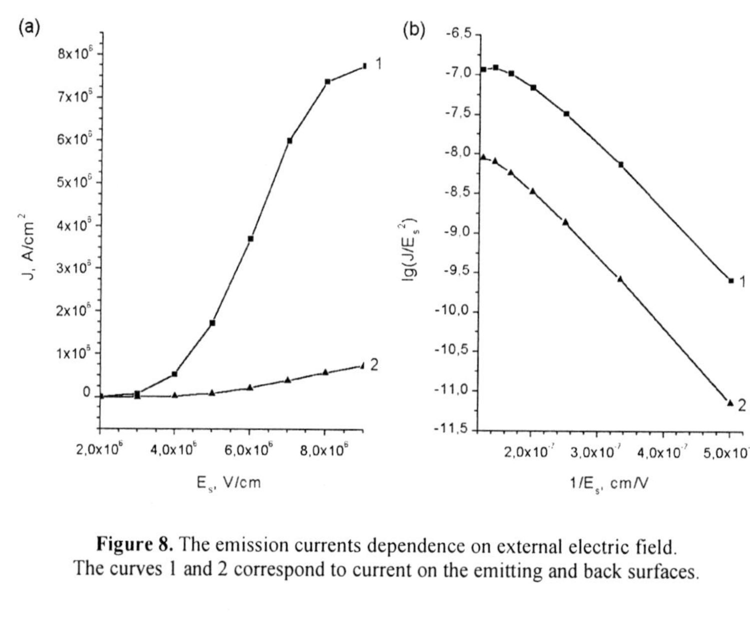

Figure 8. The emission currents dependence on external electric field. The curves 1 and 2 correspond to current on the emitting and back surfaces.

We emphasize that the cold electron emission would be negligibly small for the electric fields under consideration. Thus electron heating and impact ionization linked with it essentially affect the emission from semiconductor microcathode and current-voltage characteristics of a microcell.

4. CONCLUSION

We have considered the quasi-hydrodynamic approach to solving of semiconductor problems under the conditions of strong charge carriers heating. We have elaborated physical and mathematical model for full 3D numerical simulation of entire field emitter microcell with semiconductor microcathode and have worked out numerical algorithm for highly efficient parallel processing. We have shown that electron heating drastically affects field emission from silicon microcathode. For actual surface electric field hot electrons with the energy compared to the electron affinity dominate the emission. The impact ionization contribute markedly to electron transport and to emission current. Heavy local electron heating may also result in cathode tip instability due to intensive energy exchange between the electron gas and the lattice.

5. ACKNOWLEDGEMENT

The work is supported by the Russian Basic Research Foundation (RBRF), Projects No. 00-01-00397, 99-07-90388.

REFERENCES

1. S.Sellberherr, Analysis and simulation of semiconductor devices, Springer, Wienn, 1984.
2. R.Stratton, *Phys. Rev.*, **26**, N 6, 1962, p.2002.
3. Yu.N.Karamzin, I.G.Zakharova, New additive difference method for solving semiconductor problems, *Russ. J. Numer. Anal. Math. Modelling*, **11**, N 6, 1996, p. 477-485.
4. K.R.Shoulders and L.N.Heynick, *US patent 3*, 1966, **453**, p. 478.
5. H.F.Gray. *Techn. Dig. of the 11th IVMC'98*, 1998, p. 278.
6. V.A.Fedirko, N.A.Duzhev, and V.A.Nikolaeva, *Suppl. a la Revue "Le Vide, les Coushes Minces"* (papers from the 7th IVMC'94), N **271,** 1994, p. 158.
7. V.A.Fedirko and V.A.Nikolaeva, Mathematical Modelling (Russ.) ,**9**, N 9, 1997, p. 75.
8. V.A.Fedirko, S.V.Polyakov, Yu.N.Karamzin, and I.G.Zakharova, In: *Techn. Dig. of the 12th Intern. Vacuum Microelectronics Conf. - IVMC'99 (6 - 9 July, 1999, Darmstadt, Germany)*, 1999, p. 100-101.
9. V.Fedirko, S.Polyakov, In the book: *"Mathematical Models of Non-linear Excitation, Transfer, Dynamics and Control in Condensed systems and Other Media"* (ed. by L.A.Uvarova et al.), Kluwer Academic/Plenum Publishers, New York, 1999, p. 221-228.
10. V.Fedirko, Yu.Karamzin, S.Polyakov and I.Zakharova, In the book: *"Recent Advances in Numerical Methods and Applications"* (Proc. of 4th Intern. Conf. NMA, Sofia, Bulgaria, Aug. 19-23, 1998), World Scientific, Singapore/N-Jersey/London/H-Kong, 1999, p. 890-897.
11. S.Blakemore, Semiconductor Statistics, Pergamon Press, 1962.
12. J.L.Shaw and H.F.Gray, *Techn. Dig. of the 11th IVMC'98*, 1998, p. 146.

RELIABLE COMPUTING EXPERIMENT IN THE STUDY OF GENERALIZED CONTROLLABILITY OF LINEAR FUNCTIONAL DIFFERENTIAL SYSTEMS

Vladimir P. Maksimov and Aleksandr N. Rumyantsev*

1. INTRODUCTION

The basis of the constructive (computer-assisted) study of linear problems in the theory of FDE is a special technique of approximate description of the set of solutions to the linear FDE under study in combination with a guaranteed explicit error bound for the approximation. This technique is used in parallel with special theorems, conditions of which can be verified with use of *the reliable computing experiment* due to the modern mathematical packages (e.g., Maple, Mathematica). Presently the theoretical ground and a technology of the reliable computing experiment are worked out [1–4] as applied to studying linear boundary value problems of FDEs for the unique solvability, linear functional differential control systems for controllability, linear FDEs with delay and periodic parameters for stability. Notice that sometimes, when known traditional sufficient conditions for a property under consideration (say, the unique solvability of a boundary value problem) are inapplicable, the computer-assisted study can give the only chance to obtain the result.

The questions of theoretical validating the computer-assisted study of various classes of equations (ordinary differential, integral equations) occupy an important place in the current literature (see, for instance, [5–7]).

In this paper we consider some questions of the theoretical ground and the practical implementation of the computer-assisted study approach as applied to a generalization of the classical control problem.

* Vladimir P. Maksimov, Perm State University, Perm, Russia 614600. Aleksandr N. Rumyantsev, Perm State University, Perm, Russia 614600.

2. CONTROL PROBLEM

Consider the functional differential control system

$$(\mathrm{L}x)(t)=(Bu)(t)+v(t),\ t\in[0,T] \tag{1}$$

with linear bounded operators $\mathrm{L}:D\to L$, $B:L_2\to L$. Here L denotes the space of summable functions $z:[0,T]\to R^n$,

$$\|z\|_L=\int_0^T|z(s)|\,ds$$

($|\cdot|$ is a norm in R^n, the linear space of n-columns); D denotes the state space of absolutely continuous functions $x:[0,T]\to R^n, \|x\|_D=\|\dot{x}\|_L+|x(0)|$; L_2 is the control space of square-summable functions $u:[0,T]\to R^r$ with the inner product

$$(u_1,u_2)_{L_2}=\int_0^T u_1^{\cdot}(s)u_2(s)ds$$

($\cdot$ is the symbol of transposition). We assume the principal part of L [1-4], i.e. the operator $Q:L\to L$ defined by $Qz=\mathrm{L}(\int_0^{(\cdot)}z(s)ds)$, to be of the form

$$(Qz)(t)=z(t)-\int_0^t K(t,s)z(s)ds.$$

Here the elements $k^{ij}(t,s)$ of the kernel $K(t,s)$ are measurable on $\{(t,s):0\le s\le t\le T\}$ and satisfy the estimates $|k^{ij}(t,s)|\le m(t),\ i,j=1,\dots,n$ with an summable on [0,T] function m. In this case,

$$(Q^{-1}z)(t)=z(t)+\int_0^t H(t,s)z(s)ds,$$

where $H(t,s)$ is the resolvent kernel to the kernel $K(t,s)$.

The system Eq. (1) covers various classes of linear control systems including the systems with distributed and concentrated delay and integrodifferential system. In the classical control problem one needs to find a control u, that drives system Eq. (1) from a given initial state $x(0)=\alpha$ to a desired terminal state $x(T)=\beta$, i.e. to find $u\in L_2$ such that the boundary value problem

$$\mathrm{L}x=Bu+v,\ x(0)=\alpha,\ x(T)=\beta,$$

has a solution x_u. This solution is unique. In cases of ordinary differential systems and delay differential systems, the control problem is a subject of wide literature (see, for example, the book [8] and its references). Here we consider a more general control problem

$$\mathrm{L}x=Bu+v,\ x(0)=\alpha,\ \ell x=\beta, \tag{2}$$

where the aim of control is given by a linear bounded vector functional $\ell:D\to R^n$. The control problem of such a kind arises, in particular, in Economic Dynamics, where the aim of control can be formulated as the attainment of a given level, β, of certain characteristic of trajectory $\mathbf{x}$. For example, in the case the model Eq. (1) governs the output dynamics (in modeling the control procedure of the firm) the condition

$$\ell x \equiv \int_0^T \exp(-\mu t)dt = \beta$$

prescribes the so-called discounted integral output with a discount coefficient μ. The control problems in Economic Dynamics are studied in detail in [9].

2. CONDITION OF CONTROLLABILITY

The necessary and sufficient conditions for the solvability of problem Eq. (2) as well as the construction of the corresponding control procedures can be described in terms of the so-called Cauchy matrix, $C(t,s)$. This matrix gives the representation

$$x(t) = X(t)\alpha + \int_0^t C(t,s)f(s)ds \tag{3}$$

of the solution $x(t)$ to the Cauchy problem

$$\mathrm{L}x = f, \quad x(0) = \alpha$$

for any $f \in L,\ \alpha \in R^n$. In Eq. (3) $X(t)$ is the fundamental matrix of the solutions to the gomogeneous equation $\mathrm{L}\mathbf{x} = 0$ such that $X(0) = E,\quad E$ is the identity $n \times n$ matrix. Note, that

$$C(t,s) = E + \int_s^t H(\tau,s)d\tau$$

with the resolvent kernel $H(t,s)$, and

$$X(t) = E - \int_0^t C(t,s)(\mathrm{L}E)(s)ds\ .$$

The vector-functional ℓ in Eq. (2) has the representation

$$\ell x = \Psi x(0) + \int_0^T \Phi(\tau)\dot{x}(\tau)d\tau,$$

where $\Psi = (\ell e_1,\ldots,\ell e_n)$, e_i is the ith column of E; $n \times n$ matrix $\Phi(\tau)$ with essentially bounded measurable on [0,T] elements is defined by

$$\ell(\int_0^{(\cdot)} z(s)ds) = \int_0^T \Phi(s)z(s)ds \quad \forall z \in L\ .$$

Denote

$$\theta(s) = \Phi(s) + \int_s^T \Phi(\tau)C'_\tau(\tau,s)d\tau$$

and define $n \times n$ matrix M by the equality

$$M = \int_0^T [B^*\theta](s)[B^*\theta]^*(s)ds,$$

where $B^* : (L)^* \to (L_2)^*$ is the adjoint operator to $B : L_2 \to L$.

The invertibility of M is the necessary and sufficient condition for the solvability of

Eq. (2) for any $v \in L, \alpha, \beta \in R^n$ (ℓ-controllability of control system Eq. (1)). The control

$$\bar{u} = \left[B^*\theta\right]^* M^{-1}\gamma \tag{4}$$

with

$$\gamma = \left[\beta - \Psi\alpha - \int_0^T \Phi(\tau)X(\tau)d\tau \cdot \alpha - \int_0^T \theta(\tau)v(\tau)d\tau\right] \tag{5}$$

solves problem Eq. (2) and has the minimal norm $\|\bar{u}\|_{L_2} = \sqrt{(\bar{u},\bar{u})_{L_2}}$ among all controls that solve Eq. (2).

4. MAIN RELATIONSHIPS

The condition

$$\det \mathrm{M} \neq 0 \tag{6}$$

can be examined using computer calculations. Really, Eq. (6) holds if an invertible matrix $\tilde{M}$ is constructed such that

$$\|\mathrm{M} - \tilde{\mathrm{M}}\| < \frac{1}{\|\tilde{\mathrm{M}}^{-1}\|}. \tag{7}$$

It is appropriate to look for such an $\tilde{\mathrm{M}}$ among matrices of the form

$$\tilde{\mathrm{M}} = \int_0^T \left[B^*\tilde{\theta}\right](s)\left[B^*\tilde{\theta}\right]^*(s)\,ds,$$

where

$$\tilde{\theta}(s) = \Phi(s) + \int_s^T \Phi(\tau)\tilde{C}'_\tau(\tau,s)\,d\tau,$$

and $\tilde{C}(t,s)$ is a sufficiently accurate approximation of the Cauchy matrix $C(t,s)$. The reliable computing experiment to examine Eq. (6) consist in the construction and the successive refinement of approximations $\tilde{C}, \tilde{\theta}, \tilde{\mathrm{M}}$ with reliable error bounds, and in the immediate examination whether inequality Eq. (7) holds. As is shown in [1, 10], with the appropriate technology of constructing both $\tilde{C}(t,s)$ and its error bound, the property of ℓ-controllability can be established rigorously in the above way for any control system Eq. (1) that has this property. Here we dwell in more detail on the main part of the computing experiment. This part consists in the construction of $\tilde{C}$ and its reliable error bound. We shall describe an efficient computer-aided technique of constructing $\tilde{C}$ under the condition that the kernel $K(t,s)$ admits a piece-wise constant approximation being as accurate as we wish. This technique can be extended to more wide classes of kernels.

Split the segment [0,T] on $N+1$ equal parts by the points t_i, $0 = t_0 < t_1 < \ldots < t_{N+1} = T$, and denote $t_{i+1} - t_i = h$. Next, on every square

$$\sqcup_{ij} = (t_i, t_{i+1}) \times (t_{j-1}, t_j), \quad i = 1,\ldots,N,\ j = 1,\ldots,i$$

we replace the matrix $K(t,s)$ by the constant matrix K_{ij} and assume constant $n\times n$ matrices ΔK_{ij} to be known such that

$$\left|K(t,s)-K_{ij}\right|\le \Delta K_{ij},\ (t,s)\in\square_{ij},\ i=1,\dots,N,\ j=1,\dots,i.$$

Here the symbol $|A|$ for a matrix $A=\{a^{ij}\}$ means the matrix $\{|a^{ij}|\}$. Denote

$$\eta_i(t)=\begin{cases}1, & t\in[t_i,t_{i+1}],\\ 0, & t\notin[t_i,t_{i+1}],\end{cases}\quad i=0,1,\dots,N\ ,$$

$$\Gamma=\begin{pmatrix} E & 0 & 0 & \cdots & 0\\ -hK_{22} & E & 0 & \cdots & 0\\ -hK_{32} & -hK_{33} & E & \cdots & 0\\ \vdots & \vdots & \vdots & \ddots & \vdots\\ -hK_{N2} & -hK_{N3} & -hK_{N4} & \cdots & E\end{pmatrix},\ \Gamma^{-1}=\{P_{ij}\},\ i,j=1,\dots,N,$$

$$\tilde K(t,s)=K_{ij},\ (t,s)\in\square_{ij}, i=1,\dots,N;\ j=1,\dots,i.$$

The resolvent kernel $\tilde{\mathrm{H}}(t,s)$ for $\tilde K(t,s)$ can be found in the explicit form (see, for instance,[11]):

$$\tilde{\mathrm{H}}(t,s)=\sum_{i=1}^{N}\eta_i(t)\sum_{k=1}^{i}\sum_{j=k}^{i}P_{ij}K_{jk}\ \eta_{k-1}(s). \tag{8}$$

Define the matrices $\tilde C_t'(t,s)$ and $\tilde C(t,s)$ by the equalities

$$\tilde C_t'(t,s)=\tilde{\mathrm{H}}(t,s),\quad \tilde C(t,s)=\mathrm{E}+\int_s^t\tilde{\mathrm{H}}(\tau,s)\,d\tau.$$

Also define the linear operators $K,\tilde K,\tilde H,\ \Delta K: L\to L$ as integral Volterra operators with the kernels $K(t,s),\tilde K(t,s),\tilde H(t,s)$, and $\left[\tilde K(t,s)-K(t,s)\right]$ respectively.

In the case that the inequality

$$q\overset{def}{=}\left\|\Delta K\left(I+\tilde H\right)\right\|_{L\to L}<1 \tag{9}$$

holds, where I is the identity operator, we have the estimate

$$\left\|H-\tilde H\right\|_{L\to L}\le\frac{q}{1-q}\left\|I+\tilde H\right\|_{L\to L} \tag{10}$$

that provides the desired error bound of $\tilde C$.

Inequality Eq. (9) and estimate Eq. (10) can be adapted for computer calculating. Let us write the explicit component-wise analogs of Eqs. (9), (10). Denote the elements of the resolvent kernel $H(t,s)$ and its approximation $\tilde H(t,s)$ (see Eq. (8)) by $r^{ij}(t,s)$ and $\tilde r^{ij}(t,s)$ respectively:

$$C_t'(t,s)=\mathrm{H}(t,s)=\left\{r^{ij}(t,s)\right\}_{i,j=1,\dots,n},$$

$$\tilde C_t'(t,s)=\tilde{\mathrm{H}}(t,s)=\left\{\tilde r^{ij}(t,s)\right\}_{i,j=1,\dots,n}.$$

The estimate

$$\max_{1\le j\le n} \operatorname{vrai}\sup_{s\in[0,T]} \sum_{i=1}^{n} \int_0^T \left| r^{ij}(t,s) - \tilde{r}^{ij}(t,s) \right| dt \le \frac{\delta}{1-\delta} \left(1 + \max_{1\le j\le n} \operatorname{vrai}\sup_{s\in[0,T]} \sum_{i=1}^{n} \int_0^T \left| \tilde{r}^{ij}(t,s) \right| dt \right)$$

holds as soon as the precision of approximation $\tilde{K}(t,s) = \{k^{ij}(t,s)\}$ of $K(t,s) = \{k^{ij}(t,s)\}$ satisfies the condition

$$\delta \overset{def}{=} \max_{1\le j\le n} \left[\operatorname{vrai}\sup_{s\in[0,T]} \sum_{i=1}^{n} \int_0^T \left\{ \left| k^{ij}(t,s) - \tilde{k}^{ij}(t,s) \right| + \sum_{\nu=1}^{n} \int_0^T \left| k^{i\nu}(t,\tau) - \tilde{k}^{i\nu}(t,\tau) \right| \cdot \left| \tilde{r}^{\nu j}(\tau,s) \right| d\tau \right\} dt \right] < 1.$$

5. ON THE CHOICE OF THE CONTROL SPACE

Representation Eq. (4) provides a way to reveal several properties of $\bar{u}$, in addition to the belonging to L_2 and the minimality of its norm. Explain the aforesaid by the case when $(Bu)(t) = B(t)u(t)$ with $n \times r$ matrix B. In such a situation

$$\bar{u}(t) = B^{*}(t)\theta^{*}(t)\mathrm{M}^{-1}\gamma,$$

and thc truc smoothness of the $\bar{u}$ is defined by the smoothness of $B(\cdot), \Phi(\cdot)$, and $C'_{\tau}(\tau,\cdot)$. In applied control problems, the question on the solvability of the control problem within a class of functions of the given smoothness is of considerable importance. The properties of $C'_{\tau}(\tau,t)$ as the function of the arguments t and τ are studied in detail in [1, 2].

Another way of finding smooth controls is in connection with a special choice of the control space. Replace the space L_2 by a space U isomorphic to the direct product $L_2 \times R^r \times \ldots \times R^r$. In this case the question on the solvability of control problem Eq. (2) in the space U can be reduced efficiently to the question on the solvability of a linear algebraic system, and so can be studied by the computing experiment. Consider for short the case $U \overset{def}{=} D_2 \,\Box\, L_2 \times R^r$, i.e. the Hilbert space of absolutely continuous functions $u:[0,T] \to R^r$ with square-summable derivative and the inner product

$$(u_1, u_2)_{D_2} = (u_1(0), u_2(0))_{R^r} + (\dot{u}_1, \dot{u}_2)_{L_2}.$$

Denote

$$\mathrm{V} = \int_0^T (B^{*}\theta)(s)\,ds; \quad \mathrm{W}(s) = (\mathrm{B}^{*}\theta)(s),$$

where

$$(\mathrm{B}z)(t) = \left[B\left(\int_0^{(\cdot)} z(\tau)\,d\tau \right) \right](t).$$

A control $u \in D_2$ solves control problem Eq. (2) as soon as

$$\mathrm{V}\cdot u(0)+\int_0^T \mathrm{W}(s)\dot{u}(s)\,ds=\gamma\ , \tag{11}$$

where γ is defined by Eq. (5). Any element $u\in D_2$ can be represented in the form

$$u(t)=\mathrm{V}^*\cdot\sigma_1+g_1+\int_0^t\left[\mathrm{W}^*(s)\cdot\sigma_2+g_2(s)\right]ds, \tag{12}$$

where $\sigma_1,\sigma_2\in R^n$,

$$\left(\mathrm{V}^*\cdot\sigma_1,g_1\right)_{R^n}=0\ \forall\sigma_1\in R^n \text{ and } \left(\mathrm{W}^*\cdot\sigma_2,g_2\right)_{L_2}=0\ \forall\sigma_2\in R^n .$$

With Eqs.(11), (12) we obtain the linear algebraic system

$$\mathrm{M}_1\sigma_1+\mathrm{M}_2\sigma_2=\gamma \tag{13}$$

with respect to vectors σ_1 and σ_2 defining the control

$$\bar{u}\in D_2,\ \bar{u}(t)=\mathrm{V}^*\sigma_1+\int_0^t \mathrm{W}^*(s)\,ds\cdot\sigma_2$$

that solves Eq. (2) with $g_1=0,g_2=0$. In Eq. (13) $n\times n$ matrices M_1 and M_2 are defined by

$$\mathrm{M}_1=\mathrm{V}\cdot\mathrm{V}^*,\ \mathrm{M}_2=\int_0^T\mathrm{W}(s)\mathrm{W}^*(s)\,ds.$$

6. LINEAR CONTROL CONSTRAINTS

In conclusion consider the possibility of taking into account some additional linear constraints concerning the control. Let $\lambda: D_2\to R^r$ be a given linear bounded vector-functional with linearly independent components. The control problem with constraints can be written in the form

$$\mathrm{L}x=Bu+v,\quad x(0)=\alpha,\quad \ell x=\beta,\quad \lambda u=0. \tag{14}$$

Obtain a condition for the solvability of Eq. (14). Let $\mathrm{L}_\lambda: D_2\to L_2$ be a linear bounded operator such that the boundary value problem

$$\mathrm{L}_\lambda u=z,\quad \lambda u=0 \tag{15}$$

is uniquely solvable for every $z\in L_2$. The set of all controls $u\in D_2$ with the condition $\lambda u=0$ is governed by the equality $u(t)=(Gz)(t)$, $z\in L_2$, where $G:L_2\to\ker\lambda$ is the Green operator[1, 2] of problem Eq. (15). Using this representation with Eq. (11), come to the following equation concerning an element $z\in L_2$, which generates a control $u=Gz$ solving Eq. (14):

$$\int_0^T\theta(s)(\mathrm{B}z)(s)\,ds=\gamma .$$

Denoting $\mathrm{B}=BG$, obtain the equation

$$\int_0^T\left[\mathrm{B}^*\theta\right](s)z(s)\,ds=\gamma$$

and, next, doing again the above consideration, come to the following condition of the solvability of Eq. (14):

$$\det \int_0^T [B^*\theta](s)[B^*\theta]^* (s)\, ds \neq 0 .$$

This condition can also be checked through the computing experiment. A part of such the experiment includes the construction of a sufficiently accurate approximation to the Green operator of problem Eq. (15). As for the key relationships required to do this, we refer to [1, 3].

REFERENCES

1. N.V.Azbelev, V.P.Maksimov, and L.F.Rakhmatullina, *Introduction to the Theory of Functional Differential Equations (Russian)* (Nauka, Moscow, 1991).
2. N.V.Azbelev, V.P.Maksimov, and L.F.Rakhmatullina, *Introduction to the Theory of Linear Functional Differential Equations* (World Federation Publ., Atlanta, 1995).
3. N.V.Azbelev, V.P.Maksimov, and L.F.Rakhmatullina, *Methods of the Contemporary Theory of Linear Functional Differential Equations(Russian)* (Regular and Chaotic Dynamics, Izhevsk, 2000).
4. A.N.Rumyantsev, *The Reliable Computing Experiment in the Study of Boundary Value Problems (Russian)* (Perm State University Press, Perm, 1999).
5. E.W.Kaucher, W.L.Miranker, *Self-Validating Numerics for Function Space Problems* (Academic Press, New York, 1988).
6. M.Plum, Computer-assisted proofs for two-point boundary value problems, *Computing*, **46**, 19-34 (1991).
7. M.Plum, Existence and enclosure results for continua of solutions of parameter dependent nonlinear boundary value problems, *J. Comput. and Appl. Math.*, **60**, 187-200 (1995).
8. E.A.Andreeva, V.B.Kolmanovskii, and L.E.Shaikhet, *Control of Systems with Aftereffect (Russian)* (Nauka, Moscow, 1992).
9. V.P.Maksimov, A.N.Rumyantsev, Boundary value problems and problems of pulse control in economic dynamics. Constructive study, *Russian Mathematics(IzVUZ)*,**37**(5), 48-62 (1993).
10. A.N.Rumyantsev, Computing Experiment in the Study of Functional Differential Models: Theory and Applications. Thesis ... doct. Phys.&Math., Perm State University, Perm (unpublished) (1999).
11. V.P.Maksimov, A.N.Rumyantsev, and V.A.Shishkin, On constructing solutions of functional differential systems with guaranteed precision, *Functional Differential Equations*, Israel, **3**(1-2), 135-144 (1995).

HEAT TRANSFER IN DISPERSE SYSTEMS OF VARIOUS STRUCTURES AND CONFIGURATIONS

Marina A. Smirnova*

1. INTRODUCTION

The scientific researches of the transfer processes in disperse systems with various phisics-chemical properties assume ever greater urgency recently. It is no mere chance since with each year it is inereasedthe using of the disperse systems in the practical appendices, for example, industry, engineering, agriculture, medicine. One of the major problems of the modern physics of disperse mediums is the problem of the heat transfer, inducing problems by the thermal sources (for example, electromagnetic nature[1,2]). Such tasks are directly connected with the problem of the management by the natural phenomena and technological processes. In particular, the derosols, formed as a result of industrial activity of man may, contain valuable substances, on the one hand, and have an harmful effect on the people and environmenton. In connection with an aggravation of the ecological situation thestudy of processes occurring in such systems, becomes a more and more essential problem. Thus, the taking place research of the perculiarities features of the transfer in disperse systems homogeneous and non-uniform on structure, various on configurations and heatphysical properties, certainly, has character. In spite of the fact that the theory of transfer in dispertion systems intensively develops last years. There are some more unsolved important problems representing significant interest, both for theoretical, and for the practical applications. Still there are unsufficiently investigating processes of the trasfer occurring in the real disperse particles and assotiations the possibility of the management by such processes.

* Marina A. Smirnova, Tver Technical State University, Tver, 170026, Af. Nikitin emb., Russia

2. THE MODELLING OF THE HEAT TRANSFER IN DISPERSE TO SYSTEM, CONTAINING VARIOUS NUMBER OF PARTICLES

In the present work it is considered the heat transfer in the N particles system, induced by the thermal sources in each of the particles. In general case these thermal sources are temperature functions

$$q_i = q_{i0} f_i(T). \tag{1}$$

The present work is develoted to study of the qualitative and quantitative regularies olf the heat transfer process,l caused by both the noninearity and the collective interactions in the disperse systems.

Considering the process of the heat transfer for the dispersible medium, it is possible to offer the following mathematical model

$$\begin{aligned} & C_i \rho_i \frac{\partial T_i}{\partial t} - \vec{\nabla} \chi_i \vec{\nabla} T_i = q_i(T_i), \\ & C_e \rho_e \frac{\partial T_e}{\partial t} - \vec{\nabla} \chi_e \vec{\nabla} T_e = 0, \\ & T_i(S_i, t) = T_e(\vec{S}_i, t), \qquad i = 1 \div N \\ & -\chi_i \frac{\partial T_i}{\partial \vec{n}}(S_i, t) = -\chi_i \frac{\partial T_e}{\partial \vec{n}}(S_i, t), \\ & T_e(\infty, t) = T_\infty, \end{aligned} \tag{2}$$

where S_i - surface of i-particle, T_e - temperature outside of the particle, C_i - thermal capacity of i-particle, C_e - thermal capacity of an environment, t - is time, T_∞ - is temperature of the environment unditurbed by the presence of the particles, ρ_i - is density of substance of the particle, ρ_e - is density of substance of an environment, $\vec{n}$ - normal to a surface.

If the characteristic time of the relaxation $t_p = \max\limits_{1 \le i \le N} \{t_{pi}\}, \ t_{pi} = \dfrac{C_i \rho_i L_i^2}{\chi_i}$ (L_i - the effective size of the particles equal, to R_i for example, i.e. the radius of i-particle for a case of the spherical particles) is much less than time of the process, temperature an be defined from the stationary equations, considering $\dfrac{\partial T_i}{\partial t} \approx 0$. In the case, of the particles sizes change, the heat transfer process may be considered established (the temperature field has the time to arranged under each new configuration of system), if

$t >> t_d$ (t_d - is some time describing the dynamics of the investigated process). In this case it is possible to consider the time as parameter.

If exterior medium gas, and Knudsen's number is $Kn = \frac{l}{R_{min}}$ ($R_{min} = \min_{1 \le i \le N} \{R_i\}$, l - is average length of free run of gas molecules) noticeably differs from zero, so-called moderately large drops on Yalamov-Deryagin-Galoyan's classification, it will be necessary to ater the conditions on the boundary of the partice, introducing in them a satus of the temperature[3)]

$$T_e(S_i) = T_S + K_T l \frac{\partial T_e}{\partial \vec{n}} \mid_{S_i}, \tag{3}$$

where K_T - is saltus of the temperature, T_S - is the temperature of the surface of the particle. This is obtained boundary conditions of the third kind. Earlier the saltus of the temperature has not taken into account in researches.

If it is considered the heat transfer in limited volume or the calculation is carried out in some part of the continual medium, in which the particles were concentrated, on exterior actual or approximately specific boundary can put boundary conditions of the third kind including the heat exchange factor α.

In the system consisting of several particles, the heat transfer depends both on the configuration of the system, and from the interaction between particles. The situation is complicated also by that the heat transfer is in common case nonlinear. Now the large member of works is devoted to the of heat transfer in the particles collectives. It is considered the transfer of the electromagnetic energy, heat transfer and mass transfer proarided that the particles can evaporate.

In the present work we shall stay on the heat transfer and on the methods used for its exposition.

The "simple" case is a reviewing of collective from two particles. If particles are spherical it will be considered the system of bispherical axes and if the heat conduction equation is linear (the Poisson equation) it will be possible to receive its analytical solution.

In general case it is necessary to carry out the calculations numerically and it is caused with the several reasons: 1) if the task is linear, the analytical solution will be represented as the extremely bulkyrows; 2) the task in cjmmon case is nonlinear as the heat sourse and the thermal parameter can depend on the temperature; 3) the particles can have the various form.

In the present work we used the method of the final elements[4)].

3. THE RESULTS OF COMPUTING EXPERIMENTS AND THEIR ANALYSIS

In this part of work there are adduced the descriptions of the computing experiments carried out with the help of the developed program, the basic results of these experiments and analysis of the received results. The process of the heat transfer in dispersible systems, various on the structure, configuration and sizes was investigated which the method of the final elements. It was selected the modellingscchemes the sizes of the particles in wich are real for the dispersible systems (in particular, for aerosols).

The heat transfer process was considered at various boundary conditions:

1) boundary conditions, at which some thermal flow was set on bound of each particle (it can be caused by the management of laser radiation, heterogeneous chemical reaction, phase transition etc.), thus the system in a whole was assumed heat isolated;

2) the open system with boundary conditions of the third kind was considered sorts on external bound.

The particles of the spherical form were considered, that has allowed to calculate collective effects, to estimate dependence of temperature in the system on the size of a particle etc., however developed program allows to carry out the calculations for the particles of the any form. As the task corresponded to the conditions of the Likov-Nigmatulin's theorem, it has allowed to carry out calculations for a flat case, that has simplified computing experiment considerably.

We shall consider here concrete experiments and their basic results.
With the method of the final elements it was carried out the calculation of temperature in each unit of the system concerved a rectangular area (90 x 30 microns) with the spherical particles, placed on it, of radius 5 microns, modelling monodispersible system (further - system №1). The quantity of the particles is varied from one up to eleven, it is also varied the arrangement of particles on the square (the configuration №1 – 1 particle; №2, 3 – 2; №4, 5 – 3; №6, 7 – 4; №8, 9 – 5; №10 – 6; №11 – 7; №12 – 8; №13 – 9; №14 – 11 particles). The calculations were carried out with boundary conditions both first, and second kind, according to above mentioned classifications. The numbering of particles for monodispersible system is given in a fig. 1:

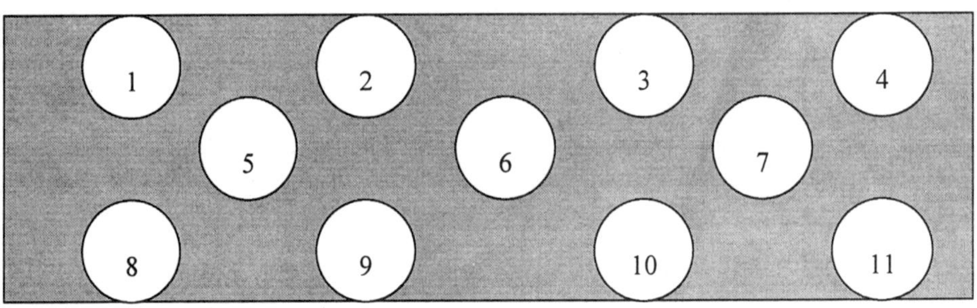

Figure 1. Numbering of particles for monodispertion of system

The diagrams, illustrated the most interesting and typieal results of the carried out accounts for systems №1 with boundary conditions of the first kind are shown on Fig. 2, 3.

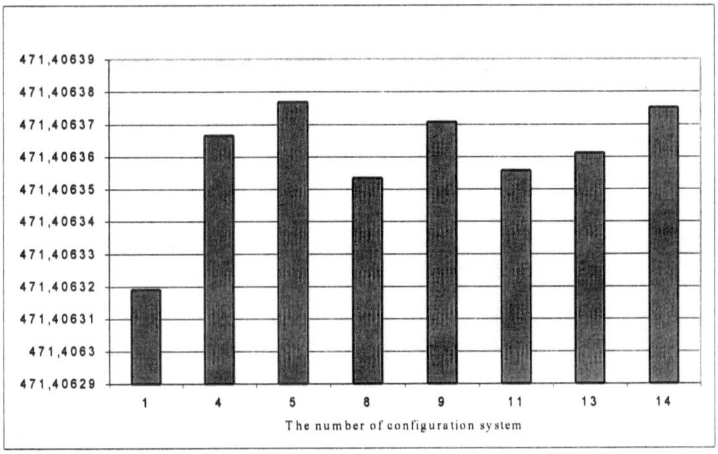

Figure 2. Temperature on the centre of the particle №6 for various configurations of system №1

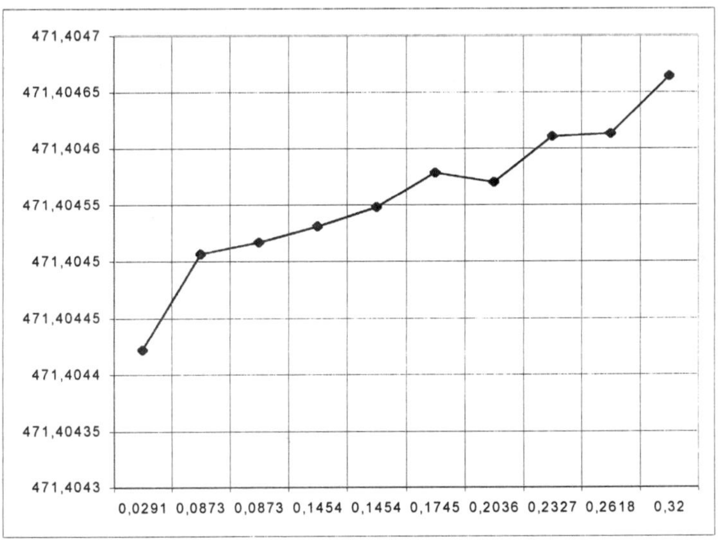

Figure 3. Diagram of dependence of average temperature for configurations 1, 4, 5, 8-14 of the system №1 on the ratio of the total area of particles of the appropriate configurations of system №1 to the area of the square

The diagrams illustrate essential nonmonotony of the dependence temperatures on number of particles for the given configurations of system. So, for example, temperature for the central particle of the configuration №8, containing five particles, appear below, than temperature of central particle for a configuration №5, containing only three particle. Fig. 3 shows, that average temperature of system depends not only on the area of particles, but also on the character of filling of space by particles.

On fig. 4, 5 are the diagrams, illustrated most interesting and characteristic results of the carried out accounts for systems №1 with boundary conditions of the second kind:

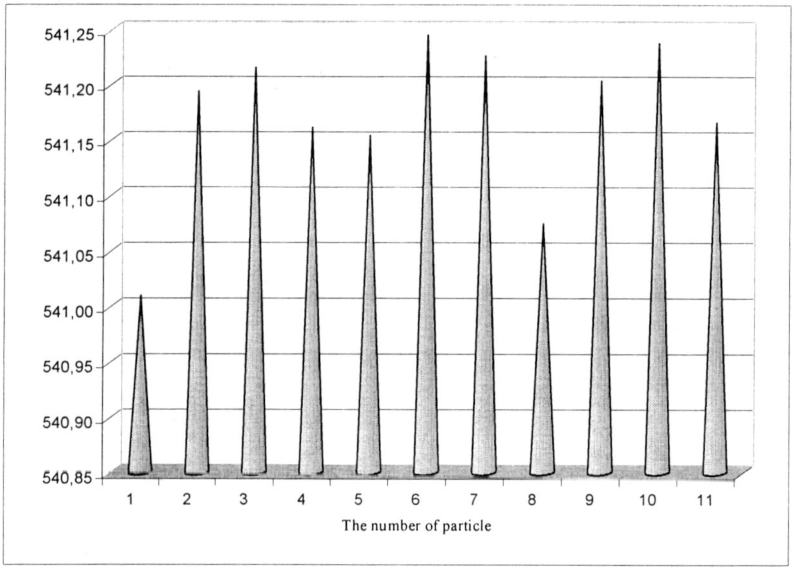

Figure 4. Average temperature of each particle for a configuration №14 of the system №1

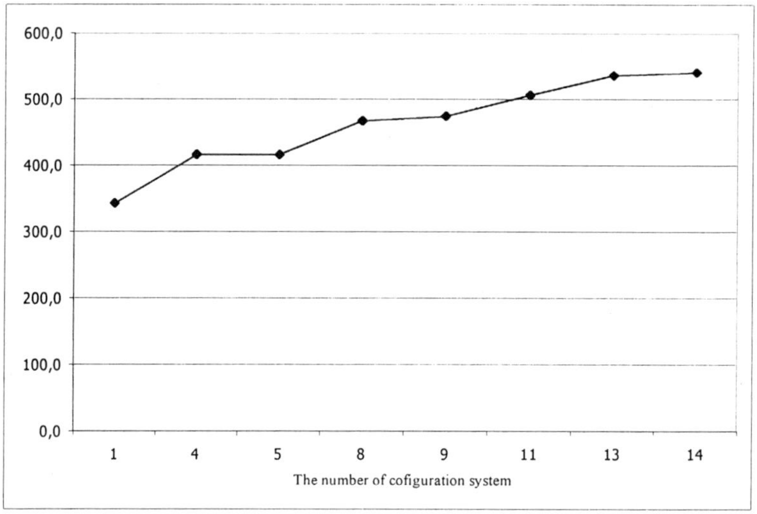

Figure 5. Temperature at the centre of particle №6 for various configurations of system №1

On diagrams are shown accounts for system " particle of water in atmosphere of air ". It is evidently, that temperature in the system depends both on a configuration of system, and on number of particles, including in the system, however, in case of boundary conditions of the first kind this dependence insignificantly influences on the temperature in the system. More brightly the reseaeches of collective effects are illustrated by this fact. Under collective in present work was understood contribution of each particle in temperature picture of system. Collective effects for "monodispersible" system were carried out in the following way: it is calculated average temperature of a single particle (T_1), average temperature of particles for various configurations of system (T_{coll}), then according to the formula (4) collective effects are calculated:

$$\varepsilon = \frac{|T_{coll} - T_1|}{T_1} \times 100\% \tag{4}$$

On Fig. 6 are shown the diagrams of the dependence of collective effects on number of particles in configuration and on their interarrangement for boundary conditions of the second kind .

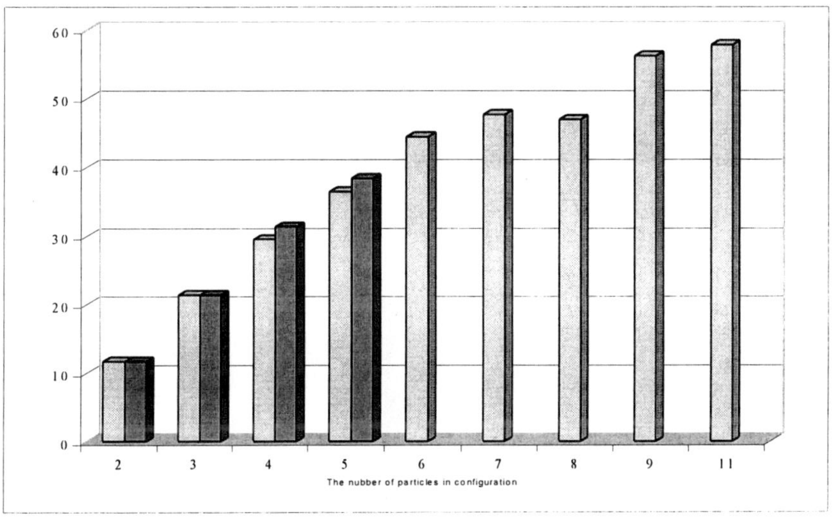

Figure 6. The diagram of dependence of collective effects on number of particles and on their interarrangement for various configurations of system №1

By thr method of final elements was carried out calculation of temperature of each unit of system representing rectangular square (90 x 70 microns) with the placed on it of spherical particles, of radii 5 and 10 microns this is the model polydispersible system (further - system №2). The quantity of particles is varied from one to nine, the arrangement of particles is varied also (the configuration №1, 11-18 – 1 particle; №2 – 4; №3, 4 – 5; №5 – 3; №6, 8 – 6; №7, 9 – 7; №10 – 9 particles).. The accounts were carried out with boundary conditions both first, and second kind, according to above mentioned classifications. The numbering of particles for polydispersible system is given on Fig. 7.

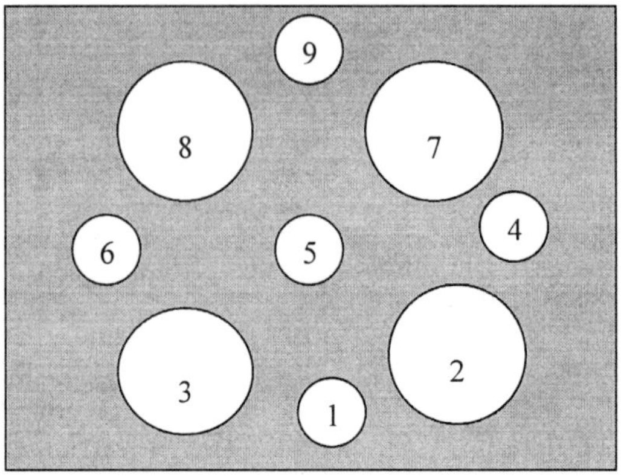

Figure 7. Numbering of particles for system №2

On Fig. 8 - 9 are the diagrams, illustrated most interesting and characteristic results of the carried out accounts for Systems №2 with boundary conditions of the first kind:

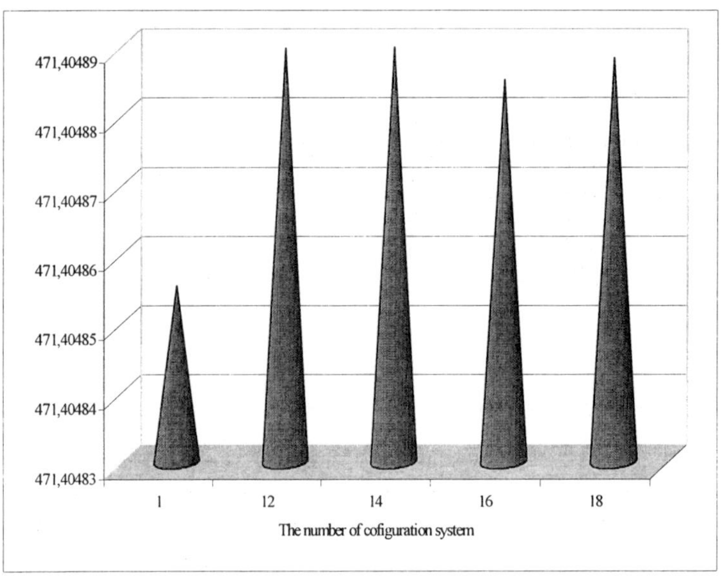

Figure 8. Average temperature of particles (radius 5 microns) for various configurations of system №2

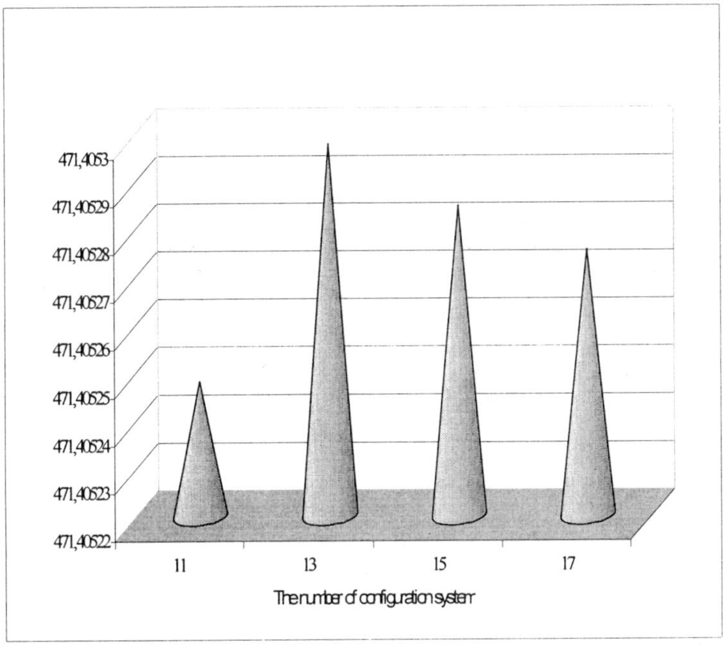

Figure 9. Average temperature of particles (radius 10 microns) for various configurations of system №2

On Fig. 10, 11 the diagrams and diagrams, illusrated most interesting and храрактерные results of the carried out accounts for systems №2 with boundary conditions of the second kind:

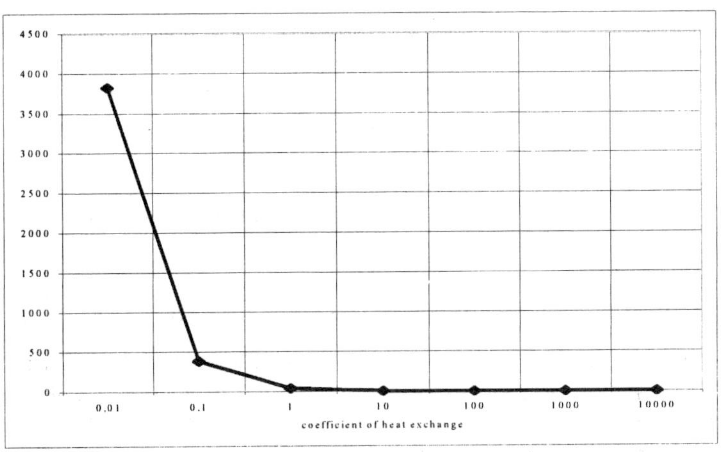

Figure 10. The diagram of temperature dependence at the centre of particles for a configuration 1 of system №2 on the coefficient of heat exchange

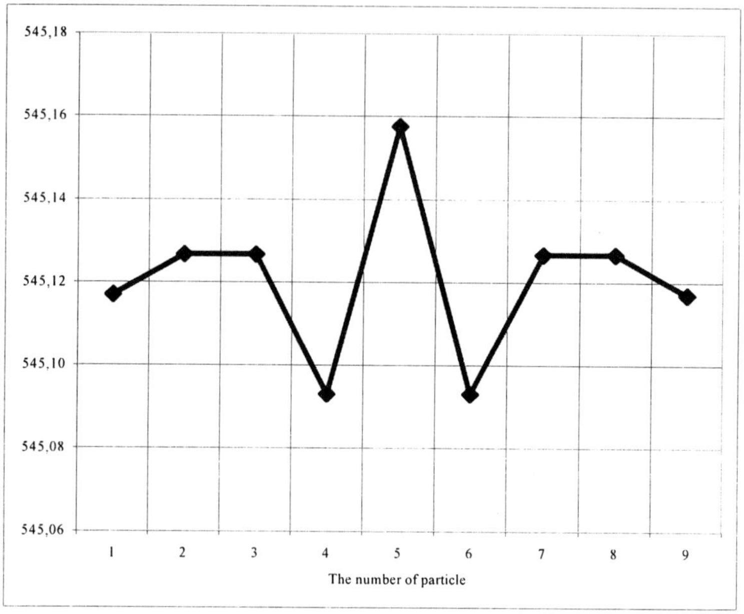

Figure 11. Average temperature for particles of a configuration 10 of system №2

From a figures it is evident, that in the polydispersible system the presence of larger particles essential influences on the temperature. For such systems the estimation of collective effects was made. The account of collective effects was carried out with the formula

$$\varepsilon = \frac{|T_{coll} - T_{ad}|}{T_{ad}} \times 100\% , \tag{5}$$

where $T_{a\partial} = \frac{\sum_{i=1}^{m} T_i S_i n_i \rho_i}{\sum_{i=1}^{m} S_i n_i \rho_i}$, T_i, $S_{i,}$ ρ_i - are the temperature, area and density of substance of i particle, n_i - number of particles of i-th grade. As it was estimated the particles of two various sizes, were the number of the system, the contribution of not single particle, but conglomerate from two particles. The received results are shown in the Table 1., and in the second column of the table are the findings - for boundary conditions of the first kind, and in third - are for second.

Table 1

Configurations of the system	Collective effect 1 (%)	Collective effect 2 (%)
11-12	0,0000037	53,047
13-14	0,0000042	53,034
15-16	0,0000018	53,032
17-18	0,0000005	53,032
13-12	0,0000042	53,034
1-11	0,0000055	53,049

It is shown, that under the boundary conditions of the third kind on external border of system (open system) the influence of collective effects changes the temperature in 1,5 and more time, under boundary conditions of the first kind this influence is not essential.

The task was decided for the following dependences of the source on the temperature: 1) $q = q_0 \left(\frac{T}{T_\infty} \right)^\sigma$, where the parameter σ is varied from -0,5 up to 2; 2) $q = q_0 e^{-\frac{E_A}{kT}}$, where the meaning of the parameter $\frac{E_A}{k}$: is equal to 0,001 - 100. The characteristic the results of such accounts for the first dependence are shown on Fig. 12, 13.

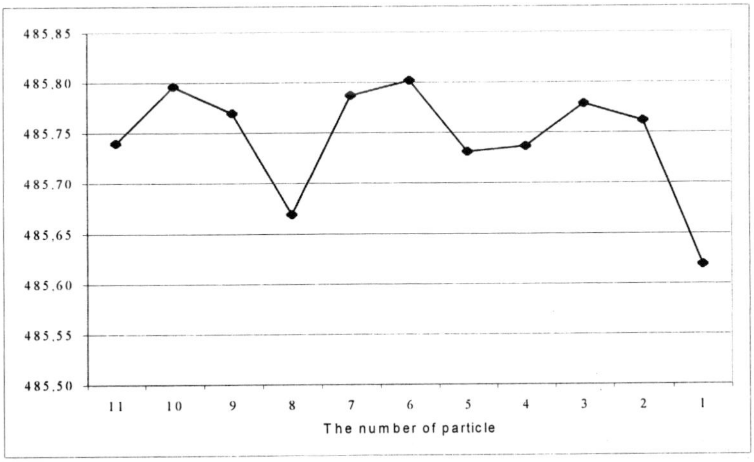

Figure 12. Temperature at the centre of each particle for the configuration 14 of the system №1 at the meaning of parameter σ = -0.5

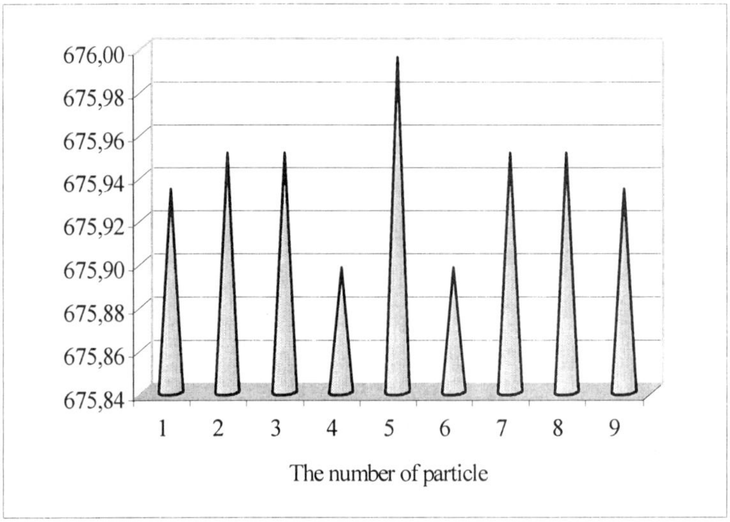

Figure 13. Temperature at the centre of each particle for the configuration 10 of the system №2 at the meaning of parameter $\sigma = 0.5$

On the Fig. 14 are shown the results of accounts for the second kind of the dependence of source on temperature.

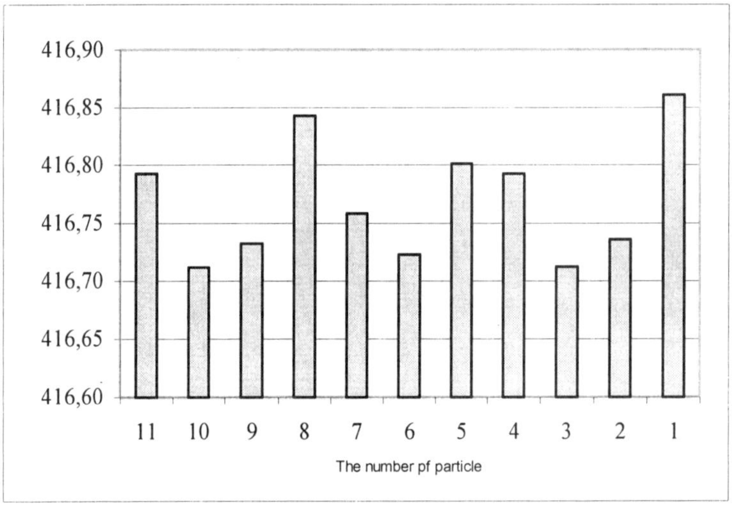

Figure 14. Temperature at the centre of each particle for configuration 14 of the system №1 at meaning of parameter σ=100

The accounting of the dependence of the thermal source on temperature results in the cosiderable overheating in the system at increase of particles number. So, for example, at σ =1, the characteristic temperature in the system for configurations, containing 11 particles, exceeds the temperature of the configuration from one particle approximately three times.

So, it is carried out the large number of computing experiments for various dispersible systems. It is shown, that the configuration of system, of the geometrical sizes, composition, kind of boundary conditions, the dependence of thermal source on temperature significally influence on the temperature in system.

REFERENCES

1. L.A Uvarova., V.K. Fedjanin *Electromagnetic and heat waves in condensed systems with non-linear characteristic (Russian).* Preprint N P17 – 96 – 379 (JINR, Dubna, 1996).
2. K. Boren., D. Khafmen *Absorption and dispersion of light by small particles.* (Mir, Moscow, 1986).
3. B.V Derjagin., U.I. Jalamov., V.S. Galojan, Reasonably large particles movement theory in the heterogeneous gas. *Report AS USSR (Russian)*, **201**, 383 – 385 (1979).
4. G.N Dulnev., V.G Parfjonov., A.V. Sigalov, *Computer usage for heat transfer problems solution.* (Vyshaya Shkola, Moscow, 1990).

3. MATHEMATICAL COMPUTER MODELS OF DISCRETE SYSTEMS

SOME NEW RESULTS IN THE THEORY OF INTELLIGENT SYSTEMS

Valery B. Kudryavtsev and Alexander S. Strogalov*

1. COMPUTER SOLVER OF MATHEMATICAL PROBLEMS

A mathematical model of intelligent system of a principally new type based on a new interpretation of the concept of induction and decision-making mechanism is developed. Professor A.S. Podkolzin invented a specific formalism for describing such systems based on a new high-level algorithmic language LSD (Logical Situation Descriptor) At the present time, this intelligent system is implemented as a Windows application in the form of a solver of mathematical problems in the symbolic form. The above-mentioned system can solve all typical problems of elementary mathematics (with the exception of geometrical and textual problems; the author is now working intensively on teaching the system to solve problems of these types) and of a number of branches of higher mathematics (differentiation, taking limits, integration, etc.) This solver demands less then 25 MB of computer memory.

It should be emphasized that similar computer solvers invented, for instance, in the USA demand enormous computational resources and are usually based on methods of computer algebra.

Note, that after an appropriate refining and tuning on the application domain, this model of intelligent system can be employed for solving problems of automatic verification of programs, of synthesis and optimization of the structure of a very large-scale integration. One of the versions of the problem solver oriented on problems of this type is protected by a series of patents (about 30) in a cooperation with the LSI Logic Corp., USA. The principle of operation of the problem solver is described in detail in Podkolzin' article[1].

* Valery B. Kudryavtsev, Moscow State University, Moscow, Russia, 119899. Alexander S. Strogalov, Moscow State University, Moscow, Russia, 119899.

2. COMPUTER-BASED TRAINING SYSTEMS

A study on computer simulation of the training process and development of real training systems on the basis of the invented models was carried out in the following directions:

1. Development of tools for the simulation of all components of the training process.
2. Problems of design of expert systems in the field of education.
3. Study of the existing types of educational systems and possibilities of their computer-based simulation.
4. Problems of development of electronic training environments.

Several papers[2, 3, 4] present the main results in this field.

Let us dwell on the main concepts of this research, which was initiated by the IDEA project[5].

Most part of the existing computer-based training systems are electronic compilations of theoretical textbooks and exercise books. Advanced versions are equipped with corresponding audio- and video effects, usually of illustrative nature. In these systems the main role is played by the user, who gets knowledge given by system, acting step by step according to the instructions. Therefore, the computer-based training system plays a passive role. Here, it is implicitly supposed that the learner can transform the educational information into the knowledge of a subject. As a rule, this transformation (or, rather, training of doing it) is the main difficulty in the training process, not to mention a self-education with the help of the computer.

The real training process contains three basic components: a teacher, a learner, and a space of their interaction having, for instance, the form of a textbook with exercises. The learning goal consists in finding an optimal sequence of interaction of the components. These components and processes of their interaction are to be simulated.

Early in the development of computer-based models and training tools these systems definitely played a positive role. At that time, the simulation of the intellectual activity at the initial stage, a better imitation the learning process better, reflecting the roles of the teacher, the learner, and the process of their interaction, could be hardly required of these systems.

In the IDEA project, the problem of designing a system, that simulates all the components of the learning process – the teacher, the learner, educational materials, and their optimal interaction – was solved and realized in practice.

It contains mechanisms of simulation of the teacher and learner; a textbook with exercises, treated as a properly organized educational material with elements of multimedia, can be designed. On this basis, the real learning process is simulated. With due account taken of its characteristic features, such as mutual tuning of the teacher and the learner, abilities of the learner, optimality of the teacher's strategy of dosing knowledge and exercises, the rate at which the learner remembers and forgets knowledge, duration and steadiness of his activity, etc.

In this approach, the teacher and the learner are interpreted as adaptive automata and the training process consists in their iterative interaction. At each step the "teacher"-automaton chooses the optimal way of giving the learning information to the "learner"-automaton depending on how well the "learner"-automaton has "learnt" such information at the previous steps of training.

The training system is rather universal for the defined class of application domains; and it is also open and can be easily supplemented with information in all its main parts.

In accordance with the above-mentioned, the problem of synthesis of an adaptive computer teacher involves the solution of the following basic problems:

1. Synthesis of the “teacher”-automaton.
2. Synthesis of the “learner”-automaton.
3. Design of an information system, similar to a textbook with exercises.
4. Elaboration of an optimal interaction strategy of components 1-3.
5. Development of a user-friendly interface.

The solution of problems 1-4 is associated with consideration of a number of questions. Among them are:

1. Development of dynamical data and knowledge bases consisting of large arrays of syntactical information with complex semantics and fuzzy logical connections. These bases should be memory compact and, at the same time, having high access speed.
2. Development of criteria space for describing the states of the "teacher" and “learner”-automata with the indication of metric-functional relations, allowing one to define the functioning of these automata.
3. Development of optimal strategies of interaction between the “teacher” and “learner”-automata, by means of both the automata theory and fuzzy logic themselves and procedures of patterns recognition, etc.

The theoretical grounding of the model is an automaton of hybrid nature and its properties[6], and some other models and methods developed in the process of investigation or borrowed from the theory of intelligent systems. By now, a number of multimedia computer-based training systems (computer science, foreign languages, humanities, medicine) are designed in the form of local versions on CD-ROM. They were developed on the basis of the software (copyright system) IDEA 3.0 for Windows Professional. At the present time, this software is distributed on the European market of computer systems by the company Link&Link Software GmbH (Dortmund, Germany).

3. ANOTHER PERSPECTIVE DIRECTIONS OF DEVELOPMENT OF INTELLIGENT SYSTEMS

1. On the base of mathematical models and methods of fuzzy logic an intelligent system of monitoring the information flow in different fields of knowledge, evaluating significance of its elements, forecasting of its development (the problem of information monitoring) was worked out. Mechanisms of information evaluation and aggregation specifically developed allows us to study the reaction of information environment on the input of new information and use the expert knowledge of elements of the environment during the decision-making process. At the present time this technology is tested in the pilot project in a cooperation with the International Agency for atomic energy (IAATE).
2. Intensively works both for developing the new mathematical models and algorithms for pattern recognition and designing real computer systems in this

field are carried out. Problems of recognition of texts (including noisy texts) in different alphabets, including Arabic character and hieroglyphs, and elements of line images from the field of problems of synthesis very large-scale integration circuits. Computer-based models of such projects and problems of analysis and synthesis of the speech, problems of correspondence "speech-text", "text-speech" were developed. The problem of recognition of the speech commands can be reduced to the solution of mathematical problem of comparison of vectors of different length by methods of dynamical programming. In this case the number of elementary comparisons linearly depends on the number of commands in the dictionary and quadratically – on the length of commands. By the work with a large dictionary (about 10 000 words) such restrictions obstruct recognition of commands in real time. Therefore for reduction of exhaustion the fast algorithms of preliminary recognition were designed, what made possible essentially accelerating of solving input problems of recognition[7].

It should be noted, that in problems of speech recognition the algorithms of developing such systems are known and clear, but an acoustic model of the language and initial teaching of automata-standards of words of the language (usually automata of Markov type) are required.

The process of primary gathering and analyzing acoustic data of native speakers is very labour-consuming: variety of voices, accents, dialects, etc. should be taken into account. Moreover, for the statistical certainly of functioning automata-standard of words every word in the dictionary must be pronounced by every representative of native speakers of current type (the announcer) several tens of times. There are over 100 000 such words. Hence, systems of speech recognition are hardly carried over from one group of languages onto another.

An application of developed mathematical problems onto solving problems of pattern recognition in the medical diagnosis, geology (searching minerals), etc. was studied. At the present time in this field the joint project with the firm INTEL (USA) has been carrying out.

A research[8] of A. Jovanovic group (mathematical department of University of Beograd, Jugoslavia) on mathematical computer-based simulation of brain-activity signals, being controlled by the man, should be mentioned explicitly. Study of this problems allows people with rough injury of motor- and speaking functions to communicate with another people, computer, etc.

3. Many applications demand developing architecture of data bases and fast retrieval in them. The problem is complicated because in the humanitarian research data bases contain hundreds of thousands textual documents, and large content of graphic information (photo archives, for instance). For the problems solution of fast retrieval in data bases of such nature the fast and superfast algorithms of retrieval and specific mechanisms of data indexing were developed. The results were published in the monograph[9]; on the base of those, in a cooperation with the Institute for Russian and soviet culture (Bochum, Germany) and Russian State archive of art and literature a CD-ROM containing lists of about 1 000 000 documents, over then 500 photographs, etc. with the realized mechanisms of the fast retrieval was released.

There are still a lot of directions of research of the MaTIS Chair, the PTC laboratory, MSC CIT and CCBTS RSUH beyond the scope of the present article.

REFERENCES

1. A. S. Podkolzin, On the organization of knowledge bases oriented on the automatic solution of problems, *Discrete Math.*, V. 3 (3), in Russian (1991).
2. V. B. Kudryavtsev, K. Waschik, A. S. Strogalov, P. A. Alisejchik, and V. V. Peretruchin, On modeling the learning process using automata, *Discrete Math.*, V. 8 (4), in Russian (1994).
3. A. S. Strogalov, S. G. Shekhovtsov, *Intellectual Activity, Learning and Education*, RSUH, Moscow, in Russian (2000).
4. A. S. Strogalov, On the problems of the theory of computer-based training systems, in: *Neural Computers and their Applications*, VI *Russian conference*, Moscow, February 16-18, 2000, in Russian.
5. K. Waschik, V. B. Kudryavtsev, A. S. Strogalov, *The IDEA Project*, Publish. Link & Link Software GmbH, Germany, Dortmund (1995).
6. V. B. Kudryavtsev, S. V. Aleshin, A. S. Podkolzin, *Introduction to the Theory of Automata*, Nauka, Moscow, in Russian (1985).
7. D. N. Babin, A. B. Kholodenko, Using lexical analyzers with pattern recognition, in: *Proceedings of the International Seminar "Dialog-99"*, V. 2, Tarusa, in Russian (1999).
8. A. Jowanowic, Computer interface using electronic brains signals, in: *Intelligent Systems*. V. 3 (1-2), Moscow, in Russian (1998).
9. E. E. Gasanov, *Net-functional Databases and Super Fast Algorithms of Information Retrieval*, RSUH, Moscow, in Russian (1997).

AN AUTOMATA APPROACH TO ANALYSIS AND SYNTHESIS OF AUDIO AND VIDEO PATTERNS

Dmitry N. Babin and Ivan L. Mazurenko*

1. INTRODUCTION

This paper describes some problems in the field of human-computer interaction, such as recognition, compression and synthesis of information in a form customary for a human. The information is represented in computer in a digital form. According to the level of complexity it can be ranked as video information, speech, handwriting texts and printed texts.

The common approach to processing of all these types of information includes the so-called "front-end", or the step of initial data representation and preprocessing.

2. IMAGE PROCESSING

The preprocessing is the main step in solving of the problems of image processing. For video and photo images it is important to represent information compactly. Here the problems of reduction of image size without loss of naturalness in human perception (image compression) and of fast reconstruction of compressed information (decompression, or image synthesis) arise. The achievements in this area are improved simultaneously with the growth of computers calculating power and have produced already such well-known technique as JPEG, MPEG, fractal compression etc. We shall not turn our attention to these algorithms.

3. SPEECH PROCESSING

Digital speech is a less complex object for the investigations. One second of non-compressed speech takes up about 10 kilobytes of the computer memory, and at the same time we need more than 10 megabytes to save one second of uncompressed video

* Dmitry N. Babin and Ivan L. Mazurenko, Moscow State University, Vorobyovy Hills, Moscow, Russia, 119899. E-mail: {dbabin, ivanmaz}@mech.math.msu.su

Mathematical Modeling: Problems, Methods, Applications
Edited by Uvarova and Latyshev, Kluwer Academic/Plenum Publishers, 2001

information with the same detail level. Moreover, it is correct to state the problem of recognition of spoken words and continuous phrases while it is not always possible for video images. As a matter of fact, spectral representation of signals in a frequency domain is usually used here (for example, with the help of Fourier transform and its modifications).

We can distinguish to directions of research in the area of automatic speech recognition (ASR).

The first one is text dictation, i.e. the recognition of the text of continuous speech produced by an arbitrary speaker (often after adopting to his/her type of voice). Here the investigators are succeeded in solving of a problem of local speech recognition almost for all known national languages. On the basis of these local solutions in the form of phoneme recognition it comes to be possible to solve the general problem of phrases recognition. Here traditionally the methods of dynamic programming[1] are used realized in the forms of dynamic time warping (DTW) and hidden Markov models (HMM)[2]. In the case of HMM approach the patterns of the recognized phrase are presented in the form of a stochastic automaton, the states of which correspond to the phonemes in the phrases transcription, and in the case of DTW the vector of a variable length is used, where each phoneme corresponds to the set of different number of vector elements, according to the duration of a phoneme. As a rule, in both cases patterns are synthesized on the basis of some set of specially recorded pronunciations of separate words or test phrases, which make up a so called speech corpus.

However, even some small local error, accumulating on a long phrase, can reduce to zero all the dignities of the methods used and the quality of speech corpora, if the sense of a pronounced text is not taken into account. Therefore, the important component of a problem of speech recognition is a problem of formal recognition of a sense (semantics) or a form (syntax) of a text, or a problem of language models construction. The case of bounded language models may be considered[3], which can be approximated by formal languages. So, it is shown that it is possible to reduce the problem of long phrases recognition to the problem of error correction in formal languages. The time of solving of this problem has an order of n^3 for regular languages and n^4 for context-free grammars, where n is the length of an analyzed word in the formal language.

For complex language models, for example for models of natural languages, the methods of frequency grammars (n-gram models) are widely used, that allow one to estimate the probability of the given word knowing n previous words in a phrase. The index of a quality of a language model here is a so-called perplexity coefficient that is an average number of different words that can be used after n known words. For English this coefficient approximately equals to 20 when n=3 and 20 000 word are used. For Russian for 150 000 word bases (that correspond to approximately 5 000 000 words) this coefficient is so huge that it is impossible to define it reliably using text databases we have.

The second problem in speech recognition is command-and-control, that is a recognition, or identification, of a bounded number of commands spoken by an arbitrary speaker, usually in hard noisy conditions. We used the following approach to solve this problem[4]. Russian phonemes were separated into some classes (such as fricatives, pauses and closures, vowels etc.) so that it found to be possible to define reliable the class which corresponds to the local fragment of digital speech (speech frame). So, the whole speech fragment is encoded by e sequence of letters of an alphabet C of these classes of sounds. The produced sequential code is than analyzed by a decoding automaton, solving the problems of noise reduction and signal compression without loss of information about phonemes order. The set of input words $\{c_i\}$ of the decoding automaton corresponding to different pronunciations of the fixed command α makes up the regular language $L(\alpha)$, which can be independently constructed directly from the text of command α and be presented in a form of a regular expression. The pronunciation of a command is recognized as α, if the result of it is coded by a sequence c of letters of alphabet C, that belongs to the language $L(\alpha)$. So, our approach allows for the given set of speech commands to estimate in advance and automatically the possibility of usage of our algorithm and the error rate of recognition. It is also possible to reduce the error rate by optimal selection of commands.

The same methods work when the additional sources of speech information are used, that allows us to increase significantly the reliability of commands recognition[5]. The usage of one microphone does not give the high reliability of robust speech recognition. Even for speech detection problem according to one microphone input signal the error rate may be more than 5%. We used the additional microphones, a photo sensor and an airflow sensor. In hard noisy conditions the signal is well detected in a closely located microphone and does not affect the signal level in a remote microphone. Comparing these two signals we increased the reliability of speech detection. The remote microphone is used also for noise level estimation.

The photo sensor is an important source of speech-information, not depending on the level of an acoustic noise. We solved a problem of maintaining the sensor stable location using the vertical line array of photo-sensors. The signal from the photo sensor as well as its derivative is used to determine the fact of speech activity since thee values are correlated with the mouth opening width and the velocity of lips movement. Here we succeeded in distinguishing the speech-related lips movements from movements of other types.

The temperature sensor was used as an airflow detector. It determines the fact of breathing reliably so that it is possible to distinguish the speech-correlated breathing from the regular breathing signals. The typical noise types for this source of information are an infra-acoustic noise connected with the external airflows (i.e. wind). The signal from this sensor along with its first derivative was used. In addition, the airflow sensor allows to detect the stop-sounds reliably that can increase the speech recognition quality.

We succeeded in increasing the reliability of speech commands recognition by uniting the results obtained from all these types of sensors. This is connected with the fact that some types of sensors can be considered as being independent from each other. Also, the typical noise for different sensors may have different nature and the probability of its simultaneous appearance is very small. In our experiments the probability of error in recognition is reduced to the level of 10^{-3}.

4. SPEECH SYNTHESIS

Speech synthesis (text-to-speech, TTS) is a problem of an automatic reading of texts by computer. On a local level the problem is successfully solved using the compilation and parametrical speech synthesis methods[6]. On a global level the problem is equivalent to the problem of text sense recognition and therefore cannot be solved satisfactorily. It is rather difficult to listen to the artificially spoken text for a long period of time.

5. HANDWRITING RECOGNITION

The next object in this hierarchy is a handwriting text. Two cases should be considered here. In the first case the information of a process of a text creation is accessible with different details level. In the second case only the final result (graphical pattern - handwriting image) is accessible. In addition, here two different cases are possible, as it is in speech recognition, - the case of "infinite" dictionary and the case of a finite set of words. We have investigated the latter problem and have analyzed the scanned images of separate hand-written words. This problem occurs in different applied areas such as an automatic recognition of some filled forms where some fields may be filled by hands with some number of words from the fixed dictionary (such as an amount of money etc.).

The known solutions of a problem of handwriting recognition are not ideal, in contrary with the problems of hand-typing and printed texts recognition. This is connected with the fact that there is no space between letters in handwriting, and in addition the number of variants of writing of different letters is huge. Therefore the algorithms developed for hand-typing or printed texts recognition cannot be transformed to be useful in this case. There exist some systems, realized generally on portable ("pocket") PCs that solve the problem of handwriting recognition with an acceptable reliability, that use additional parameters, mostly the time. The problem of recognition of handwriting texts presented only by their images is actual up to now. The one way of its solution is to try to recover the time parameter by analyzing the curves in the handwritten image. The second approach is based on some algorithms automatically calculating some features of an image. We used the second approach and utilized our own method of image preprocessing the result of which is a vector of typical image parameters.

The formalization of a problem is as follows. Let $\{A_1,A_2,\ldots,A_M\}$ be a dictionary of words of some natural language. Let $\{P_1,P_2,\ldots,P_L\}$, $L\geq M$, be a set of matrices with elements in $\{0,1\}$. Let $p_{ij}=1$ if the point with coordinates (i,j) belongs to the image of a word, and $p_{ij}=0$ if (i,j) defines the point in the background. Let us know some correspondence between images of words and the words themselves. We need to construct the automaton on the basis of the training material, which will recognize the images of new words. The word image is considered as some structural unit, that is we do not try do separate it into images of letters, because the letter problem is connected with some additional difficulties covered below.

The feature vector for matrix P_i is a vector $W_i=(w_{i1},w_{i2},\ldots,w_{ik})$, where w_{ij} equals to the number of columns of ones in j-th column or in j-th row of a matrix P_i, separated by the sets of zeros. On a physical level it corresponds to the number of intersections of the corresponding horizontal or vertical line with our word. It is also possible to set some angle of the intersecting line. In any case we get some vector $W_i=(w_{i1},w_{i2},\ldots,w_{ik})$, where $w_{ij}\leq M$ and M is the maximal allowable number of intersections. After that, we can note that the human may shrink or stretch some elements the words during handwriting in horizontal direction. Therefore we do not need to consider the repeating elements and it is important only to know the sequence of different elements in a training vector of features.

So we come to the problem of finite automaton synthesis representing the set of training vectors.

As an application of the described technique the computer demo program was created, which recognizes the handwritten images of Russian words after preliminary training. The program uses the feature vectors obtained by an intersection of an image by horizontal and vertical lines as well as by separate intersection of an upper and a lower parts of an image by vertical lines and of left and right parts of a word by horizontal lines. The program testing showed the following results. After training procedure on the base of written words {one, two, ..., twenty, rubles} by one person, each word once, the reliability of recognition of the same words by the same person was found to be 90%, by another person – from 50 to 90%. After training on the base of two variants of writing of each word the reliability did not increase significantly. After the 10th iteration of the training process the reliability of recognition of words written by the native person was close to 100% and by another person – from 70% to 90%. Finally, after training of system on the material, which included writing of each word 10 times by 10 different persons, the reliability of recognition of an 11th one was about 95%.

6. PRINTED TEXT RECOGNITION

The simplest problem from the point of view of a dimension is a problem of printed text recognition (also known as optical character recognition, OCR). These images can be obtained from a scanner, a photo or a video camera. Not taking into account the problems of preliminary image processing in all of these cases, we can find out the common regularity connected with the complexity of the problem. In sequential images recognition, such as ASR, handwriting recognition, OCR, the problem of separation of a signal into elementary units such as phonemes or letters is most complex. Exhaustive solving of these problems results in the algorithms the time of work of which depends on the length of the sequence exponentially. For the problem of separating of a printed text distorted by noise we succeeded in developing the algorithm with polynomial dependency on the length of the sequence, where the degree of a polynomial is defined by the number of different types of width of characters in the alphabet. Applying this algorithm for solving the problem of separating of a handwritten text into letters we have got the exponential time again[7].

REFERENCES

1. R. Bellman. Dynamic Programming (Princeton University Press, Princeton, N.J., 1957).

2. L.R. Rabiner. A Tutorial om Hidden Markov Models and Selected Applications in Speech Recognition. *Proc. of IEEE*, v.77, No 2 (1989), pp.257-286.

3. A. B. Kholodenko. Using Lexical Analyzers with Pattern Recognition for Natural Languages, in: Intelligent Suystems, v.4 (1-2), in Russian (Moscow, 1999).

4 I. L. Mazurenko. On Reduction of an Exhaustive Search in Dictionaries of Speech Commands in ASR Systems, in: *Intelligent Systems*. V. 2 (1-4), in Russian (Moscow, 1997).

5 I. L. Mazurenko. Multi-channel Speech Recognition System, in: *Proceedings of VI All-Russia Conference "Neurocomputers and Applications"*, in Russian (Moscow, February 16-18, 2000).

6 V. N. Sorokin. Speech Synthesis, in Russian (Nauka Publishing, Moscow, 1992).

7. S. V. Kovatsenko. On Reconstruction of Segmentation of Dynamic Patterns in: Discrete Mathematics, v.8 (4) , in Russian (Moscow, 1996), p.143-149.

4. MATHEMATICAL MODELS IN ECONOMICS

A MATHEMATICAL MODEL OF CONTROLLING THE PORTFOLIO OF A COMMERCIAL BANK

Elena M. Krasavina, Aleksey P. Kolchanov and Aleksandr N. Rumyantsev*

1. INTRODUCTION

Nowadays methods of mathematical modeling are widely used in Economics. In this regard banking is one of the most promising field. The necessity of using methods of mathematical modeling in banking stems from the fact that the present-day Russia banking system has very constrained possibilities for effective lend-borrow operating (in view of the unstable resource base, stringent credit limitations, tendency for a decrease of interest rates and a decline of profit margins of bank operation). In this connection there is growing in importance the scientific approach to solving the bank resources control problem. This approach enables one to take into account complicated economic interrelations, the multitude of external and internal factors in the activity of a bank. In this fields, a crucial role is played by mathematical modeling [1, 2] with the use of the theory of functional differential equations [3, 4]. As proposed in what follows, mathematical models in the form of boundary value problems for impulse functional differential systems can be studied using the reliable computer experiment (in sense of [5]).

2. PRELIMINARIES FROM THE THEORY OF FUNCTIONAL DIFFERENTIAL SYSTEM

Here we follow the books [3, 4]. Let R^n be the linear space of columns with $\alpha = \mathrm{col}\{\alpha_1,\dots,\alpha_n\}$, $\|\alpha\| = \max_{1\le i\le n}|\alpha_i|$; L be the space of integrable functions $z:[0,T]\to R^n$,

* Elena M. Krasavina, MIREA, Technical University, Moscow, Russia 117454, Aleksey P. Kolchanov, Aleksandr N. Rumyantsev, Perm State University, Perm, Russia 614600.

Mathematical Modeling: Problems, Methods, Applications
Edited by Uvarova and Latyshev, Kluwer Academic/Plenum Publishers, 2001

$$\| z \|_L = \int_0^T \|z(s)\| ds .$$

Fix a system of rational points $t_1, \ldots, t_m$, $0 = t_0 < t_1 < \ldots < t_m < t_{m+1} = T$. Denote by $DS(m)$ the space of functions $y : [0,T] \to R^n$, $\dot{y} \in L$, of the form

$$y(t) = y(0) + \int_0^t \dot{y}(s) ds + \sum_{q=1}^{m} \chi_{[t_q,T]}(t) \Delta y(t_q) ,$$

where $\Delta y(t_q) = y(t_q) - y(t_q - 0)$; $\chi_{[t_q,T]}(t)$ is characteristic function of $[t_q, T]$:

$$\chi_{[t_q,T]}(t) = \begin{cases} 1, & t \in [t_q, T] \\ 0, & t \notin [t_q, T] \end{cases}, \quad q = 1, \ldots, m .$$

This space equipped with the norm $\|y\|_{DS(m)} = \|\dot{y}\|_L + \|\Delta y\|_{mn+n}$, is a Banach space. Below Δy stands for $\operatorname{col}\{y(0), \Delta y(t_1), \ldots, \Delta y(t_m)\}$.

Consider the equation

$$(\mathrm{L}y)(t) = f(t), \ t \in [0,T], \tag{1}$$

with a linear bounded operator $\mathrm{L} : DS(m) \to L$, $f \in L$. We assume that the principal part of L, i.e. the operator $Q : L \to L$, $Qz = \mathrm{L}(\int_0^{(\cdot)} z(s) ds)$, is invertible. In this case the space of all solutions to the homogeneous equation

$$(\mathrm{L}y)(t) = 0 , \ t \in [0,T] \tag{2}$$

is finite-dimensional, its dimension equals $mn + n$. Let $\{y_1, \ldots, y_{mn+n}\}$ be a basis in this space. A matrix $Y(\cdot) = \{y_1(\cdot), \ldots, y_{mn+n}(\cdot)\}$ called to be a fundamental matrix to Eq. (2). We deal with Y such that $\Delta Y = E_{mn+n}$ (the identity $(mn+n) \times (mn+n)$ matrix). The so-called principal boundary value problem

$$\begin{gathered} (\mathrm{L}y)(t) = f(t), \ t \in [0,T], \\ \Delta y = \alpha , \end{gathered} \tag{3}$$

is uniquely solvable for any $f \in L$, $\alpha \in R^{mn+n}$.

Let $\ell : DS(m) \to R^N$ be a linear bounded vector functional. There takes place the representation

$$\ell y = \int_0^T \Phi(s) \dot{y}(s) ds + \Psi_0 y(0) + \sum_{q=1}^{m} \Psi_q \Delta y(t_q),$$

where the elements of $N \times n$ matrix Φ are essentially bounded and Ψ_q, $q = 0, \ldots, m+1$, are $N \times n$ matrix with real elements. The system

$$(\mathrm{L}y)(t) = f(t), \ t \in [0,T], \tag{4}$$

$$\ell y = \beta ,$$

is called the linear boundary value problem.

The solvability (the unique solvability) of Eq. (4) is equivalent to the solvability (the unique solvability) with respect to $\alpha \in R^n$ of the linear algebraic system

$$\ell Y \cdot \alpha = \beta - \ell g , \tag{5}$$

where ℓg is the value of the vector functional ℓ over g, the solution of the principal boundary value problem

$$\begin{gathered}(\mathrm{L}y)(t) = f(t),\ t \in [0,T], \\ \Delta y = 0 .\end{gathered} \tag{6}$$

In case $N = mn + n$, a necessary and sufficient condition of the unique solvability of Eq. (4) is the invertibility of the matrix ℓY:

$$\det \ell Y \neq 0 . \tag{7}$$

The system

$$\begin{gathered}(\mathrm{L}y)(t) = f(t),\ t \in [0,T], \\ \ell y \leq \beta ,\end{gathered} \tag{8}$$

is called the boundary value problem with inequalities. Problem Eq. (8) is solvable when there is solvable the linear system of inequalities

$$\ell Y \cdot \alpha \leq \beta - \ell g . \tag{9}$$

3. A RELIABLE COMPUTING TECHNOLOGY IN THE STUDY OF BOUNDARY VALUE PROBLEMS FOR THE SOLVABILITY

It is clear that the condition Eq. (7) can not be examined by direct calculations since the fundamental matrix $Y(t)$ and the matrix ℓY can not be calculated in explicit form (generally speaking). A computerized technique of reliable examining Eq. (7) is based on the known invertible operator theorem which states that the invertibility of $M = \ell Y$, $M = \{m_{ij}\}_{i,j=1}^{mn+n}$, is established if there is constructed an invertible matrix M_a, $M_a = \{m_{ij}^a\}_{i,j=1}^{mn+n}$, such that

$$\|M - M_a\| \leq \frac{1}{\|M_a^{-1}\|} . \tag{10}$$

As is shown in [5, 4], such an M_a with rational elements can be found (of course, for M being invertible) among matrices $\{\overline{\ell}\overline{Y}\}$, where $\overline{\ell} : DS(m) \to R^{mn+n}$ is a linear bounded vector functional closed to ℓ and $\overline{Y}$ is a fundamental matrix to $\overline{\mathrm{L}}y = 0$ with a linear bounded operators $\overline{\mathrm{L}} : DS(m) \to L$ closed to L. There are worked out algorithms of constructing $\overline{\ell}$ and $\overline{\mathrm{L}}$ such that in parallel with this there is being constructed an error bound matrix M_{v} with rational elements m_{ij}^{v} such that

$$\left| m_{ij} - m_{ij}^a \right| \leq m_{ij}^{\mathrm{v}},\ i,j = 1,\ldots,mn+n . \tag{11}$$

Since M_a и M_a^{-1} are exactly calculated, the condition

$$\|M_v\| \le \frac{1}{\|M_a^{-1}\|} \tag{12}$$

can be examined by direct calculations. It is clear that inequality implies Eq. (10).

As for problem Eq. (8), in [6] there is proposed the following way of examining this problem for the solvability. Define the matrices $\underline{M}$ and $\bar{M}$ by

$$\underline{M} = M_a - M_v, \quad \bar{M} = M_a + M_v.$$

Obviously that $\underline{M} \le M \le \bar{M}$. In a similar way the estimate $\underline{d} \le \ell g \le \bar{d}$ is constructed where g is the solution of Eq. (6). The system Eq. (9) has a solution if the system

$$\bar{M}\alpha \le \beta - \bar{d} \tag{13}$$

is solvable. For $\bar{\alpha} \in R^{mn+n}$ being a solution of Eq. (13) a solution $\bar{y}$ of principal boundary value problem

$$\begin{gathered}(\mathrm{L}y)(t) = f(t),\ t \in [0,T], \\ \Delta y = \bar{\alpha}\end{gathered} \tag{14}$$

is also a solution of Eq. (8).

4. A MODEL OF CONTROLLING THE BANK PORTFOLIO

The goal of the control problem for the commercial bank portfolio is the construction of optimal (with respect to a given criterium) volumes, terms and rates of interest in the operations over the period of time under consideration. Contrary [1, 2], we consider here a model with the continuous time-variable. Such an approach is reasonable in the case of a big bank with a wide variety of assets and liabilities.

Denote by y the vector of bank resources (cash assets, liquid securities, interest-earning long-term assets, investment portfolio, inter-bank credits, deposits, bills, and so on); $y = \mathrm{col}\{y_1, \dots, y_n\}$, y_i is the i-th kind of resources. Let us consider $y \in DS(m)$ with t_q, $q = 1, \cdots, m$, being the points in time at which control is being brought into operation. In such the points a jump-like change of a bank resources is possible. The equation governing the resource dynamics can be written in the form

$$\dot{y}(t) = (Wy)(t) - (Fy)(t),\ t \in [0,T]. \tag{15}$$

Here an operator W describes an increase of the resources and F does a decrease of ones. It is to be noted that W and F are the so-called Volterra operators and they enable one to take into account the prehistory of the resources dynamics. It is clear that in practical banking there takes place an aftereffect (time delay). Volterra operators (memory operators) enable one to model the aftereffect in detail.

Consider the case of linear equation Eq. (15). In this case we can write Eq. (15) in the form

$$\dot{y}_i(t) - \sum_{j=1}^{n} \sum_{k=1}^{n_{ij}} p_{ij}^k(t) y_j\left[h_{ij}^k(t)\right] = f_i(t),\ t \in [0,T]. \tag{16}$$

$$y_j(\xi) = \varphi_j(\xi),\ \xi < 0,\ i,j = 1,\dots,n.$$

Here $f = \mathrm{col}\{f_1,\ldots,f_n\} \in L$, function $h_{ij}^k : [0,T] \to R$ is piece-wise continuous, $h_{ij}^k(t) \le t$. The delayed argument $h_{ij}^k(t)$ and the prehistory of the resources dynamics φ_j are assumed to be given.

There are three groups of restrictions with respect to the desired trajectory:

- the qualitative restrictions;
- the restrictions imposed by the Central Bank;
- the market restrictions.

The qualitative restrictions are generated primarily due to the bank risks. To bound the risk of loss in the liquidity, the restrictions to a minimal value of the liquid assets are entered. Due to similar reasoning with respect to the credit risk and the interest risk, some corresponding restrictions are added too. Another restriction arises as a goal of control. For example, the minimal value of the net profits (overall return) can be prescribed. In case of linear model the restrictions mentioned above can be written in the form $\ell_s y \le \gamma_s$, $s = 1,\ldots,n_1$, with ℓ_s, linear functionals.

The restrictions imposed by the Central Bank include

- the balance equations with taking into account obligatory reserves at prescribed points in time: $\ell_s y = \gamma_s$, $s = n_1,\ldots,n_2$;
- some limitations by the Central Bank, which can also be written in the for $\ell_s y \le \gamma_s$, $s = n_2,\ldots,n_3$.

The market restrictions are entered to take into account practical possibilities of the bank resources operation in financial markets. In particular, there are imposed requirements on a maximum value of investments securities under operation. Besides, some markets impose a lower bound of the rate of a bargain (say, the bargains in the inter-bank market are prescribed to be gross in value). These restrictions can also be written in the form $\ell_s y \le \gamma_s$, $s = n_3,\ldots,n_4$.

Thus all of the above restrictions allow the form

$$\ell_s y \le \gamma_s, \quad s = 1,\ldots,\tilde{n}. \tag{17}$$

Note that, in general $\tilde{n} \ne mn + n$.

The final setting of the control problem is as follows: find a trajectory $y \in DS(m)$ that is a solution to Eq. (16), satisfies the system of inequalities Eq. (17) and a given optimum criterion. Using the computerized technique mentioned above, the control problem is transacted to a linear programming problem in $c = \mathrm{col}\{c_1,\ldots,c_{mn+n}\}$:

$$\left\{ \sum_{i=1}^{mn+n} a_i c_i \,:\, \Lambda c \le b \right\} \to \min\,(\max). \tag{18}$$

A solution $\tilde{c}$ of this problem enables one to define a solution of the control problem as the solution $\tilde{y}$ of Eq. (16) with the condition $\Delta\tilde{y} = \tilde{c}$.

REFERENCES

1. A.P.Kolchanov, A model of optimal controlling the bank portfolio (Russian), *Mathematics and Applied Mathematics. Bulletin of Perm State Techn.Univ.*, Perm, 52-47 (1999).

2. A.S.Lapushkin, A.N.Rumyantsev, A Mathematical model of controlling the portfolio of a commercial bank, *Russian Proceeding of the IV Intern. Conference on Math. Modeling*, STANKIN,Moscow, 289-294 (2001).
3. N.V.Azbelev, V.P.Maksimov, and L.F.Rakhmatullina, *Introduction to the Theory of Linear Functional Differential Equations* (World Federation Publ., Atlanta, 1995).
4. N.V.Azbelev, V.P.Maksimov, and L.F.Rakhmatullina, *Methods of the Contemporary Theory of Linear Functional Differential Equations(Russian)* (Regular and Chaotic Dynamics, Izhevsk, 2000).
5. A.N.Rumyantsev, *The Reliable Computing Experiment in the Study of Boundary Value Problems (Russian)* (Perm State University Press, Perm, 1999).
6. V.P.Maksimov, A.N.Rumyantsev, Boundary value problems and problems of pulse control in economic dynamics. Constructive study, *Russian Mathematics(IzVUZ)*, **37**(5), 48-62 (1993).

TUTORING PROCESS AS OBJECT FOR SITUATIONAL CONTROL

Victor I. Miheev, MariaV. Massalitina, Igor L.Tolmachev *

1. INTRODUCTION

At present the great variety of computer tutoring systems exists, although generally the principle of information transfer from the computer to the Student and certain testing procedure, checking some skills of Student, form the basis of the system. But in reality the tutoring process some closer interactions between the Tutor and the Student is implying [1]. The paper presents an analysis of these interactions and the automation means used.

2. ESSENTIAL ASPECTS OF ITS

Let examine the essential aspects of Intellectual Tutoring System (ITS) that can be considered as system with situational control over tutoring process, where Student is an object of control and the control system simulates tutor's actions.

The automation subject is a tutor's activity, and the object of control, i.e. the Student, has several degrees of freedom, which provide him with maximal possibilities for his activity within the framework of the studied data domain.

Now an interaction model "Tutor-Student" in more detailed form is described. It is based on two aspects (see Figure 1):

1. Formalization of profuse theoretical pedagogical experience (observation and interpretation) on Student conduct diagnostics.

2. Revealing external manifestations of characteristics of Student's thinking.

In the paper the first aspect is regarded.

Note that most of proposed principles can be applied to a model, where the Student's actions can be evaluated on "Question-Answer" principle. Statistical methods suitable for metric scales of measurement to resolve a problem of pedagogical measurements and evaluations can be used.

However in this situation much qualitative information is appeared to be lost; consequently suitability of interpretation of the situation is considerably reduced. Defects aroused can be diminished by study of Student's task-deciding process.

---

* Victor I. Miheev, MariaV. Massalitina, Igor L.Tolmachev, Peoples' Friendship University of Russia, Moscow, Russia

Mathematical Modeling: Problems, Methods, Applications
Edited by Uvarova and Latyshev, Kluwer Academic/Plenum Publishers, 2001

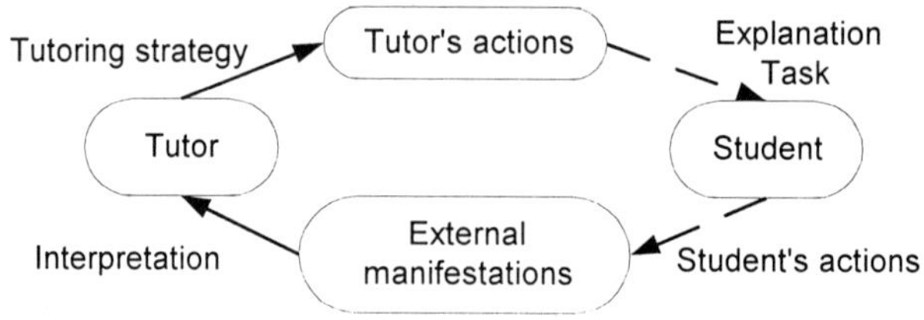

Figure 1. Tutoring model with a feedback

Using the stated approach of tutoring process modeling the conception based on the control system, which simulate tutor activity, is proposed. Subsequently, let examine a question related to semantic data domain description. That have relevance to pedagogics, and then, let analyze the problem of formation of control decisions sequences to determine the goal to be sought implying teaching objectives in this situation.

3. CONTROL PARADIGM OVER TUTORING PROCESS

Generally, it is assumed to see a knowledge representation in the form of theme spaces or concept spaces. We mean that the knowledge in content domain is not a static construction or substance that can be transmitted to the Student. Human knowledge is composed in real time through some cross scale relationship between memory substructures and the characteristics of the present moment. If human memory and our awareness of the contents of mental events are at two separated time scales, then a static mono-level representation of memory and knowledge is inappropriate. The dynamic real time emergence of knowledge representation is also required.

In general the training can be represented as the control over Student's knowledge and skills with considering the Student's individual characteristics. Note that we use the Leontiev's theory of activity as methodological device for analysis of Student's behavior. Considering Student's individual characteristics requires a solution of a diagnostic problem i.e. it is necessary to estimate the Student's psychological characteristics and cognitive activities.

However, the training theories are not sufficiently investigated and are weak-formalizable therefore the Tutor's system situation is characterized by uniqueness and dynamics of an object of control, incompleteness of knowledge, importance of subjective factors. Thus any control will not be thorough without the situation semantic revising.

The Tutor's model is also necessary Quasi Axiomatic Theory (QAT). In QAT, the set of quasi axioms is a semiotically open set: a set whose membership rules are non-stationary. To see the notion of a semiotically open set, let O be a set of interpretations of the perceived signs of an object. If ω is a measurement device, then the collection of pre-experiences, $\omega^{-1}(O)$, must be regarded as composed from some theoretical construct consisting of unmeasured states. The construct is modeled around some theoretical notions about unmeasured quantum mechanical states. The set of states must be considered as open set where, at any time, a new class of states may be measured from this construct [2].

As control paradigm we use the theory of situational control. Traditional control methods are based on some formal model, while situational control method is based on the semiotic model. In the semiotic systems the ontologies can be created using the extensive logical theory and heuristic methods developed by Pospelov.

Following Pospelov [3], we use the term "formal system" to refer to a four-term expression:

$$M = \langle T, P, A, R \rangle \tag{1}$$

where T = set of basic elements, P = syntactic rules, A = set of axioms, and R = semantic rules. This notation describes a first order cybernetic system. The rule modification transforms, of a second order cybernetic system, apply to the set of basic elements, to the syntactic rules, to the axioms, and to the semantic rules. In Pospelov's notation, a second order cybernetic system has the following form:

$$C = \langle M, C(T), X(P), X(A), X(R) \rangle \tag{2}$$

where $X(\cdot)$ is an operator capable of introducing change to the associated set. In comparing Eq. (1) and Eq. (2) the time has been introduced. Thus in C the basic elements properties (syntax, semantics, pragmatics) may be revised.

4. SEMANTIC DATA DOMAIN DESCRIPTION

In situational control a language is needed that is sufficient for the description of situations that need to be controlled, and has formal properties that allow the inference engines to perform automated situational analysis. The imposition of a grammar on the elements of a bag of tokens, in the language, is one methodology that results in generative situational analysis.

Any situation on the object of control or in the control system can be described on above-mentioned language, with some lexical units of data domain dictionary usage. In this case the dictionary of the basic lexical elements can be represented as follows: the basic lexicon dictionary contains subsequent base concepts (and each of these concepts is in turn a set of concepts): Instructor, Student, subject skill, generalized skill (individual characteristic), pedagogical principle, pedagogical rule, task, problem, explanation etc.

The dictionary contains following operations-decisions assumptions of control system: to give a task, to give an explanation, to inform about correctness/incorrectness of certain operation, operation's refinement requirement etc.

Finally, this dictionary contains following base relationships: assume the name, assume an attribute, attribute-measure, measure-value, cause-effect etc.

Evidently, the concepts and actions dictionaries are wholly defined by problem domain semantics, and the nature of relationship dictionary is broader. A syntagmatic unit of the situational control language is an ordered triple in the form (X R Z) where X and Z are concepts and R is a relational variable.

The following example illustrates the pedagogical rule's formalizing.

Example. "The estimation of Student's capability to explain his operations in detail enables to estimate such knowledge property as comprehension."

Notions:	Relationships:
- Skill (p_1)	- Assume the name (r_1)
- Explanation of operations in detail (p_2)	- Assume the attribute (r_2)
- Estimation (p_3)	- Attribute-measure (r_3)
- Knowledge (p_4)	- Cause-effect (r_4)
- Property (p_5)	
- Comprehension (p_6)	

$$((p_1r_1p_2)r_2p_3)r_4(p_4r_2(p_5r_1p_6))r_2p_3)$$

Considering the semantics of a simulated activity sphere concerned to pedagogics, a quantitative information evaluation, contained in qualitative primary evaluations, turn out as one of the key problems. Therefore the basic notions, which must be formalized in process of translation from natural language into the situational-control language, are qualitative notions. To overcome the formalization difficulties we use the fuzzy sets device. In this case each attribute in the situation description can be considered as linguistic variable with a set of terms, which are qualitative data estimations aroused from the Student's actions.

5. FORMATION OF CONTROL DECISIONS SEQUENCES

Let examine formation of control-decisions sequence by transition from a "current situation" to a "goal situation". As a type of planning procedure we chose a planning on the "positions": each "position" represents the object of control state, environment "position" and "position" of control system. The plan represents in this case a certain path in the "position" space and it can be named as a "control strategy". In examined data domain a plan building corresponds to a choice of certain pedagogics methods and principles.

The closing of initial set of facts generates each system "position". The synchronous and diachronous transition rules are used here [4]. The synchronous transition rules change the estimation of characteristics (for example, Student's knowledge and skills) that can be estimated after each Student's action, and the diachronous transition rules change estimation of characteristics obtained. It can be estimated only by means of accumulation of information.

To build the plan the following mode is used. Let envisage a network where each node is symbolizing a current situation. This situation represents a standard fuzzy situation describing some state of Student's knowledges and skills. Each control decision is based on the rule of the rule-based system. The control decisions for the fuzzy input situation are based on analyzing of permitted transactions from a current situation to a goal situation [5]. A goal situation in the network is defined by analysis of preference degrees of control decisions. To determine the goal situation it is necessary to construct the fuzzy situational net (FSN) presenting the fuzzy oriented weighted graph. The nods of FSN correspond to the standard fuzzy situations; the arcs are marked at control decision and are weighted by selected degrees of decisions. The preference degrees depend upon the situation and demonstrate the Tutor's strategy choice in current situation. The control decision in current situation is the decisions sequence that transits from current "position" to goal "position" on the FSN path that represent control strategy.

Note that in present model a notion of tutoring strategy is very close to control strategy.

6. CONCLUSION

The paper presents an overall approach to simulation and formalization of a tutoring process as applied to construction of a Intellectual Tutoring System with a feedback. Tutoring system is considered here as system with situational control over tutoring process. The use of fuzzy sets is proposed to formalize the data domain.

The main advantage of presented model is possibility of selection of tasks system to develop the new Student's knowledge and skills with basing on an acquired material, but with guiding mainly on his individual characteristics.

REFERENCES

1. V.I. Miheev, I.L. Iolmachev, Modeling of process of training, *Proceeding of the 4th International Seminar on Applied Semiotics, Semiotic and Intellectual Control*, Moscow, 1999, pp.143-147. (in Russian)
2. Paul S. Prueitt, The Autonomous Organization of Data through Semiotic Methods (unpublished); http://www.bcngroup.org/area2/review.htm
3. D.A. Pospelov, Situational Control: theory and practice, Moscow, 1986 (in Russian)
4. G.S. Osipov, Dynamics in Integrated Knowledge-Based Systems, *Proceedings of the 1998 IEEE International Symposium on Intelligent Control*,Gaithersburg, Maryland, USA, September, 1998, pp.199-203.
5. A.N. Melihov, L.S. Bershtein, C.I. Korovin, Situational systems with fuzzy logic, Moscow, 1990 (in Russian)

5. NONLINEAR MODELS IN CHEMICAL PHYSICS AND PHYSICAL CHEMISTRY

NONLINEAR DYNAMICS OF STRONGLY NON-HOMOGENEOUS CHAINS WITH SYMMETRIC CHARACTERISTICS

D.V. Gorlov, L.I. Manevitch*

1. INTRODUCTION

Homogeneous chains of coupled nonlinear oscillators have become the subject of great interest in connection with different physical and mechanical problems[1,2]. One of the most efficient techniques for study of corresponding mathematical problems implies the use of complex variables and further multiple–scale expansions[2].

If the inertial and elastic characteristics of coupled oscillators are weakly non–homogeneous, the solution for homogeneous chain can be considered as appropriate first approximations. However, in many cases we deal with strongly non–homogeneous chains. Such a inhomogeneity can be caused, e.g. by significant difference of masses or elastic characteristics. In this case, alongside with small parameters reflecting a weak nonlinearity or coupling between oscillators it is necessary to take into account new small parameters characterized isotopic or elastic inhomogeneity of the chain.

We present a general asymptotic approach to nonlinear dynamics of non–homogeneous chains based on generalization of complex representations for homogeneous chains proposed in[2,3].

2. INITIAL SYSTEM

Let us consider an infinite system of coupled nonlinear oscillators with strong difference between alternating inertial or elastic characteristics. The dynamics of this system is described by the infinite system of nonlinear equations (Eqs. 1). We use the dimensionless variables $u_{j,k} = U_{j,k}/r_0$ where r_0 is the distance between particles, $U_{j,k}$ are displacements of corresponding oscillators.

* D. V. Gorlov, Russian State University of Oil and Gas, Leninsky prosp., 65, Moscow, L. I. Manevitch, Institute of Chemical Physics, Kosygin str., 4, Moscow, Russia.

$$m_1 \frac{d^2 u_{j,1}}{dt^2} + c_1 u_{j,1} + c_{13} r_0^2 u_{j,1}^3 + c(2u_{j,1} - u_{j-1,2} - u_{j,2}) = 0$$
$$-\infty < j < \infty \qquad (1)$$
$$m_2 \frac{d^2 u_{j,2}}{dt^2} + c_2 u_{j,2} + c_{23} r_0^2 u_{j,2}^3 + c(2u_{j,2} - u_{j,1} - u_{j+1,1}) = 0$$

Parameter c characterizes the linear coupling between oscillators, m_k are masses, c_k. and c_{k3} specify elastic properties of k oscillator (k =1,2).

One can consider different cases depending on the values of coefficients.

3. INERTIAL INHOMOGENEITY

Considering the case of inertial inhomogeneity we mean that $m_2/m_1 = \varepsilon << 1$. This relation gives us a small parameter for the system under consideration.

3.1. Acoustic branch

Acoustic branch corresponds to low frequency modes, and the order of chain oscillation period (with respect to small parameter ε) is the same as for low–frequency oscillator. That is why we introduce dimensionless time variable

$$\tilde{\tau} = \omega_0 t \qquad (2)$$

where $\omega_0 = \sqrt{c_1/m_1}$ is the frequency of the linearized low–frequency oscillator.

So, the initial system is reformulated as

$$\frac{d^2 u_{j,1}}{d\tilde{\tau}^2} + u_{j,1} + \frac{c_{13} r_0^2}{c_1} u_{j,1}^3 + \frac{c}{c_1}(2u_{j,1} - u_{j-1,2} - u_{j,2}) = 0$$
$$\varepsilon \frac{d^2 u_{j,2}}{d\tilde{\tau}^2} + \frac{c_2}{c_1} u_{j,2} + \frac{c_{23} r_0^2}{c_1} u_{j,2}^3 + \frac{c}{c_1}(2u_{j,2} - u_{j,1} - u_{j+1,1}) = 0 \qquad (3)$$

The most interesting situation for acoustic branch corresponds to estimations

$$\frac{c_2}{c_1} = \tilde{\nu} \quad \frac{c_{23} r_0^2}{c_2} = \varepsilon^2 \tilde{\nu}_3 \quad \frac{c_{13} r_0^2}{c_2} = \varepsilon^2 \tilde{\xi}_3 \quad \frac{c}{c_2} = \tilde{\eta} \qquad (4)$$

where $\tilde{\nu}$, $\tilde{\nu}_3$, $\tilde{\xi}_3$, $\tilde{\eta} \approx O(1)$.

3.1.1. *Acoustic Branch, Long Wavelength*

One can consider a continuum function $y_k(\tilde{\tau}, \chi)$, χ being a new variable for distance measuring. It is expedient to assume that wavelength is unity, then the distance between particles is $\rho\varepsilon/2r_0$, $\rho = O(1)$. We can use series expansion of $y_k(\tilde{\tau}, \chi)$ for identical oscillators of each type:

$$u_{j,1}=y_1(\tilde{\tau},\chi)\quad u_{j+1,1}=y_1(\tilde{\tau},\chi)+\varepsilon\rho\frac{\partial y_1(\tilde{\tau},\chi)}{\partial\chi}+\frac{\varepsilon^2\rho^2}{2}\frac{\partial^2 y_1(\tilde{\tau},\chi)}{\partial\chi^2}+\dots$$
$$u_{j,2}=y_2(\tilde{\tau},\chi)\quad u_{j-1,2}=y_2(\tilde{\tau},\chi)-\varepsilon\rho\frac{\partial y_2(\tilde{\tau},\chi)}{\partial\chi}+\frac{\varepsilon^2\rho^2}{2}\frac{\partial^2 y_2(\tilde{\tau},\chi)}{\partial\chi^2}-\dots \tag{5}$$

These expansions lead us to the system

$$\frac{\partial^2 y_1}{\partial\tilde{\tau}^2}+y_1+\varepsilon^2\tilde{\xi}_3 y_1^3+2\tilde{\eta}(y_1-y_2)-\tilde{\eta}\left(-\varepsilon\rho\frac{\partial y_2}{\partial\chi}+\frac{\varepsilon^2\rho^2}{2}\frac{\partial^2 y_2}{\partial\chi^2}-\dots\right)=0$$
$$\varepsilon\frac{\partial^2 y_2}{\partial\tilde{\tau}^2}+\tilde{\nu}y_2+\varepsilon^2\tilde{\nu}_3 y_2^3+2\tilde{\eta}(y_2-y_1)-\tilde{\eta}\left(\varepsilon\rho\frac{\partial y_1}{\partial\chi}+\frac{\varepsilon^2\rho^2}{2}\frac{\partial^2 y_1}{\partial\chi^2}+\dots\right)=0 \tag{6}$$

We have to express y_2 from the second equation as follows:

$$y_2=\frac{2\tilde{\eta}}{2\tilde{\eta}+\tilde{\nu}}y_1-\frac{\varepsilon\Gamma}{2\tilde{\eta}+\tilde{\nu}}\qquad \Gamma=\frac{\partial^2 y_2}{\partial\tilde{\tau}^2}+\varepsilon\tilde{\nu}_3 y_2^3-\tilde{\eta}\left(\rho\frac{\partial y_1}{\partial\chi}+\frac{\varepsilon\rho^2}{2}\frac{\partial^2 y_1}{\partial\chi^2}+\dots\right) \tag{7}$$

and introduce a new change of time variable and make some changes in parameters to simplify system

$$\tau=\sqrt{\gamma}\tilde{\tau}\qquad \gamma=\frac{2\tilde{\eta}\tilde{\nu}+2\tilde{\eta}+\tilde{\nu}}{2\tilde{\eta}+\tilde{\nu}}\qquad \eta=\frac{\tilde{\eta}}{\gamma}\qquad \nu=\frac{\tilde{\nu}+2\tilde{\eta}}{\gamma}\qquad \nu_3=\frac{\tilde{\nu}_3}{8\gamma}\qquad \xi_3=\frac{\tilde{\xi}_3}{8\gamma} \tag{8}$$

After dividing by γ we come to a set of equations

$$\frac{\partial^2 y_1}{\partial\tau^2}+y_1+\varepsilon^2 8\xi_3 y_1^3+\varepsilon^2\frac{2\eta}{\nu}8\nu_3 y_2^3+\varepsilon\frac{2\eta}{\nu}\frac{\partial^2 y_2}{\partial\tau^2}+\varepsilon^2\frac{\eta}{\nu}\left(-\rho\frac{\partial\Gamma}{\partial\chi}+\frac{\varepsilon\rho^2}{2}\frac{\partial^2\Gamma}{\partial\chi^2}-\dots\right)=0$$
$$\varepsilon\frac{\partial^2 y_2}{\partial\tau^2}+\nu y_2+\varepsilon^2 8\nu_3 y_2^3-2\eta y_1-\eta\left(\varepsilon\rho\frac{\partial y_1}{\partial\chi}+\frac{\varepsilon^2\rho^2}{2}\frac{\partial^2 y_1}{\partial\chi^2}+\dots\right)=0 \tag{9}$$
$$\Gamma=\frac{\partial^2 y_2}{\partial\tau^2}+\varepsilon 8\nu_3 y_2^3-\eta\left(\rho\frac{\partial y_1}{\partial\chi}+\frac{\varepsilon\rho^2}{2}\frac{\partial^2 y_1}{\partial\chi^2}+\dots\right)$$

Then one can apply an approach given in[2] :

1. We use complex representation of unknown functions

$$\psi_k=\frac{\partial y_k}{\partial\tau}+iy_k\quad \psi_j^*=\frac{\partial y_k}{\partial\tau}-iy_k\quad i=\sqrt{-1} \tag{10}$$

2. After introducing of the new variables and "slow" times

$$\varphi_k=e^{i\tau}\psi_k\quad \tau_1=\varepsilon\tau_0\quad \tau_2=\varepsilon^2\tau_0\dots\text{ where }\quad \tau_0=\tau \tag{11}$$

   we consider φ_k as functions of many times τ_0, τ_1, τ_2, ...

3. Then we present a solution in a form of power expansion as follows

$$\varphi_k(\tau_0,\tau_1,\tau_2,\ldots)=\varphi_{k,0}+\varepsilon\varphi_{k,1}+\varepsilon^2\varphi_{k,2}+\ldots \tag{12}$$

4. After substitutions (Eqs. 10–12) both equations are divided by $e^{i\tau_0}$.

Scheme (10)–(12) transforms the terms in (9) in such way

$$\begin{gathered}
\frac{\partial^2 y_k}{\partial\tau^2}\longrightarrow\frac{\partial\varphi_{k,0}}{\partial\tau_0}+\varepsilon\left(\frac{\partial\varphi_{k,1}}{\partial\tau_0}+\frac{\partial\varphi_{k,0}}{\partial\tau_1}\right)+\varepsilon^2\left(\frac{\partial\varphi_{k,2}}{\partial\tau_0}+\frac{\partial\varphi_{k,1}}{\partial\tau_1}+\frac{\partial\varphi_{k,0}}{\partial\tau_2}\right)+\ldots+\frac{i}{2}E_k\\
y_k\longrightarrow-\frac{i}{2}E_k\qquad y_k^3\longrightarrow\frac{i}{8}E_k^3e^{2i\tau_0}\\
E_k=(\varphi_{k,0}+\varepsilon\varphi_{k,1}+\ldots)-(\varphi^*_{k,0}+\varepsilon\varphi^*_{k,1}+\ldots)e^{-2i\tau_0}
\end{gathered}\tag{13}$$

System (Eq. 9) after applying Eq. (13)

$$\begin{gathered}
\frac{\partial\varphi_{1,0}}{\partial\tau_0}+\varepsilon\left(\frac{\partial\varphi_{1,1}}{\partial\tau_0}+\frac{\partial\varphi_{1,0}}{\partial\tau_1}\right)+\varepsilon^2\left(\frac{\partial\varphi_{1,2}}{\partial\tau_0}+\frac{\partial\varphi_{1,1}}{\partial\tau_1}+\frac{\partial\varphi_{1,0}}{\partial\tau_2}\right)+\ldots+\varepsilon^2 i\xi_3E_1^3e^{2i\tau_0}+\\
\varepsilon^2 i\frac{2\eta\nu_3}{\nu}E_2^3e^{2i\tau_0}+\varepsilon\frac{2\eta}{\nu}\left[\frac{\partial\varphi_{2,0}}{\partial\tau_0}+\varepsilon\left(\frac{\partial\varphi_{2,1}}{\partial\tau_0}+\frac{\partial\varphi_{2,0}}{\partial\tau_1}\right)+\ldots\right]+\\
+\varepsilon^2\frac{\eta}{\nu}\left(-\rho\frac{\partial\Gamma}{\partial\chi}+\frac{\varepsilon\rho^2}{2}\frac{\partial^2\Gamma}{\partial\chi^2}-\ldots\right)=0
\end{gathered}\tag{14}$$

$$\begin{gathered}
\varepsilon\left[\frac{\partial\varphi_{2,0}}{\partial\tau_0}+\varepsilon\left(\frac{\partial\varphi_{2,1}}{\partial\tau_0}+\frac{\partial\varphi_{2,0}}{\partial\tau_1}\right)+\ldots\right]-\frac{i}{2}\nu E_2+\varepsilon^2 i\nu_3E_2^3e^{2i\tau_0}+\eta iE_1+\\
+\frac{i\eta}{2}\left(\varepsilon\rho\frac{\partial E_1}{\partial\chi}+\frac{\varepsilon^2\rho^2}{2}\frac{\partial^2E_1}{\partial\chi^2}+\ldots\right)=0
\end{gathered}$$

$$\Gamma=\frac{\partial\varphi_{2,0}}{\partial\tau_0}+\varepsilon\left(\frac{\partial\varphi_{2,1}}{\partial\tau_0}+\frac{\partial\varphi_{2,0}}{\partial\tau_1}\right)+\ldots+\frac{\partial^2y_2}{\partial\tau^2}+\varepsilon i\nu_3E_k^3e^{2i\tau_0}+\frac{i\eta}{2}\left(\rho\frac{\partial E_1}{\partial\chi}+\frac{\varepsilon\rho^2}{2}\frac{\partial^2E_1}{\partial\chi^2}+\ldots\right)$$

is ready for following asymptotic analysis. Then we equate coefficients at each from growing powers of the parameter ε:

ε^0 The first equation gives

$$\frac{\partial\varphi_{1,0}}{\partial\tau_0}=0 \tag{15}$$

Therefore $\varphi_{1,0}=\varphi_{1,0}(\tau_1,\tau_2,\ldots,\chi)$.
Second equation is algebraic one

$$-\frac{i}{2}\left((2\eta+1)\varphi_{2,0}-2\eta\varphi_{1,0}\right)+\frac{i}{2}\left((2\eta+1)\varphi^*_{2,0}-2\eta\varphi^*_{1,0}\right)e^{-2i\tau_0}=0 \tag{16}$$

It means that

$$\varphi_{2,0} = \frac{2\eta + \alpha_1 \nu}{\nu}\varphi_{1,0} + \alpha_1 e^{-2i\tau_0}\varphi^*_{1,0} \tag{17}$$

α_1 being an undetermined coefficient, which will be found later.

ε^1 First equation with taking into account (17) is a following:

$$\frac{\partial \varphi_{1,1}}{\partial \tau_0} + \frac{\partial \varphi_{1,0}}{\partial \tau_1} + \frac{2\eta^2 i}{\nu^2}\varphi_{1,0} - \frac{2\eta(2\alpha_1\nu + \eta)i}{\nu^2} e^{-2i\tau_0}\varphi^*_{1,0} = 0 \tag{18}$$

The condition of secular terms absence leads to equation

$$\frac{\partial \varphi_{1,0}}{\partial \tau_1} + \frac{2\eta^2 i}{\nu^2}\varphi_{1,0} = 0 \tag{19}$$

therefore

$$\varphi_{1,0} = e^{iF_1\tau_1}\Phi_{1,0}(\tau_2,\ldots,\chi) \quad F_1 = -\frac{2\eta^2}{\nu^2} \tag{20}$$

The solution of Eqs. (18), (19) is a following:

$$\varphi_{1,1} = -\frac{\eta(2\alpha_1\nu + \eta)}{\nu^2} e^{-iF_1\tau_1} e^{-2i\tau_0}\Phi^*_{1,0} \tag{21}$$

Then the second equation with taking into account Eqs. (17), (20), (21) gives us

$$-\frac{\nu i}{2}\left(\varphi_{2,1} - e^{-2i\tau_0}\varphi^*_{2,1}\right) + Aie^{iF_1\tau_1}\Phi_{1,0} - Bie^{-iF_1\tau_1}e^{-2i\tau_0}\Phi^*_{1,0} + $$
$$+\frac{\eta\rho i}{2}\left(e^{iF_1\tau_1}\frac{\partial \Phi_{1,0}}{\partial \chi} - e^{-iF_1\tau_1}e^{-2i\tau_0}\frac{\partial \Phi^*_{1,0}}{\partial \chi}\right) = 0 \tag{22}$$
$$A = \frac{\eta(\eta^2 + 2\eta\nu\alpha_1 + \nu)}{\nu^2} \quad B = \frac{(\eta^2+\nu)(2\nu\alpha_1 + \eta)}{\nu^2}$$

This equation has a solution if $A=B$, that is $\alpha_1=0$ and $A=B=\eta(\eta^2+\nu)/\nu^2$. The function $\varphi_{2,1}$ is determined from Eq. (22)

$$\varphi_{2,1} = Ae^{iF_1\tau_1}\Phi_{1,0} + \alpha_2 e^{-iF_1\tau_1}e^{-2i\tau_0}\Phi^*_{1,0} + \alpha_3\rho e^{iF_1\tau_1}\frac{\partial \Phi_{1,0}}{\partial \chi} + Be^{-iF_1\tau_1}e^{-2i\tau_0}\frac{\partial \Phi^*_{1,0}}{\partial \chi}$$
$$A = \frac{2\eta^3 + 2\eta\nu + \nu^3\alpha_2}{\nu^3} \quad B = \frac{(\alpha_3\nu - \eta)\rho}{\nu} \tag{23}$$

α_2 and α_3 being undetermined coefficients.

ε^2: The first equation gives us

$$\left(\frac{\partial\varphi_{1,2}}{\partial\tau_0}+e^{iF_1\tau_1}\frac{\partial\Phi_{1,0}}{\partial\tau_2}\right)+Aie^{iF_1\tau_1}\Phi_{1,0}-Bie^{-iF_1\tau_1}e^{-2i\tau_0}\Phi_{1,0}^{*}+$$
$$+\frac{\nu^4\xi_3+16\eta^4\nu_3}{\nu^4}i\left[e^{3iF_1\tau_1}e^{2i\tau_0}\Phi_{1,0}^{\;3}-3e^{iF_1\tau_1}\left|\Phi_{1,0}\right|^2\Phi_{1,0}+3e^{-iF_1\tau_1}e^{-2i\tau_0}\left|\Phi_{1,0}\right|^2\Phi_{1,0}^{*}-\right.$$
$$\left.-e^{-3iF_1\tau_1}e^{-4i\tau_0}\Phi_{1,0}^{*\;3}\right]-\frac{4\eta\rho(\nu\alpha_3-\eta)}{\nu^2}ie^{-iF_1\tau_1}e^{-2i\tau_0}\frac{\partial\Phi_{1,0}^{*}}{\partial\chi}+\tag{24}$$
$$+\frac{\eta^2\rho^2 i}{2\nu}\left(e^{iF_1\tau_1}\frac{\partial^2\Phi_{1,0}}{\partial\chi^2}-e^{-iF_1\tau_1}e^{-2i\tau_0}\frac{\partial^2\Phi_{1,0}^{*}}{\partial\chi^2}\right)=0$$
$$A=\frac{2\eta^2\left(\nu-3\eta^2\right)}{\nu^4}\qquad B=\frac{2\eta\left(2\eta^3+\eta\nu+2\nu^3\alpha_2\right)}{\nu^4}$$

The condition of the secular terms absence gives the principal approximation for the case under consideration

$$\frac{\partial\Phi_{1,0}}{\partial\tau_2}-F_2 i\Phi_{1,0}+Ci\left|\Phi_{1,0}\right|^2\Phi_{1,0}+Di\frac{\partial^2\Phi_{1,0}}{\partial\chi^2}=0,\tag{25}$$
$$F_2=\frac{2\eta^2\left(3\eta^2-\nu\right)}{\nu^4}\qquad C=-\frac{3\nu^4\xi_3+48\eta^4\nu_3}{\nu^4}\qquad D=\frac{\eta^2\rho^2}{2\nu}$$

After change $\Theta_{1,0}=e^{iF_2\tau_2}\Phi_{1,0}$ we come to a principal equation for considered system in the form

$$\frac{\partial\Theta_{1,0}}{\partial\tau_2}+Ci\left|\Theta_{1,0}\right|^2\Theta_{1,0}+Di\frac{\partial^2\Theta_{1,0}}{\partial\chi^2}=0\tag{26}$$

3.1.2. Acoustic Branch, Short Wavelength

One can consider a continuum function $y_k(\tilde{\tau},\chi)$ as an envelope wave, χ being a new variable for distance measuring. When assuming that envelope wave length is unity the distance between particles is $\rho\varepsilon/2r_0$, $\rho=O(1)$. We can use series expansion of $y_k(\tilde{\tau},\chi)$ similar to Eq. (5):

$$\begin{aligned}u_{j,1}&=y_1(\tilde{\tau},\chi)\quad u_{j+1,1}=-\left(y_1(\tilde{\tau},\chi)+\varepsilon\rho\frac{\partial y_1(\tilde{\tau},\chi)}{\partial\chi}+\frac{\varepsilon^2\rho^2}{2}\frac{\partial^2 y_1(\tilde{\tau},\chi)}{\partial\chi^2}+\ldots\right)\\ u_{j,2}&=y_2(\tilde{\tau},\chi)\quad u_{j-1,2}=-\left(y_2(\tilde{\tau},\chi)-\varepsilon\rho\frac{\partial y_2(\tilde{\tau},\chi)}{\partial\chi}+\frac{\varepsilon^2\rho^2}{2}\frac{\partial^2 y_2(\tilde{\tau},\chi)}{\partial\chi^2}-\ldots\right)\end{aligned}\tag{27}$$

After it we obtain

$$\begin{aligned}
&\frac{\partial^2 u_1}{\partial\tilde{\tau}^2} + u_1 + \varepsilon^2\tilde{\xi}_3 u_1^3 + 2\tilde{\eta}u_1 + \tilde{\eta}\left(-\varepsilon\rho\frac{\partial u_2}{\partial\chi} + \frac{\varepsilon^2\rho^2}{2}\frac{\partial^2 u_2}{\partial\chi^2} - \ldots\right) = 0 \\
&\varepsilon\frac{\partial^2 u_2}{\partial\tilde{\tau}^2} + \tilde{\nu}u_2 + \varepsilon^2\tilde{\nu}_3 u_2^3 + 2\tilde{\eta}u_2 + \tilde{\eta}\left(\varepsilon\rho\frac{\partial u_1}{\partial\chi} + \frac{\varepsilon^2\rho^2}{2}\frac{\partial^2 u_1}{\partial\chi^2} + \ldots\right) = 0
\end{aligned} \tag{28}$$

After the change of variable

$$\tau = \sqrt{\gamma}\tilde{\tau} \qquad \gamma = 2\tilde{\eta} + 1 \quad \eta = \frac{\tilde{\eta}}{\gamma} \quad \nu = \frac{\tilde{\nu} + 2\tilde{\eta}}{\gamma} \quad \nu_3 = \frac{\tilde{\nu}_3}{8\gamma} \quad \xi_3 = \frac{\tilde{\xi}_3}{8\gamma} \tag{29}$$

we can use our scheme to perform a further asymptotic analysis. The result is

$$\begin{aligned}
&\varphi_{1,0} = \varphi_{1,0}(\tau_2, \ldots, x) \qquad \varphi_1 = \varphi_{1,1} + o(\varepsilon) \\
&\varphi_2 = -\varepsilon^2\left(\frac{(\eta - \alpha_1\nu)\rho}{\nu}\frac{\partial\varphi_{1,0}}{\partial\chi} + \alpha_1\rho e^{-2i\tau_0}\frac{\partial\varphi_{1,0}^*}{\partial\chi}\right) + o(\varepsilon^2)
\end{aligned} \tag{30}$$

$$\frac{\partial\varphi_{1,0}}{\partial\tau_2} + Ci|\varphi_{1,0}|^2\varphi_{1,0} + Di\frac{\partial^2\varphi_{1,0}}{\partial\chi^2} = 0 \quad C = -\frac{3\xi_3}{\nu} \quad D = -\frac{\eta^2\rho^2}{2\nu} \tag{31}$$

Equation (31) is the principal approximation for high-frequency acoustic oscillations and waves.

3.2. Optical branch

Optical branch corresponds to high–frequency modes, and chain oscillation period order is the same as for high–frequency oscillator. Therefore we introduce a new time variable $\tilde{\tau} = \omega_0 t$, where $\omega_0 = \sqrt{c_2/m_2}$ is the frequency of the linearized high–frequency oscillator.

Then the initial system is presented as follows

$$\begin{aligned}
&\varepsilon^{-1}\frac{d^2 u_{j,1}}{d\tilde{\tau}^2} + \frac{c_1}{c_2}u_{j,1} + \frac{c_{13}r_0^2}{c_2}u_{j,1}^3 + \frac{c}{c_2}\left(2u_{j,1} - u_{j-1,2} - u_{j,2}\right) = 0 \\
&\frac{d^2 u_{j,2}}{d\tilde{\tau}^2} + u_{j,2} + \frac{c_{23}r_0^2}{c_2}u_{j,2}^3 + \frac{c}{c_2}\left(2u_{j,2} - u_{j,1} - u_{j+1,1}\right) = 0
\end{aligned} \tag{32}$$

The most interesting situation for optical branch corresponds to estimations

$$\frac{c_1}{c_2} = \tilde{\xi} \quad \frac{c_{23}r_0^2}{c_2} = \varepsilon^3\tilde{\nu}_3 \quad \frac{c_{13}r_0^2}{c_2} = \varepsilon^3\tilde{\xi}_3 \quad \frac{c}{c_2} = \tilde{\eta} \tag{33}$$

where $\tilde{\xi},\ \tilde{\nu}_3,\ \tilde{\xi}_3, \tilde{\eta} \approx O(1)$.

3.2.1. *Optical Branch, Long Wavelength*

When using the expansions for long-length wave (Eq. 5), we introduce a new time variable $\tau = \sqrt{2\tilde{\eta}+1}\tilde{\tau}$, then make parameters change

$$\eta = \frac{\tilde{\eta}}{\gamma} \quad \xi = \frac{\tilde{\xi}+2\tilde{\eta}}{\gamma} \quad \nu_3 = \frac{\tilde{\nu}_3}{8\gamma} \quad \xi_3 = \frac{\tilde{\xi}_3}{8\gamma} \tag{34}$$

The results of analysis are given below

$$\begin{gathered} \varphi_{2,0} = e^{iF_1\tau_1} e^{iF_2\tau_2} e^{iF_3\tau_3} \Theta_{2,0}(\tau_3,\ldots,\chi) \\ F_1 = 2\eta^2 \;\; F_2 = -2\eta^2(5\eta^2-\xi) \;\; F_3 = 2\eta^2(\xi^2+42\eta^4-14\xi\eta^2) \end{gathered} \tag{35}$$

$$\begin{gathered} \varphi_2 = \varphi_{2,0} + \varepsilon\eta^2 e^{-2i\tau_0}\varphi_{2,0}^* - \varepsilon^2\eta^2(6\eta^2-\xi)e^{-2i\tau_0}\varphi_{2,0}^* + o(\varepsilon^2) \\ \varphi_1 = -\varepsilon\;(2\eta\varphi_{2,0}) + \varepsilon^2\left((8\eta^3-2\xi\eta)\varphi_{2,0} - 2\eta^3 e^{-2i\tau_0}\varphi_{2,0}^* + \eta\rho\frac{\partial\varphi_{2,0}}{\partial x}\right) + o(\varepsilon^2) \end{gathered} \tag{36}$$

The principal equation can be written as follows:

$$\frac{\partial\Theta_{2,0}}{\partial\tau_3} + C\,i\left|\Theta_{2,0}\right|^2\Theta_{2,0} + D\,i\frac{\partial^2\Theta_{2,0}}{\partial\chi^2} = 0 \quad C = -3\nu_3 \quad D = -\frac{\eta^2\rho^2}{2} \tag{37}$$

3.2.2. *Optical Branch, Short Wavelength*

Using the relation (Eq. 27), we introduce a new time variable and then make the parameters change

$$\tau = \sqrt{2\tilde{\eta}+1}\tilde{\tau} \quad \eta = \frac{\tilde{\eta}}{\gamma} \quad \xi = \frac{\tilde{\xi}+2\tilde{\eta}}{\gamma} \quad \nu_3 = \frac{\tilde{\nu}_3}{8\gamma} \quad \xi_3 = \frac{\tilde{\xi}_3}{8\gamma} \tag{38}$$

Taking into account the relations (Eq. 13) we obtain after asymptotic analysis

$$\varphi_{2,0} = \varphi_{2,0}(\tau_3,\ldots,\chi) \quad \varphi_2 = \varphi_{2,0} + o(\varepsilon^2) \quad \varphi_1 = -\varepsilon^2\eta\rho\frac{\partial\varphi_{2,0}}{\partial\chi} + o(\varepsilon^2) \tag{39}$$

The condition of the secular terms absence in second equation gives the principal approximation for short wavelength optic oscillations and waves

$$\frac{\partial\varphi_{2,0}}{\partial\tau_3} + Ci\left|\varphi_{2,0}\right|^2\varphi_{2,0} + Di\frac{\partial^2\varphi_{2,0}}{\partial\chi^2} = 0 \quad C = -3\nu_3 \quad D = \frac{\eta^2\rho^2}{2} \tag{40}$$

4. ELASTIC INHOMOGENEITY

Considering the case of elastic inhomogeneity we mean that $c_1/c_2 = \varepsilon << 1$. This relation gives us small parameter for the system under consideration.

4.1. Acoustic branch

Acoustic branch corresponds to low frequency modes, and chain oscillation period order is the same as for linearized low-frequency oscillator. That is why we introduce a new time variable like in Eq. (3)

The most interesting situation for acoustic branch corresponds to the estimations

$$\frac{m_2}{m_1} = \mu \qquad \frac{c_{13} r_0^2}{c_1} = \varepsilon^3 \tilde{\xi}_3 \qquad \frac{c_{23} r_0^2}{c_1} = \varepsilon^3 \tilde{\nu}_3 \qquad \frac{c}{c_1} = \tilde{\eta} \tag{41}$$

where $\mu,\ \tilde{\nu}_3,\ \tilde{\xi}_3,\ \tilde{\eta} \approx O(1)$. Then the initial system is rewrited as

$$\begin{gathered} \frac{d^2 u_{j,1}}{d\tilde{\tau}^2} + u_{j,1} + \varepsilon^3 \tilde{\xi}_3 u_{j,1}^3 + \tilde{\eta}\left(2u_{j,1} - u_{j-1,2} - u_{j,2}\right) = 0 \\ \mu \frac{d^2 u_{j,2}}{d\tilde{\tau}^2} + \varepsilon^{-1} u_{j,2} + \varepsilon^3 \tilde{\nu}_3 u_{j,2}^3 + \tilde{\eta}\left(2u_{j,2} - u_{j,1} - u_{j+1,1}\right) = 0 \end{gathered} \tag{42}$$

4.1.1. *Acoustic Branch, Long Wavelength*

After using of Eq. (5), we introduce a new time variable and make parameters change

$$\tau = \sqrt{2\tilde{\eta}+1}\,\tilde{\tau} \qquad \eta = \frac{\tilde{\eta}}{\gamma} \qquad \nu = \frac{1}{\gamma} \qquad \nu_3 = \frac{\tilde{\nu}_3}{8\gamma} \qquad \xi_3 = \frac{\tilde{\xi}_3}{8\gamma} \tag{43}$$

We obtain as well as earlier

$$\begin{gathered} \varphi_{1,0} = e^{iF_1\tau_1} e^{iF_2\tau_2} e^{iF_3\tau_3} \Theta_{1,0}(\tau_3,\ldots,\chi) \qquad F_1 = -\frac{2\eta^2}{\nu} \\ F_2 = -\frac{2\eta^2(\eta^2 - 2\eta + \mu)}{\nu^2} \quad F_3 = -2\eta^2 \frac{(2\eta^4 - 4\eta^3 - 2\mu\eta^2 + \mu^2 - 4\mu\eta + 4\eta^2)}{\nu^3} \end{gathered} \tag{44}$$

$$\begin{gathered} \varphi_1 = \varphi_{1,0} - \varepsilon\left(\frac{\eta^2}{\nu} e^{-2i\tau_0} \varphi_{1,0}^*\right) - \varepsilon^2\left(\eta^2 \frac{2\eta^2 - 2\eta + \mu}{\nu^2} e^{-2i\tau_0} \varphi_{1,0}^*\right) - o(\varepsilon^2) \\ \varphi_2 = \varepsilon\left(\frac{2\eta}{\nu}\varphi_{1,0}\right) + \varepsilon^2\left(\frac{2\mu\eta - 4\eta^2}{\nu^2}\varphi_{1,0} - 2\frac{\eta^3}{\nu^2} e^{-2i\tau_0}\varphi_{1,0}^* + \frac{\eta}{\nu}\rho\frac{\partial\varphi_{1,0}}{\partial\chi}\right) + o(\varepsilon^2) \end{gathered} \tag{45}$$

and come to the equation corresponding to principal approximation

$$\frac{\partial\Theta_{1,0}}{\partial\tau_3}+Ci\left|\Theta_{1,0}\right|^2\Theta_{1,0}+Di\frac{\partial^2\Theta_{1,0}}{\partial\chi^2}=0 \qquad C=-3\xi_3 \quad D=\frac{\eta^2\rho^2}{2\nu} \tag{46}$$

4.1.2. Acoustic Branch, Short Wavelength

After use of Eq. (27) we introduce a new time variable and make parameters change

$$\tau=\sqrt{2\tilde{\eta}+1}\tilde{\tau} \qquad \eta=\frac{\tilde{\eta}}{\gamma} \qquad \nu=\frac{1}{\gamma} \qquad \nu_3=\frac{\tilde{\nu}_3}{8\gamma} \qquad \xi_3=\frac{\tilde{\xi}_3}{8\gamma} \tag{47}$$

Scheme (Eq. 15) and asymptotic analysis lead to expressions

$$\varphi_{1,0}=\varphi_{1,0}(\tau_3,\dots,\chi) \quad \varphi_1=\varphi_{1,0}+o(\varepsilon^2) \quad \varphi_2=-\varepsilon^2\left(\frac{\eta}{\nu}\rho\frac{\partial\varphi_{1,0}}{\partial\chi}\right)+o(\varepsilon^2) \tag{48}$$

The condition of the secular terms absence gives a principal equation for short wavelength acoustic oscillations and waves

$$\frac{\partial\varphi_{1,0}}{\partial\tau_3}+Ci\left|\varphi_{1,0}\right|^2\varphi_{1,0}+Di\frac{\partial^2\varphi_{1,0}}{\partial\chi^2}=0 \qquad C=-3\xi_3 \qquad D=-\frac{\eta^2L^2}{2\nu} \tag{49}$$

4.2. Optical branch

Optical branch corresponds to high-frequency modes, and the order of chain oscillation period is the same as for high-frequency oscillator. That is why we introduce a new time variable $\tilde{\tau}=\omega_0 t$, where ω_0 is the frequency of the linearized high–frequency oscillator (just as in item 4.2).

Then the initial system is reformulated now as

$$\begin{aligned}&\mu^{-1}\frac{d^2u_{j,1}}{d\tilde{\tau}^2}+\varepsilon u_{j,1}+\frac{c_{13}r_0^2}{c_2}u_{j,1}^3+\frac{c}{c_2}\left(2u_{j,1}-u_{j-1,2}-u_{j,2}\right)=0\\&\frac{d^2u_{j,2}}{d\tilde{\tau}^2}+u_{j,2}+\frac{c_{23}r_0^2}{c_2}u_{j,2}^3+\frac{c}{c_2}\left(2u_{j,2}-u_{j,1}-u_{j+1,1}\right)=0\end{aligned} \tag{50}$$

The most interesting situation for optical branch corresponds to estimations

$$\frac{c_{23}r_0^2}{c_2}=\varepsilon^4\tilde{\nu}_3 \qquad \frac{c_{13}r_0^2}{c_2}=\varepsilon^4\tilde{\xi}_3 \qquad \frac{c}{c_2}=\tilde{\eta}\varepsilon \tag{51}$$

where $\tilde{\nu}_3,\ \tilde{\xi}_3,\ \tilde{\eta}\approx O(1)$.

4.2.1. Optical Branch, Long Wavelength

After use of Eq. (5), we introduce, as earlier, a new time variable, but now it is same as initial one. We would like to introduce as well as earlier, but $\tau=\tilde{\tau}$. Then we make the parameters change

$$\eta = \tilde{\eta} \quad \xi = 1 + 2\tilde{\eta} \quad \nu_3 = \frac{\tilde{\nu}_3}{8} \quad \xi_3 = \frac{\tilde{\xi}_3}{8} \tag{52}$$

and obtain

$$\varphi_{2,0} = e^{iF_1\tau_1} e^{iF_2\tau_2} e^{iF_3\tau_3} e^{iF_4\tau_4} \Theta_{2,0}(\tau_4, \ldots, \chi) \tag{53}$$

$$F_1 = \eta \;\; F_2 = \frac{(4\mu - 1)\eta^2}{2} \;\; F_3 = \frac{(4\mu^2\xi + \eta - 12\eta\mu)\eta^2}{2} \;\; F_4 = \frac{\eta^2(16\xi^2 + 35\eta^2 - 80\xi\eta)}{8}$$

$$\varphi_2 = \varphi_{2,0} + \varepsilon\left(\frac{\eta}{2} e^{-2i\tau_0}\varphi^*_{2,0}\right) - \varepsilon^2\left(\frac{2\mu - 1}{2}\eta^2 e^{-2i\tau_0}\varphi^*_{2,0}\right) + \\ + \varepsilon^3\left(\frac{8\mu^2\xi - 32\eta\mu + 5\eta}{8}\eta^2 e^{-2i\tau_0}\varphi^*_{2,0}\right) + o(\varepsilon^3) \tag{54}$$

$$\varphi_1 = -\varepsilon\,(2\eta\mu\varphi_{2,0}) + \varepsilon^2\left(2\eta(2\mu\eta - \mu^2\xi)\varphi_{2,0} - \mu\eta^2 e^{-2i\tau_0}\varphi^*_{2,0} + \mu\eta\rho\frac{\partial\varphi_{2,0}}{\partial\chi}\right) + o(\varepsilon^2)$$

The principal approximation for low-frequency optical oscillations and waves is written below

$$\frac{\partial\Theta_{1,0}}{\partial\tau_4} + Ci\left|\Theta_{1,0}\right|^2\Theta_{1,0} + Di\frac{\partial^2\Theta_{1,0}}{\partial\chi^2} = 0 \quad C = -3\nu_3 \quad D = -\frac{\eta^2\rho^2\mu}{2} \tag{55}$$

4.2.2. *Optical Branch, Short Wavelength*

After using Eq. (27) all computations are the same as in item 4.2.1 Their results are

$$\varphi_{2,0} = e^{iF_1\tau_1} e^{iF_2\tau_2} e^{iF_3\tau_3} e^{iF_4\tau_4} \Theta_{2,0}(\tau_4, \ldots, \chi) \\ F_1 = \eta \;\; F_2 = -\frac{\eta^2}{2} \;\; F_3 = \frac{\eta^3}{2} \;\; F_4 = -\frac{5\eta^4}{8} \tag{56}$$

$$\varphi_2 = \varphi_{2,0} + \varepsilon\left(\frac{\eta}{2} e^{-2i\tau_0}\varphi^*_{2,0}\right) - \varepsilon^2\left(\frac{\eta^2}{2} e^{-2i\tau_0}\varphi^*_{2,0}\right) + \varepsilon^3\left(\frac{5\eta^3}{8} e^{-2i\tau_0}\varphi^*_{2,0}\right) + o(\varepsilon^3)$$

$$\varphi_1 = -\varepsilon^2\left(\mu\eta\rho\frac{\partial\varphi_{2,0}}{\partial\chi}\right) + o(\varepsilon^2) \tag{57}$$

and we obtain the principal approximation for short-wave optic oscillations and waves

$$\frac{\partial\Theta_{1,0}}{\partial\tau_4} + Ci\left|\Theta_{1,0}\right|^2\Theta_{1,0} + Di\frac{\partial^2\Theta_{1,0}}{\partial\chi^2} = 0 \quad C = -3\nu_3 \quad D = \frac{\eta^2\rho^2\mu}{2} \tag{58}$$

5. CONCLUSION

We see, that in all cases considered (for both acoustic and optical branches) the final equation corresponding to main asymptotic approach is Nonlinear Shroedinger Equation (NSE). This cases are the most interesting ones because for other relations between parameters of the initial system we come in final approximation to uncoupled oscillators with effective characteristics or to the linearized coupled system.

In our approach the signs of nonlinear and elastic coefficients for the final NSE can be different depending on signs of unharmonic coefficients and the branch under consideration.

When these signs are similar ("soft" nonlinearity in items 3.1.2, 3.2.1, 4.1.2, 4.2.1, "strong" nonlinearity in 3.1.1, 3.2.2, 4.1.1, 4.2.2) NSE admits localized nonlinear waves which are envelope solitons.

E.g. in the case of mass inhomogeneity for the short wavelength corresponding to acoustic branch we have got Eq. (31). For $\xi_3 > 0$ the parameters C and D have equal signs and

$$\varphi_{1,0}(\chi,\tau_2) = \left(2S\frac{D}{C}\right)^{1/2} e^{i(k\chi-\omega\tau_2)}\,\mathrm{sech}\left[S^{1/2}(\chi - \nu\tau_2)\right] \tag{59}$$

where $k = \nu/2D$, $\omega = \nu^2/4D^2 - S$.

In all cases mentioned above nonlocalized, cooperative waves are unstable. On the contrary, in all other cases where signs of C and second derivatives D are different the cooperative harmonic-like motions are stable.

6. ACKNOWLEDGEMENTS

This work is supported by RFBR.

REFERENCES

1. Kosevitch A.M. and Kovalyov A.S., 1974. Introduction to nonlinear physical mechanics (in Russian). Naukova dumka, Kiev, pp. 57–71, 107–119.

2. Manevitch L.I. 1999, Complex representation of dynamics of coupled nonlinear oscillators. In: Mathematical models of non–linear excitations, transfer, dynamics, and control in condensed systems and other media. Kluwer Academic/Plenum publishers.

3. Manevitch L.I. 2001, The description of localized normal modes in a chain of nonlinear coupled oscillators: using of comlex variables. Nonlinear Dynamics. (to be published)

MODELS OF DIRECTED SELF-AVOIDING WALKS AND STATISTICS OF RIGID POLYMER MOLECULES

Arkadii E. Arinstein*

1. INTRODUCTION

Many physical properties of polymer systems are determined by the conformational properties of macromolecules, i.e., the set of their possible conformations and their conformational mobility. The progress achieved in the description of the statistics of the conformations of linear polymer chains is largely due to the fact that the conformational properties of polymer chains as a whole depend very little on the characteristic features of their chemical structure, i.e., they reflect certain general, fundamental properties of polymers. In addition, the fact that the number N of links in a polymer chain is large makes it possible to pass, when necessary, to the asymptotic limit ($N \to \infty$). The simplest model that makes it possible to take into account the conformational properties of long linear molecules is the model of an ideal (phantom) flexible polymer chain consisting of freely-jointed immaterial links [1-3]. Even though this model is extreme idealized it is extremely helpful for understanding many features of polymer systems. Furthermore, according to Flory theorem, the model of ideal (phantom) flexible polymer chain describes the properties of real polymer chains in dilute solutions in θ-solvents and polymer melts, where the interaction of the units of one chain is compensated by their interaction with the environment chains [1, 2].

The orientations of two neighboring units in an ideal freely jointed chain are independent of one another. However, if the polymer molecule possesses some interlink rigidity and the orientations of two neighboring units are correlated, then the concept of a Kuhn segment (or persistence length) is used to describe a chain of such kind [1, 2]. A Kuhn segment is introduced as a chain section of length Δl in which the memory of the orientation of its starting section is completely lost, so that the orientations of neighboring Kuhn segments are uncorrelated. The application of Kuhn segment as a structureless object makes possible to describe the statistics of the conformational states of a finite rigid

---

* N.N. Semenov Institute of Chemical Physics of Russian Academy of Sciences, Moscow, Russia 117334; Moscow State University of Technology "STANKIN", Moscow, Russia 101472.

Mathematical Modeling: Problems, Methods, Applications
Edited by Uvarova and Latyshev, Kluwer Academic/Plenum Publishers, 2001

molecule using the description of a freely jointed chain with a unit equal to the Kuhn segment. Really, in general case the structure of the Khun segments which is determined by the characteristic features of the chemical structure of specific polymer molecules essentially doesn't have to influence on the conformational properties of a single long polymer chain as a whole.

The introducing of Kuhn segment as a structureless object makes possible to describe the statistics of the conformational states of a molecule with a finite interunits stiffness using the description of a freely-jointed chain with a unit equal to the Kuhn segment.

In more general (idealized) models a Kuhn segment is treated as a structureless object, since it is assumed that if a large number of Kuhn segments fits within the length of a molecule, then the structure of the segments, as determined by the characteristic features of the chemical structure of specific polymer molecules, essentially has no effect on the conformational properties of a long polymer chain as a whole. Thus, a molecule with a finite stiffness can be treated as freely jointed only if a large number of Kuhn segments fits within the length of the molecule.

If the stiffness of a polymer molecule is large, i.e., the number of Kuhn segments in the chain is small, the asymptotic representation of a polymer molecule as a freely-jointed chain is unsuitable and more complicated models must be used to describe its statistical properties. It is certainly of interest to study such rigid-chain systems as well as systems consisting of chains of limited length theoretically, since this study will make it possible to explain a number of experimentally observed unusual properties of so-called oligomer systems, which fall between low- and high-molecular chemical compounds.

But even if a polymer molecule contains a sufficiently large number of Kuhn segments, in some cases treating each Kuhn segment as a structureless (isotropic) object, is a too coarse approximation. Obviously, the structure of a Kuhn segment is determined by the peculiarities of chemical structure of a polymer chain, and at first glance it appears that taking account of the internal structure of a Kuhn segment will result in different consequences for different homological series. However, for all types of linear polymers the Kuhn segments possess a very important common property: since, by definition, a Kuhn segment is an almost rectilinear section of a chain, it is, first and foremost, a strongly anisotropic object. This means that a rigid-chain polymer consists of strongly anisotropic elements, and it is this local anisotropy (the presence of quite extended rectilinear sections) of a polymer molecule that can result in the appearance of supramolecular structures, influencing the macroscopic properties of the polymer system and completely determining its state in some cases. It turns out that the anisotropy of Kuhn segments results in the appearance of a local anisotropy of a rigid-chain polymer melt, and under certain conditions it can lead to liquid-crystal ordering of the entire system as a whole. For example, anomalies in the rheological properties are observed in liquid oligomeric systems [4], a liquid-crystal state arises in melts of rigid-chain polymers (polyethylene type) [5], and so on. Scaling analysis of the appearance of anisotropy in a rigid-chain polymeric system for various ratios of the lengths of a chain and its Kuhn segment which was performed by us not long ago [6] in framework of Onsager model [7] modified for high density system [8], showed that if $\left|\ln(1-\Theta)\right|^{\frac{-4}{3\nu+1}} N^{1-\frac{2}{3\nu+1}} \le n_K \le \left|\ln(1-\Theta)\right|^{\frac{-2}{\nu+1}} N^{1-\frac{1}{\nu+1}}$ (here n_K is the monomer number in Kuhn segment, N is the monomer number in polymer chain and Θ is the volume density of polymer melt) the molecule properties are neither a low-molecular ones nor a freely-jointed polymer chain ones.

Thus, under certain conditions the properties of rigid-chain polymers are determined by their small-scale structure (on scales less than or the order of the persistence length). It is obvious that when these scales are taken into account the statistics of the conformational states of rigid-chain molecules is different from that of long flexible polymer chains, and the asymptotic description employed for the latter is inapplicable for rigid-chain molecules, since it is too coarse.

The properties of long rigid molecule were investigated alse by Bresler–Frenkel [9], who received, seemingly for the first time, well-known equation specifying an average squared size of a rigid chain versus its length

$$\left\langle \mathbf{R}^2 \right\rangle = 2(a/T)^2 \left[LT/a - 1 + \exp(-LT/a) \right]. \tag{1}$$

Until recently the properties of rigid polymer chain have been examined in frameworks of Kratky-Porod model [10, 11] and its modifications based on the methods of field theory, in particular, the path-integral technique. The use of field theory methods has appeared extremely productive for a problem under consideration. Furthermore, taking into account the properties of boson and fermion propagators the Dirac equation has been successfully proposed for description of properties of rigid polymer chains [12] (see, also [13] and refs. there). This approach is based on the observation that the Brownian path with its size R and length L in bosonic case is connected according to the diffusion low: $\langle \mathbf{R}^2 \rangle \sim L$, while in the fermion case this dependence principally differs, namely, $\langle \mathbf{R}^2 \rangle \sim L^2$. So the ultra-relativistic limit corresponds to strong rigid polymer chain, while the non-relativistic (classical) limit corresponds to flexible polymer chain. It seems quite reasonable to assume that the intervening asymptotic form correctly describes the properties of polymer molecule with finite stiffness. This assumption is confirmed by computer simulations [14, 15], although the rigorous proof of this fact is necessary.

Some variants of the description of the conformational statistics of polymer molecule with finite stiffness of arbitrary length proposed recently by authors by means of model of directed self-avoiding walks (DSAW) [6] are presented in the present paper. This model makes it possible to take into account more fully all the above-enumerated features of the conformational statistics of rigid chains and to construct the generating function for the spatial distribution function of a polymeric chain. Noting that all outcomes are the exact ones. In particular, in framework of DSAW models the conditions under which the Dirac propagator describes exactly the polymer molecule with finite stiffness, and the necessary proof has been demonstrated in this paper.

2. DSAW ON REGULAR LATTICES AS A MODEL OF A LINEAR CHAIN WITH ARBITRARY STIFFNESS

We shall use the random-walk model to describe the statistics of the conformations of chains possessing a finite inter-units bending stiffness. We shall consider random walks of a particle on a regular lattice for which the probability of the direction of each subsequent hop depends on the direction of the hop at the preceding step. We shall call such walks as directed self-avoiding as well as the trajectories of such a random walk correspond to the conformations of a linear chain with finite stiffness.

The stiffness of the chain (the probability of choosing a direction of the walk at the next step in the lattice as a function of the direction of the preceding step) is taken into account as follows:

- the probability that the motion realized in the same direction is maximum and equals α_+;
- the probability that motion occurs in the perpendicular direction is $\alpha_\perp$;
- the probability of a 180° turn is α_-.

It is obvious that the probabilities α_i introduced in this manner satisfy to inequality $\alpha_- < \alpha_\perp < \alpha_+$, and normalization requirement also $\alpha_+ + \alpha_- + 2(d-1)\alpha_\perp = 1$ (here $d = 1, 2, 3$ is the dimension of space).

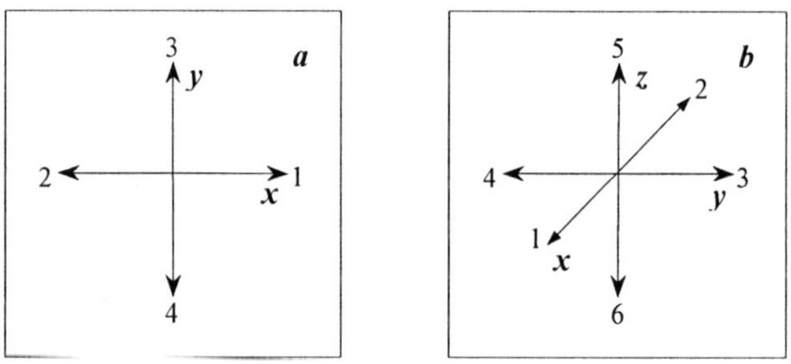

Figure 1. The enumeration of lattice directions.

We shall enumerate the directions in the plane ($d = 2$, see Fig. 1*a*) and in the space ($d = 3$, see Fig. 1*b*), as well as introduce the column vector of the distribution of probabilities of the direction of motion at a step n:

$$|P(n)\rangle_d = (P_i(n)), \qquad i = 1, 2, \ldots, 2d. \tag{2}$$

The column vector of the probability distribution at step $n+1$ is expressed by the one at step n by means of a transition matrix:

$$|P(n+1)\rangle_d = \mathbf{T}_d |P(n)\rangle_d, \qquad \mathbf{T}_2 = \begin{pmatrix} \mathbf{A} & \mathbf{B} \\ \mathbf{B} & \mathbf{A} \end{pmatrix}, \qquad \mathbf{T}_3 = \begin{pmatrix} \mathbf{A} & \mathbf{B} & \mathbf{B} \\ \mathbf{B} & \mathbf{A} & \mathbf{B} \\ \mathbf{B} & \mathbf{B} & \mathbf{A} \end{pmatrix}, \tag{3}$$

where $\mathbf{A}$ and $\mathbf{B}$ are 2×2 square matrices: $\mathbf{A} = \begin{pmatrix} \alpha_+ & \alpha_- \\ \alpha_- & \alpha_+ \end{pmatrix}$, $\mathbf{B} = \begin{pmatrix} \alpha_\perp & \alpha_\perp \\ \alpha_\perp & \alpha_\perp \end{pmatrix}$, moreover the matrices $\mathbf{A}$ and $\mathbf{B}$ commute with one another. This fact allows easily to diagonalize the matrices $\mathbf{T}_d$ and to calculate, for example, an average squared size of a rigid chain depending on its length.

Indeed, considering the matrices $\mathbf{A}$ and $\mathbf{B}$ as c-numbers we can write:

$$\mathbf{T}_2 = \begin{pmatrix} \mathbf{O}/\sqrt{2} & \mathbf{O}/\sqrt{2} \\ -\mathbf{O}/\sqrt{2} & \mathbf{O}/\sqrt{2} \end{pmatrix} \begin{pmatrix} \mathbf{d}_{2-} & 0 \\ 0 & \mathbf{d}_{2+} \end{pmatrix} \begin{pmatrix} \mathbf{O}^{-1}/\sqrt{2} & -\mathbf{O}^{-1}/\sqrt{2} \\ \mathbf{O}^{-1}/\sqrt{2} & \mathbf{O}^{-1}/\sqrt{2} \end{pmatrix}. \tag{4}$$

$$\mathbf{T}_3=\begin{pmatrix} \mathrm{O}/\sqrt{3} & \mathrm{O}/\sqrt{3} & \mathrm{O}/\sqrt{3} \\ 0 & -\mathrm{O}/\sqrt{3} & \mathrm{O}/\sqrt{3} \\ -\mathrm{O}/\sqrt{3} & 0 & \mathrm{O}/\sqrt{3} \end{pmatrix}\begin{pmatrix} \mathbf{d}_{3-} & 0 & 0 \\ 0 & \mathbf{d}_{3-} & 0 \\ 0 & 0 & \mathbf{d}_{3+} \end{pmatrix}\begin{pmatrix} \mathrm{O}^{-1}/\sqrt{3} & \mathrm{O}^{-1}/\sqrt{3} & -2\mathrm{O}^{-1}/\sqrt{3} \\ \mathrm{O}^{-1}/\sqrt{3} & -2\mathrm{O}^{-1}/\sqrt{3} & \mathrm{O}^{-1}/\sqrt{3} \\ \mathrm{O}^{-1}/\sqrt{3} & \mathrm{O}^{-1}/\sqrt{3} & \mathrm{O}^{-1}/\sqrt{3} \end{pmatrix}. \quad (5)$$

where $\mathbf{d}_{2-}=\begin{pmatrix} \alpha_+-\alpha_- & 0 \\ 0 & \alpha_+-\alpha_--2\alpha_\perp \end{pmatrix}$, $\mathbf{d}_{2+}=\begin{pmatrix} \alpha_+-\alpha_- & 0 \\ 0 & \alpha_++\alpha_-+2\alpha_\perp \end{pmatrix}=\begin{pmatrix} \alpha_+-\alpha_- & 0 \\ 0 & 1 \end{pmatrix}$,

$$\mathbf{d}_{3-}=\mathbf{d}_{2-} \quad \mathbf{d}_{3+}=\begin{pmatrix} \alpha_+-\alpha_- & 0 \\ 0 & \alpha_++\alpha_-+4\alpha_\perp \end{pmatrix}=\begin{pmatrix} \alpha_+-\alpha_- & 0 \\ 0 & 1 \end{pmatrix}.$$

Let the vector $\mathbf{e}(n)$ correspond to an edge in the lattice along which a hop occurs at the n step of the walks. The same vector $\mathbf{e}(n) \equiv \mathbf{e}(l)$ corresponds to the n-th link in the chain ($l = l_0\, n$, l_0 is the length of a lattice edge, corresponding to the length of one link in the chain). We shall represent the vector connecting the beginning and end of the chain using the vectors $\mathbf{e}(n)$ introduced:

$$\mathbf{R}(N)=l_0\sum_{n}^{N}\mathbf{e}(n). \qquad (6)$$

Using *eq*. (6) we get the following expression for the squared size of a molecule:

$$\left\langle \mathbf{R}^2(N)\right\rangle = l_0^2\left\{N+2\sum_{n_1=1}^{N}\sum_{n_2=1}^{N-n_1}\left\langle \mathbf{e}(n_1)\cdot\mathbf{e}(n_1+n_2)\right\rangle\right\} = l_0^2\left(\frac{1+\theta}{1-\theta}N-\frac{2\theta}{1-\theta}\cdot\frac{1-\theta^N}{1-\theta}\right), \qquad (7)$$

here $g(m)=\langle \mathbf{e}(n)\cdot\mathbf{e}(n+m)\rangle=\left(\mathbf{T}_d^m\right)_{1,1}-\left(\mathbf{T}_d^m\right)_{1,2}=(\alpha_+-\alpha_-)^m=\theta^m$, that coincides with well known expressions (see *eq*. (1), for example).

3. SPATIAL DISTRIBUTION OF THE POLYMER CHAIN UNITS

The complete information about the state of a chain is contained in the distribution function $G(\mathbf{R}_n, n)$. The last one gives the probability that the n-th link of the chain (more accurately, the end of the chain) starting at the origin of the coordinates falls to a lattice site with the coordinates $\mathbf{R}_n \equiv \{R_{nx} = l_0 m_{nx},\ R_{ny} = l_0 m_{ny},\ R_{nz} = l_0 m_{nz}\}$. For $n = N$ the distribution function $G(\mathbf{R}_N, N)$ gives the probability that the end of the chain has moved relatively to its beginning by the vector $\mathbf{R}_N$.

The probability of a wandering particle reaching some point in the random-walk model at a particular step can depend only on the location of the particle at the preceding steps and is independent completely from the future location of the particle. Consequently, the distribution function $G(\mathbf{R}_n, n)$ does not depend on the total number of steps. In relation to linear molecule this fact means that the distribution function $G(\mathbf{R}_n, n)$ determining specifically the probability that the distance between the n links of a chain is $\mathbf{R}_n$ does not depend on the length of the molecule. Knowledge of the distribution function $G(\mathbf{R}_n, n)$ gives us a complete picture of the spatial distribution of the molecule mass.

$(2d)^N$ different trajectories can be realized in N steps in the walk of a particle (both random and directed self-avoiding). We find the desired distribution function $G(\mathbf{R}_n, N)$ determining the fraction of trajectories which correspond to the displacement $\mathbf{R}_n$. Therefore, singling out the corresponding trajectories, which we specify using *eq*. (6), and averaging over all possible realizations of the walk of a particle, the distribution function $G(\mathbf{R}_n, N)$ can be written as

$$G(\mathbf{R}_N, N) = \left\langle \delta\left(\mathbf{R}_N - l_0 \sum_n^N \mathbf{e}(n)\right)\right\rangle = \left\langle \prod_j \delta\left(m_{Nj} - \sum_{n=1}^N \sigma_n^j\right)\right\rangle, \qquad j = x, y, z, \quad (8)$$

where the brackets signify averaging of all possible trajectories and the analog of the Dirac delta-function with a vector argument is used – the Kronecker symbol with vector argument – $\delta(\mathbf{R}) \equiv \delta(m_x)\delta(m_y)\delta(m_z)$: $\delta(m) \equiv \delta_{m,0}$, $\delta_{m,0} = 0$ if $m \neq 0$ and $\delta_{m,0} = 1$ if $m = 0$; the numbers σ_n^x, σ_n^y, and σ_n^z assume the values ± 1 or 0 and averaging is performed over all possible values of σ_n^x, σ_n^y, and σ_n^z satisfying the condition $(\sigma_n^x(^2 + (\sigma_n^y)^2 + (\sigma_n^z)^2 = 1$. In *eq*. (10) was taken into account that the vectors $\mathbf{e}(n)$ can possess components 1, 0, or –1.

In two-dimensional space this expression simplifies somewhat:

$$G(\mathbf{R}_N, N) = \left\langle \delta\left(\mathbf{R}_N - l_0 \sum_n^N \mathbf{e}(n)\right)\right\rangle = \left\langle \delta\left(m_{Nx} - \sum_{n=1}^N \sigma_n^x\right)\delta\left(m_{Ny} - \sum_{n=1}^N \sigma_n^y\right)\right\rangle, \quad (9)$$

where the numbers σ_n^x and σ_n^y assume the values ± 1 or 0, and the averaging is performed over all possible values of and satisfying the condition $(\sigma_n^x(^2 + (\sigma_n^y)^2 = 1$.

An even simpler expression is obtained in the one-dimensional case:

$$G(x_N, N) = \left\langle \delta\left(m_N - \sum_{n=1}^N \sigma_n\right)\right\rangle, \quad (10)$$

where the averaging extends only over two values, $\sigma_n = \pm 1$.

The construction of the distribution function $G(\mathbf{R}_N, N)$ in accordance with *eqs*. (8)–(10) is a quite difficult problem, as the probability of direction choice at the each step depends on the direction at previous step. This fact is taken into account at the averaging by means of matrix $\mathbf{T}_d$, the matrix elements of which are the probability distribution for values of vectors σ_{n+1} at given value of vector σ_n.

In the calculations below we shall use the integral representation of the Kronecker symbol (more accurately, its Fourier transform) $\delta(m) = \frac{1}{2\pi}\int_{-\pi}^{\pi} \exp(imq)dq$. Substituting this representation of the Kronecker symbol into the *eqs*. (8)–(10) we obtain for the Fourier transform of the distribution function $G(\mathbf{R}_N, N)$:

$$K_3(\mathbf{q}, N) = \left\langle \exp\left[i \sum_{n=1}^N \left(\sigma_n^x q_x + \sigma_n^y q_y + \sigma_n^z q_z\right)\right]\right\rangle, \qquad d = 3; \quad (11)$$

$$K_2(\mathbf{q},N) = \left\langle \exp\left[i\sum_{n=1}^{N} \left(\sigma_n^x q_x + \sigma_n^y q_y\right)\right]\right\rangle, \qquad d=2; \tag{12}$$

$$K_1(q,N) = \left\langle \exp\left[i\sum_{n=1}^{N} \left(\sigma_n q\right)\right]\right\rangle, \qquad d=1; \tag{13}$$

where the vectors $\mathbf{m}_N = \{m_{Nx}, m_{Ny}, m_{Nz}\}$ and $\mathbf{q} = \{q_x, q_y, q_z\}$, and

$$G_d(\mathbf{R}_N,N) = \frac{1}{(2\pi)^d}\int \exp(i\mathbf{m}_N\cdot\mathbf{q})K_d(\mathbf{q},N)d\mathbf{q}\,. \tag{14}$$

After averaging of *eqs*. (14)–(16) we have

$$K_d(\mathbf{q},N) = \sum_{j,k}\left[(\hat{\mathbf{q}}_d\cdot \mathrm{T}_d)^{N-1}\,\hat{\mathbf{q}}_d\right]_{j,k}, \tag{15}$$

where $\hat{\mathbf{q}}_3 = \begin{pmatrix} \hat{\mathbf{q}}_x & 0 & 0 \\ 0 & \hat{\mathbf{q}}_y & 0 \\ 0 & 0 & \hat{\mathbf{q}}_z \end{pmatrix}$, $\quad \hat{\mathbf{q}}_2 = \begin{pmatrix} \hat{\mathbf{q}}_x & 0 \\ 0 & \hat{\mathbf{q}}_y \end{pmatrix}$, $\quad \hat{\mathbf{q}}_1 = \begin{pmatrix} e^{iq} & 0 \\ 0 & e^{-iq} \end{pmatrix}$, $\quad \mathbf{T}_1 = \mathbf{A}$,

$$\hat{\mathbf{q}}_x = \begin{pmatrix} e^{iq_x} & 0 \\ 0 & e^{-iq_x} \end{pmatrix}, \quad \hat{\mathbf{q}}_y = \begin{pmatrix} e^{iq_y} & 0 \\ 0 & e^{-iq_y} \end{pmatrix}, \quad \hat{\mathbf{q}}_z = \begin{pmatrix} e^{iq_z} & 0 \\ 0 & e^{-iq_z} \end{pmatrix}.$$

4. THE DISTRIBUTION FUNCTION AND STATISTICAL SUMS OF ONE-DIMENTIONAL STATISTICAL MODELS

We pay attention especially to the one-dimensional variant, when the problem is reducing to one-dimensional Ising model. The exact solution of this model, which has the enough simple form [16], gives the key to the problem in general. However the one-dimensional variant is of more then methodological interest. Actually, the one-dimensional case is of little physical interest for the problem of the conformations of a rigid chain: it can describe only the folded structure of oriented (i.e., possessing liquid-crystal ordering) chains, which are subjected to stretching. However, for describing the diffusion of impurities in a condensed phase even the one-dimensional variant of the problem of DSAW has an important physical meaning [17].

The calculation of the distribution function (15) in one-dimensional case is reduced to diagonalization of 2×2 matrix $\mathbf{q}_1\mathbf{A}$, that is easily to realize:

$$\hat{\mathbf{q}}\,\mathbf{A} = \hat{\mathbf{O}}\begin{pmatrix} \lambda_+ & 0 \\ 0 & \lambda_- \end{pmatrix}\hat{\mathbf{O}}^{-1}, \tag{16}$$

where $\lambda_{\pm}=\alpha_{+}\left[\cos q\pm\sqrt{\left(\alpha_{-}/\alpha_{+}\right)^{2}-\sin^{2}q}\right]$, $\hat{\mathbf{O}}=\frac{1}{\sqrt{\Delta}}\begin{pmatrix}\alpha_{-} & \alpha_{0}e^{iq}\\ -\alpha_{0}e^{-iq} & \alpha_{-}\end{pmatrix}$,

$\Delta=\alpha_{-}^{2}+\alpha_{0}^{2}$, and $\alpha_{0}=\alpha_{+}\left[i\sin q-\sqrt{\left(\alpha_{-}/\alpha_{+}\right)^{2}-\sin^{2}q}\right]$.

Substituting (16) into *eq.* (15) we obtain

$$K_{1}(q,N)=\frac{1}{\Delta}\left[\left(\alpha_{-}^{2}\lambda_{+}^{N-1}+\alpha_{0}^{2}\lambda_{-}^{N-1}\right)\cos q+\alpha_{0}\alpha_{-}\left(\lambda_{-}^{N-1}-\lambda_{+}^{N-1}\right)\right] \tag{17}$$

The *eq.* (17) can be used to calculate the generating function for the distribution function:

$$\Gamma^{(1)}(q,\xi)=1+\sum_{N=1}^{\infty}\xi^{N}K_{1}(q,N)=\frac{1-(\alpha_{+}-\alpha_{-})\xi\cos q}{1-2\alpha_{+}\xi\cos q+(\alpha_{+}-\alpha_{-})\xi^{2}}\,. \tag{18}$$

However the possibility to represent the one-dimensional distribution function (18) as the statistical sum of one-dimensional Ising model is of most interest for us

$$K_{1}(q,N)=\left(\alpha_{+}\alpha_{-}\right)^{(N-1)/2}\sum_{\sigma_{1}\sigma_{N}}\left[Z(q,N)\right]_{\sigma_{1}\sigma_{N}}, \tag{19}$$

where $\left[Z(q,N)\right]_{\sigma_{1}\sigma_{N}}=\sum_{\sigma_{2},\ldots,\sigma_{N-1}=\pm1}\exp\left(iq\sum_{n=1}^{N}\sigma_{n}+h\sum_{n=1}^{N-1}\sigma_{n}\sigma_{n+1}\right)$ and $h=0{,}5\ln(\alpha_{+}/\alpha_{-})$.

This fact namely allows for the first time to construct the distribution function in continuous limit as the solution of Dirac equation [12, 13]:

$$K_{1}(q,l)=\frac{1+E}{2E}\exp[-ml(1-E)]-\frac{1-E}{2E}\exp[-ml(1+E)] \tag{20}$$

where $E^{2}=1-p^{2}/m^{2}$, $p=q/l_{0}$, $l=l_{0}N$ and m is a parameter characterizing the chain rigidity.

If we neglect the normalizing factor $\exp(-ml)$ *eq.* (19) has a form [13] after the Laplace transform:

$$\widetilde{K}_{1}(q,s)=\frac{s+m}{s^{2}+p^{2}-m^{2}}\,. \tag{21}$$

To proceed to many-dimensional space it was supposed that in two- and three-dimensional variants it is sufficient to replace in *eq.* (21) $p^{2}=l_{0}^{2}q^{2}$ by the squared scalar of vector $\mathbf{p}^{2}=l_{0}^{2}\mathbf{q}^{2}$ [13]. As a result, for $K_{2,3}(\mathbf{q},l)$ we can use *eq.* (20) in which $E^{2}=1-\mathbf{p}^{2}/m^{2}$, and for $\widetilde{K}_{2,3}(\mathbf{q},s)$ we can use *eq.* (21) in denominator of which p^{2} is replaced by $\mathbf{p}^{2}$.

The *eqs.* (20) and (21) which are related to each other by means of Laplace transform, are accurately the trace of the Euclidean version of Dirac propagator, are two- and three-dimensional distribution functions of polymer chain units in continuous limit. As it has been above mentioned there is no rigorous proof of this conclusion. After calculation of generating function for the chain distribution function in two- and three-dimensional cases we will come back to this point.

The diagonalization of matrices 4×4 (for two-dimensional variant) is a very complicated and tedious problem. And in a three-dimensional variant the diagonalization of matrices 6×6, in general case, is impossible. However, we can find the generating function for the chain distribution function without diagonalization of matrix using the analogy with statistical model of a one-dimensional system. Indeed, introducing the notation $p_{\pm} = \frac{1}{2}\left(q_x \pm q_y\right)$, $g = \frac{1}{2}\ln\left(\alpha_+\alpha_-/\alpha_\perp^2\right)$, $h = \frac{1}{2}\ln\left(\alpha_+/\alpha_-\right)$, it is possible to represent $K_2(\mathbf{q}, N)$ as a statistical sum of a one-dimensional chain of two types of "spin-1/2" particles:

$$K_2(\mathbf{q}, N) = \alpha_\perp^{N-1} \sum_{\substack{\sigma_2,\ldots,\sigma_N=\pm 1 \\ \nu_2,\ldots,\nu_N=\pm 1}} \exp\left[i\sum_{n=1}^{N} \sigma_n\left(p_+ + p_-\nu_n\right) + \sum_{n=1}^{N-1}\frac{1}{2}\left(1+\nu_n\nu_{n+1}\right)\left(g + h\sigma_n\sigma_{n+1}\right)\right]. \quad (22)$$

The structure of the expression obtained for the statistical sum (22) is such that if the nearest neighbors are the different kinds of particles ($\nu_n \neq \nu_n + 1$) they do not interact. This means that the chain consists of two types of alternating noninteracting chains, and the calculation of the partition function (22) is reduced to averaging procedure over the lengths of these chains and (n_j^x and n_j^y) and their number (m) under the condition that the total length of the entire chain is fixed: $\sum n_j^x + \sum n_j^y = N$.

The partition of the entire chain into a collection of two types of alternating noninteracting chains means in the language of walk trajectories, that first n_1^x steps are taken in the direction of x, n_1^y steps are taken in the direction of y, n_2^x steps are taken in the direction of x, n_2^y steps are taken in the direction of y, and so on.

Tedious but not complicated calculations lead to simple expression for the generating function

$$\Gamma^{(2)}(\mathbf{q},\xi) = 1 + \frac{P_1(q_x,\xi) + P_1(q_y,\xi) + 2P_1(q_x,\xi)P_1(q_y,\xi)}{4\alpha_\perp[1 - P_1(q_x,\xi)P_1(q_y,\xi)]}, \quad (23)$$

where $P_1(q,\xi) = \Gamma^{(1)}(q,\xi) - 1 = \dfrac{2\alpha_\perp\xi\left[\cos q - (\alpha_+ - \alpha_-)\xi\right]}{1 - 2\alpha_+\xi\cos q + (\alpha_+^2 - \alpha_-^2)\xi^2}$.

The construction of a generating function for the distribution function in three-dimensional space is entirely analogous. The only difference is that we average over walk trajectories of the following type: first n_1^x steps are taken in the one-dimensional subspace x, then n_1^{yz} steps in the plane yz, then n_2^x steps once again in the one-dimensional subspace x, then n_2^{yz} steps once again in the plane yz, and so on.

After averaging over trajectories of such type we obtain that the generating function $\Gamma^{(3)}(q, \xi)$ for the distribution function in three-dimensional space has a form which is similar to the two-dimensional function $\Gamma^{(2)}(q, \xi)$:

$$\Gamma^{(3)}(\mathbf{q},\xi) = 1 + \frac{1}{6\alpha_\perp}\left\{\frac{[1+P_2(q_y,q_z,\xi)]P_1(q_x,\xi)}{1-P_2(q_y,q_z,\xi)P_1(q_x,\xi)} + \frac{[1+P_2(q_z,q_x,\xi)]P_1(q_y,\xi)}{1-P_2(q_z,q_x,\xi)P_1(q_y,\xi)} + \frac{[1+P_2(q_x,q_y,\xi)]P_1(q_z,\xi)}{1-P_2(q_x,q_y,\xi)P_1(q_z,\xi)}\right\}, \tag{24}$$

where $P_2(q_1,q_2,\xi) = 4\alpha_\perp\left[\Gamma^{(2)}(\mathbf{q},\xi)-1\right] = \dfrac{P_1(q_1,\xi)+P_1(q_2,\xi)+2P_1(q_1,\xi)P_1(q_2,\xi)}{1-P_1(q_1,\xi)P_1(q_2,\xi)}$.

The fundamental difference of obtained generating function in comparison with the generating function for isotropic random walks consists in the presence of additional poles on complex ξ-plane: $2d$ poles instead of one pole as it takes place in classic generating function [3]. These poles arise in infinity as a result of symmetry breaking, and the number of these poles is equal to lattice coordination number.

5. THE GENERATING FUNCTION FOR THE DISTRIBUTION FUNCTION IN TWO- AND THREE-DIMENSIONAL SPACES AND DIRAC PROPAGATOR

To assess the reasonableness of transition from one-dimensional Dirac propagator (20) to two- and three-dimensional Dirac propagator which has been obtained on the base of statistical sum of one-dimensional Ising model related to generating function (18) (which has two poles whereas the generating function (23) has four poles and the one (24) has six poles) we have to pay our attention to the fact that the degree of symmetry breaking in these two cases is different. Indeed, in Dirac propagators (21) it is assumed that the probability to save the motion direction α_+ and the probability to turn to the perpendicular direction $\alpha_\perp$ are equal ($\alpha_+ = \alpha_\perp$) and only one probability (the probability of a 180° turn) α_- differs from they. On this reason we assign in *eq.* (23) that

$$\Gamma^{(1)}(q,\xi) = 1 + P_1(q_x,\xi) = \frac{1-(\alpha_+-\alpha_-)^2\xi^2}{1-2\alpha_+\xi\cos q+(\alpha_+^2-\alpha_-^2)\xi^2}, \tag{25}$$

and as a result for two-dimensional generating function we have

$$\begin{aligned}\Gamma^{(2)}(\mathbf{q},\xi) &= 1 + \frac{P_1(q_x,\xi)+P_1(q_y,\xi)+2P_1(q_x,\xi)P_1(q_y,\xi)}{4\alpha_+[1-P_1(q_x,\xi)P_1(q_y,\xi)]} = \\ &= 1 - \frac{\Gamma^{(1)}(q_x,\xi)+\Gamma^{(1)}(q_y,\xi)-2\Gamma^{(1)}(q_x,\xi)\Gamma^{(1)}(q_y,\xi)}{4\alpha_+[\Gamma^{(1)}(q_x,\xi)+\Gamma^{(1)}(q_y,\xi)-\Gamma^{(1)}(q_x,\xi)\Gamma^{(1)}(q_y,\xi)]} = \frac{1+\theta\xi\Sigma_2}{1-(1+\theta)\xi\Sigma_2+\theta\xi^2}\end{aligned}, \tag{26}$$

where $\Sigma_2 = (\cos q_x + \cos q_y)/2$.

For three-dimensional case we obtain the similar expression, so, in common, we have

$$\Gamma^{(d)}(\mathbf{q},\xi)=\frac{1+\theta\xi\Sigma_d}{1-(1+\theta)\xi\Sigma_d+\theta\xi^2},\tag{27}$$

where $\Sigma_3 = (\cos q_x + \cos q_y + \cos q_z)/3$, ($\Sigma_2$ see above); $\theta = \alpha_+ - \alpha_-$, $2(d-1)\alpha_+ + \alpha_- = 1$.

Now we can derive difference evolutionary equation for distribution functions $K_d(\mathbf{q}, n)$, representing *eq.* (27) as:

$$\left[\theta(1/\xi+\xi-2)+(1-\theta)(1/\xi-1)+(1+\theta)(1-\Sigma_d)\right]\sum_{n=0}^{\infty}\xi^n K_d(\mathbf{q},n)=1/\xi+\theta\Sigma_d.\tag{28}$$

Equating the addends for the same exponents, we obtain the difference equation for $K_d(\mathbf{q}, n)$

$$\begin{aligned}&\theta\left[K_d(\mathbf{q},n+1)+K_d(\mathbf{q},n-1)-2K_d(\mathbf{q},n)\right]++(1-\theta)\left[K_d(\mathbf{q},n+1)-K_d(\mathbf{q},n)\right]+\\&+K_d(\mathbf{q},n)(1+\theta)(1-\Sigma_d)=0,\qquad n\ge 2,\qquad K_d(\mathbf{q},0)=1,\qquad K_d(\mathbf{q},1)=\Sigma_d.\end{aligned}\tag{29}$$

Proceeding to continuous limit (taking into account that $\cos q \approx 1 - q^2/2$ and $\mathbf{q} = l_0\mathbf{p}$) we come to the requisite equation

$$\left(\theta\frac{\partial^2}{\partial l^2}+\frac{1-\theta}{l_0}\frac{\partial}{\partial l}+(1+\theta)\frac{\mathbf{p}^2}{\mathbf{2d}}\right)K_d(\mathbf{p},l)=0,\qquad K_d(\mathbf{p},0)=1,\qquad \left.\frac{\partial K_d(\mathbf{p},l)}{\partial l}\right|_{l=0}=0.\tag{30}$$

Executing the Laplace transform we find that

$$\widetilde{K}_d(\mathbf{q},s)=\frac{s+(1-\theta)/l_0\theta}{s^2+s(1-\theta)/l_0\theta+\mathbf{p}^2(1+\theta)/2d\theta},\tag{31}$$

that gives after inverse Laplace transform

$$K_d(\mathbf{q},l)=\frac{1+E}{2E}\exp\left[-\frac{1-\theta}{2\theta}(1-E)l/l_0\right]-\frac{1-E}{2E}\exp\left[-\frac{1-\theta}{2\theta}(1+E)l/l_0\right],\tag{32}$$

where $E=\sqrt{1-l_0^2\mathbf{p}^2 2\theta(1+\theta)/d(1-\theta)^2}$.

The comparison of *eq.* (23) and (32) demonstrates that indeed, the expressions like Dirac propagator are exact distribution functions for rigid polymer chain. However, it's necessary to specify the conditions under which it comes true. The matter is that the polymer chain stiffness in *eq.* (20) is represented by parameter m. If m tends to zero we deal with flexible freely jointed polymer chain, while the contrary limit, when m tends to infinity, corresponds to the absolutely rigid, rod-like polymer molecule. However, according to *eq.* (32) $m = l_0^2 2\theta(1+\theta)/(1-\theta)^2$, where $0 \le \theta \le (1 - 2d\alpha_-)/(2d-1) \le 1$, i.e. $m \to 0$, if $\theta \to 0$, and $m \to \infty$, if $\theta \to 1$. If θ tends to zero we deal with isotropic random walks which describe the flexible freely jointed polymer chain. But the contrary limit

$(m \to \infty)$ is impossible as the parameter θ is restricted by valuation $1/(2d-1) < 1$. So, the maximal valuation of parameter m (if $\alpha_- = 0$) is $l_0^2 d/(d-1)^2 \ll \infty$, that signifies that the polymer chain which is describing by distribution function like Dirac propagator (20), or (32), can't be too rigid. Thus, the distribution function like Dirac propagator describes indeed the rigid polymer chain, but the stiffness of this chain is restricted.

On the other hand, this distribution function corresponds to the system, which is very valuable from physical point of view. Indeed, as has been noted above, the model of ideal (phantom) chain in which the self-crossings of polymer chain are not forbidden, well describes the polymer melt. The reason of this fact is that the intramolecular interactions (between remote units of one polymer chain) in polymer melt are compensated exactly by the intermolecular interactions (between various chains of polymer melt). However, this compensation can't take place if we consider two nearest neighbor units of chain. Therefore, even in the polymer melt this interaction between two nearest neighbor units of chain is to be taken into account, and in frameworks of DSAW model at $\alpha_- \equiv 0$ this problem is solved as it was demonstrated above.

6. CONCLUSIONS

Briefly summing up our outcomes it is possible to note that DSDAW model permits to describe main peculiarities of conformational statistical properties of polymer chain possessing arbitrary stiffness, at that time the form of final mathematical expressions is facile enough and well-behaved. Unfortunately, because of limitation of article size, we had no possibility to discuss many problems being of great interest, which can be solved in frameworks of DSAW model, such as the accounting of self-crossings forbiddingness, scale renormalization, self-organization and formation of supramolecular structures etc.

7. ACKNOWLEDGMENTS

In conclusion, I would like to express thanks to my colleagues L. Manevich and S. Timman for many helpful discussions of this problem, also to my namesake A. Kholodenko for the kindly given useful materials and explanation of a number of fundamental problems.

REFERENCES

1. P. G. De Gennes, *Scaling Concepts in the Physics of Polymers* (Cornell Univ. Press, Ithaca, 1979).
2. A. Yu. Grosberg and A. R. Khokhlov, *Statistical Physics of Macromolecules* (Nauka, M.,1989) (in Russian).
3. J. M. Ziman, *Models of Disorder: the Theoretical Physics of Homogeneously Disordered Systems* (Cambridge Univ. Press, Cambridge, 1979).
4. A. E. Arinstein and S. M. Mezikovskii, Polym. Eng. Sci. v. **37**, 1339 (1997).
5. A. N. Semenov and P. R. Khokhlov, Usp. Fiz. Nauk v. **156**, 417 (1988) (in Russian).
6. A. E. Arinstein, J. Exp. and Theor. Phys. v. **91**, No. 1, 206 (2000).
7. L. Onsager, Ann. N. Y. Acad. Sci. v. **51**, 627 (1949).
8. A. R. Khokhlov and A. N. Semenov, J. Stat. Phys. v. **38**, 161 (1985).
9. L. D. Landau and E. M. Lifshitz, *Statistical Physics*, Part 1 (Nauka, M., 1995); Pergamon, Oxford, 1980.
10. O. Kratky, G. Porod. Rec. Trav. Chim. v. **68**, 1106 (1949).
11. G. Porod, Monatsh. Chem. v. **80**, 251 (1949).
12. A. L. Kholodenko, Ann. Phys. v. **202**, 186 (1990).
13. A. L. Kholodenko, Physica A v. **260**, 267 (1998).
14. A. L. Kholodenko, Macromolecules v. **26**, No 16, 1479 (1993).
15. D. Pötschke, P. Hickl and all, Macromol. Theory Simul., v. **9**, 345 (2000).
16. R. P. Feynman, *Statistical Mechanics: a Set of Lectures* (Benjamin, Reading, Mass., 1972).
17. A. E. Arinshtein, Dokl. Akad. Nauk v. **358**, 350 (1998) (in Russian).

POSTULATE OF THE ARITHMETICAL MEAN AND NONBONDED INTERACTIONS

Yurii.G. Papulov, Marina.G. Vinogradova and M.N. Saltykova*

1.INTRODUCTION

The postulate of the arithmetical mean for binary interactions of particles is formulated as the relation

$$p_{HX} = (1/2)(p_{HH} + p_{XX}) \tag{1}$$

(interaction of unlike particles H and X is equal to one-half of the sum of the interactions of the like particles). Analogously, for ternary interactions

$$p_{HHX} = (1/3)(2p_{HHH} + p_{XXX}), \quad p_{HXX} = (1/3)(p_{HHH} + 2p_{XXX}); \tag{2}$$

and, for quaternary interactions

$$\begin{aligned} &p_{HHHX} = (1/4)(3p_{HHHH} + p_{XXXX}), \quad p_{HXXX} = (1/4)(p_{HHHH} + 3p_{XXXX}), \\ &p_{HHXX} = (1/4)(2p_{HHHH} + 2p_{XXXX}) \end{aligned} \tag{3}$$

and so on.

For bonded atom-atomic interactions in diatomic molecules the condition (1) is introduced by L. Pauling[1], who has put forward on the basis (1) the well-known concept of electronegativity[1,2].

The conditions (1)-(3) for nonbonded atom-atomic interactions in substituted methanes and their analogs are elucidated in the papers[3,4]. And also, the condition (1) is discussed for nonbonded interactions of atoms in substituted ethanes[5,6], ethylenes[5,7], benzenes[5,6,8], cyclopropanes[6,9], etc[5,10].

* Tver State University, Tver, Sadovy per., 35, 170002, Russia

Our aim here is to examine the action of the postulate of the arithmetical mean for nonbonded interactions in alkanes by means of the additive schemes of calculation and prediction of physicochemical properties.

2.STARTING POINT

From the phenomenological point of view, molecule is an interacting atom unity (physical model). Some molecular property (P) may be presented as sum of properties of atom-atomic interactions: one-body (p_{α}), two-body ($p_{\alpha\beta}$), three-body ($p_{\alpha\beta\gamma}$), etc[11,12]

$$P = \sum_{\alpha} p_{\alpha} + \sum_{\alpha,\beta} p_{\alpha\beta} + \sum_{\alpha,\beta,\gamma} p_{\alpha\beta\gamma} + \dots \tag{3}$$

(mathematical model). The summation in (3) is over all atoms, all pairs of atoms, all triads of atoms, etc.

From (3), there follow various additive schemes[13,14] which are discussed below (for alkanes).

3.INTRAMOLECULAR INTERACTIONS

Two partner atom-atomic interactions in molecule[see13-15] may be divided into bonded interactions (ξ) and nonbonded interactions, separated by one skeletal atom along the molecular chain (η-interactions), two such atoms (ζ-interactions), three such atoms (ϑ-interactions), four atoms (κ-interactions) and so on.

Let us introduce the *effective bond property terms* for alkanes

$$\xi^*_{CC} = \xi_{CC} + (1/2)\varepsilon_C, \; \xi^*_{CH} = \xi_{CH} + (1/4)\varepsilon_C + \varepsilon_H , \tag{4}$$

in which ε_C and ε_H are contributions of atoms.

The trio η-interactions of nonlinked atoms are taken into account also[13,14].

In staggered (stable) conformations of alkanes[14,15] ζ-interactions are divided into folowing kinds: ζ^t (*trans*) and ζ^g (*gauche*); ϑ – interactions are divided into ϑ^{tt}, ϑ^{tg}, ϑ^{gt}, $\vartheta^{gg'}$, $\vartheta^{gg''}$; κ-interactions are divided into κ^{ttt}, κ^{ttg}, κ^{tgt} and so on.

Here we assume that $\zeta^t \approx \zeta^g \approx \zeta$, and $\vartheta^{tt} \approx \vartheta^{tg} \approx \vartheta^{gt} \approx \vartheta^{gg'} \approx \vartheta^{gg''} \approx \vartheta$, etc.

4.SIMPLE SCHEMES

Simple additive schemes do not take into consideration the mutial influence between nonlinked atoms. For alkanes, equation (3) gives Fajans scheme[16]

$$P(C_nH_{2n+2}) = (n-1)p_{C-C} + (2n+2)p_{C-H} . \tag{5}$$

Here, $p_{C-C} = \xi^*_{CC}$, $p_{C-H} = \xi^*_{CH}$ are given by (4)

Simple schemes are crude: they do not reflect the effect of isomerism.

5.THE FIRST-ORDER APPROXIMATION

In this approximation we consider the bonded interactions and nonbonded interactions, separated by one skeletal atom along the molecular chain. Taking into account the binary interactions, from (3) we obtain Zahn's scheme[17]

$$P(C_nH_{2n+2}) = (n-1)p'_{C-C} + (2n+2)p_{C-H} + x_{CC}\Gamma_{CC}\,, \qquad (6)$$

where

$$\left.\begin{aligned} &p'_{C-C} = \xi^*_{CC} + 6\eta_{CH} - 3\eta_{HH}\,,\quad p'_{C-H} = \xi^*_{CH} + (3/2)\eta_{HH}\,, \\ &\Gamma_{CC} = \eta_{CC} - 2\eta_{CH} + \eta_{HH}\;; \end{aligned}\right\} \qquad (7)$$

Γ_{CC} is effective interaction of nonlinked pair of carbon atoms attached to the same atom, x_{CC} is the number of such pairs.

The scheme (6) reflect partly the structural isomers, for example, *n*-butane and 2-methylpropane; but it do not distinguish between such isomers as 2-methylpentane and 3-methylpentane, or 2-methylhexane, 3-methylhexane and 3-ethylpentane, etc.

If we postulate the arithmetical mean (2) for binary nonbonded interactions via one carbon atom, i.e.

$$\eta_{CH} = (1/2)(\eta_{CC} + \eta_{HH})\,, \qquad (8)$$

then, as can be seen from (7), $\Gamma_{CC} = 0$, and we return to the simple scheme (5).

Taking into account the ternary interactions of nonlinked atoms, we obtain Allen's scheme[18]

$$P(C_nH_{2n+2}) = (n-1)p'_{C-C} + (2n+2)p_{C-H} + x_{CC_1}\Gamma_{CC} + x_{CCC_1}\Delta_{CCC}\,, \qquad (9)$$

where

$$\left.\begin{aligned} &p'_{C-C} = \xi^*_{CC} + 6\eta_{CH} - 3\eta_{HH} + 6\eta_{CHH} - 4\eta_{HHH}\,, \\ &p_{C-H} = \xi^*_{CH} + (3/2)\eta_{HH} + \eta_{HHH}\,, \\ &\Gamma_{CC} = \eta_{CC} - 2\eta_{CH} + \eta_{HH} + 2\eta_{CCH} - 4\eta_{CHH} + 2\eta_{HHH}\,, \end{aligned}\right\} \qquad (10)$$

$$\Delta_{CCC} = \eta_{CCC} - 3\eta_{CCH} + 3\eta_{CHH} - \eta_{HHH} \;;$$

Δ_{CCC} is effective interaction of nonlinked trio of carbon atoms attached to the same atom, x_{CCC} is the number of such trios.

Also the scheme (9) do not distinguish between such isomers as 2-methylpentane and 3-methylpentane, or 2-methylhexane, 3-methylhexane and 3-ethylpentane, or 2,3-dimethylpentane and 2,4-dimethylpentane, etc.

If we postulate the arithmetical mean (3) for ternary nonbonded interactions via one carbon atom

$$\eta_{CCH} = (1/3)(2\eta_{CCC} + \eta_{HHH}),\ \eta_{CHH} = (1/3)(\eta_{CCC} + 2\eta_{HHH}), \tag{11}$$

then, as can be seen from (10), $\Delta_{CCC} = 0$, and we return to the scheme (6).

6.THE SECOND-ORDER APPROXIMATION

In this approximation we consider the bonded interactions and nonbonded interactions, separated by one and two skeletal atoms along the molecular chain. Taking into account the binary interactions, from (3) we obtain the following expression:

$$\begin{aligned} P(C_nH_{2n+2}) = (n-1)p''_{C-C} + (2n+2)p_{C-H} + x_{CC}\Gamma'_{CC} + \\ + x_{CCC}\Delta_{CCC} + y_{CC}\tau_{CC}, \end{aligned} \tag{12}$$

in which p_{C-H} and Δ_{CCC} are the same as in (10), and

$$\begin{aligned} p''_{C-C} = p'_{C-C} + 9\varsigma_{HH},\ \Gamma'_{CC} = \\ = \Gamma_{CC} + 6\varsigma_{CH} - 6\varsigma_{HH},\ \tau_{CC} = \varsigma_{CC} - 2\varsigma_{CH} + \varsigma_{HH}\;; \end{aligned} \tag{13}$$

τ_{CC} is effective interaction of nonlinked pair of carbon atoms across two atoms, y_{CC} is the number of such pairs.

The second-order appoximation schemes are more suitable for monosemantic description of the structural isomers. These schemes do not distinguish between such isomers as 3-methylheptane and 4-methylheptane, etc.

If we postulate the (1) for nonbonded interactions via two carbon atoms

$$\varsigma_{CH} = (1/2)(\varsigma_{CC} + \varsigma_{HH}), \tag{14}$$

then, as can be seen from (13), $\tau_{CC} = 0$, and we return to the additive schemes of the first-order approximation.

7.THE THIRD-ORDER APPROXIMATION

In this approximation we consider the bonded interactions and nonbonded interactions, separated by one, two and three skeletal atoms along the molecular chain. Taking into account the binary interactions, from (3) we obtain the following expression:

$$P(C_nH_{2n+2}) = (n-1)p''_{C-C} + (2n+2)p_{C-H} + x_{CC}\Gamma''_{CC} + \\ + x_{CCC}\Delta_{CCC} + y_{CC}\tau'_{CC} + z_{CC}\omega_{CC}, \tag{15}$$

in which p_{C-H}, Δ_{CCC} and p''_{C-C} are the same as in (10) and (13), and

$$\Gamma''_{CC} = \Gamma'_{CC} + 9\vartheta_{HH}, \tau'_{CC} = \tau_{CC} + 6\vartheta_{CH} - 6\vartheta_{HH}, \\ \omega_{CC} = \vartheta_{CC} - 2\vartheta_{CH} + \vartheta_{HH}; \tag{16}$$

ω_{CC} is effective interaction of nonlinked pair of carbon atoms across three atoms, z_{CC} is the number of such pairs.

If we postulate the (1) for nonbonded interactions via three carbon atoms

$$\vartheta_{CH} = (1/2)(\vartheta_{CC} + \vartheta_{HH}), \tag{17}$$

then, as can be seen from (16), ω_{CC} = 0, and we return to the additive schemes of the second-order approximation.

8.DISTANT INTERACTIONS

In the fouth-order approximation we consider the bonded interactions and nonbonded interactions, separated by one, two, three and four skeletal atoms along the molecular chain. Taking into account the binary interactions, from (3) we obtain the following expression:

$$P(C_nH_{2n+2}) = (n-1)p''_{C-C} + (2n+2)p_{C-H} + x_{CC}\Gamma''_{CC} + \\ + x_{CCC}\Delta_{CCC} + y_{CC}\tau''_{CC} + z_{CC}\omega'_{CC} + u_{CC}\nu_{CC}, \tag{18}$$

in which p_{C-H}, Δ_{CCC}, p''_{C-C}, Γ''_{CC} are the same as in (10), (13) and (16), and

$$\tau''_{CC} = \tau'_{CC} + \kappa_{HH}, \omega'_{CC} = \omega_{CC} + 6\kappa_{CH} - 6\kappa_{HH}, \nu_{CC} = \\ = \kappa_{CC} - 2\kappa_{CH} + \kappa_{HH}; \tag{19}$$

ν_{CC} is effective interaction of nonlinked pair of carbon atoms across four atoms, u_{CC} is the number of such pairs.

If we postulate the (1) for nonbonded interactions via four carbon atoms

$$\kappa_{CH} = (1/2)(\kappa_{CC} + \kappa_{HH}); \tag{20}$$

then, the parameter ν_{CC} , as would be seen from (19), vanishes; so we return to the additive schemes of the three-order approximation.

Additive schemes may be extended to higher-order approximations, including the bonded interactions and nonbonded interactions, separated through one, two, three, four, five,... skeletal atoms along the molecular chain.

9.GRAPH-THEORETICAL ASPECTS

In the language of graph theory structural formula of chemical compound appear as *molecular graph* (MG), their vertices correspond to atoms and the edges are associated with chemical bonds.

Let p_l be the numbers of paths of length l = 1, 2, 3, ... in MG; and R be the number of trio adjacency edges with common vertex. As it is seen,

$$p_2 = x_{CC},\ R = x_{CCC},\ p_3 = y_{CC},\ p_4 = z_{CC},\ p_5 = u_{CC},\ \dots .$$

Then, schemes (5), (6), (9), (12), (15), (18), ... may be rewritten (omitting the superscripts) as follows[14,15,19]:

$$\begin{aligned} P(C_nH_{2n+2}) = {} & a + bn + p_2\Gamma_{CC} + R\Delta_{CCC} + p_3\tau_{CC} + \\ & + p_4\omega_{CC} + p_5\nu_{CC} + p_6\mu_{CC} + p_7 o_{CC} + \dots . \end{aligned} \tag{21}$$

Here $a = -p_{C-C} + 2p_{C-H}$, $b = p_{C-C} + 2p_{C-H}$ (bonded terms); $\Gamma_{CC}, \Delta_{CCC}, \tau_{CC}, \omega_{CC}, \nu_{CC}$ are the same as before; μ_{CC}, o_{CC} are effective interactions of nonlinked pairs of carbon atoms across five and six atoms respectively; p_l and R were defined above.

10.NUMERICAL CALCULATIONS

The parameters of additive schemes may be estimated from some experimental values of the property P by using least-squares procedures. Then, formulae (5), (6), (9), (12), (15), (18), ... , or else (21) may be used for calculation and prediction of the properties of alkanes.

This technique may be applied to majority of properties of alkanes such as enthalpy of formation, entropy, Gibbs free energy, molal volume, polarizability and molal refraction, diamagnetic susceptibility, heat of vaporization, logarithm of saturated vapor pressure, boiling point (at different pressures) and so on[13,20].

The formula (21) is tested here for the standard enthalpy of formation alkanes of gas phase[21] $\Delta_f H^0$ (g, 298 K) with:

- 2 parameters (a, b),
- 3 parameters (a, b, Γ_{CC}),
- 4 parameters (a, b, Γ_{CC}, Δ_{CCC}),
- 5 parameters (a, b, Γ_{CC}, Δ_{CCC}, τ_{CC}),
- 6 parameters (a, b, Γ_{CC}, Δ_{CCC}, τ_{CC}, ω_{CC}),
- 7 parameters (a, b, Γ_{CC}, Δ_{CCC}, τ_{CC}, ω_{CC}, ν_{CC}),
- 8 parameters (a, b, Γ_{CC}, Δ_{CCC}, τ_{CC}, ω_{CC}, ν_{CC}, μ_{CC}),
- 9 parameters (a, b, Γ_{CC}, Δ_{CCC}, τ_{CC}, ω_{CC}, ν_{CC}, μ_{CC}, o_{CC}).

The average calculation deviation ($|\bar{\varepsilon}|$) are 5.2; 2.9; 2.9; 1.4; 0.8; 0.8; 0.7; 0.7 kJ/mol respectively, and the maximum deviation ($\varepsilon_{max} = \Delta_f H^0_{calc.} - \Delta_f H^0_{exptl.}$) are 14.6; 7.8; 8.1; -5.6; -4.5; -3.9; -3.8; -3.8 kJ/mol respectively.

As it is seen, the calculations generally agree with the results of experiment. A good agreements between observed and estimated values are found at second-order and other approximations. Nine enthalpy parameters were used for prediction of the unknown enthalpy of formation of higher alkanes[14] (nonanes and and decanes).

11.CONCLUSION

In the phenomenological approach the extensive physicochemical property of alkanes may be expressed as sum of bonded terms (e.g. a + bn) and nonbonded terms from the effective interactions of nonlinked CC pairs (Γ_{CC}) and CCC trios (Δ_{CCC}), separated by the same carbon atom, plus the effective interactions of nonlinked CC pairs (τ_{CC}, ω_{CC}, ν_{CC},...), separated through two, three, four,... carbon atoms respectively.

The action of the postulate of the arithmetical mean for nonbonded interactions is very interesting. If the conditions (8), (11), (14), (17), (20),... are fulfilled, then the parameters (Γ_{CC}, Δ_{CCC}, τ_{CC}, ω_{CC}, ν_{CC},...) in the corresponding formulae vanish, and the property P of series C_nH_{2n+2} becomes *a linear function* of the number of carbon atoms.

The observed dependences "Property P of alkanes versus the carbon number n" generally are non-linearity[13,14].

12.ACKNOWLEDGMENTS

This work was supported by the Russian Foundation for Basic Research (project code № 00-03-32982) and the Scientific Programme "Russian University – Fundamental Research".

REFERENCES

1. L. Pauling, *The Nature of the Chemical Bond* (Oxford University Press, London, 1940).
2. L. Pauling, *General Chemistry* (W.H. Freeman and company, San-Francisco, 1970).
3. M.G. Vinogradova, in: *Properties of Substances and Structure of Molecules* (Tver Univ. Press, Tver. 1998), pp. 3-19.
4. Yu.G. Papulov and M.G. Vinogradova, in: *Mathematical models of non-linear exitations, transfer, dynamics, and control in condensed systems and other media.*, edited by L. Uvarova, A. Arinstein and A. latyshev (Plenum Press, New York, 1999), pp. 399-408.
5. Yu.G. Papulov, in: *Calculation Methods in Physical Chemistry* (Kalinin Univ. Press, Kalinin, 1983), pp. 3-15.
6. Yu.G. Papulov, V.M. Smolyakov and A.N. Leshina, in: *Phenomenological and quantumchemical methods of prediction of the thermodynamic properties of organic compounds* (IVTAN Press, Moskva, 1989), pp. 33-42.
7. M.G. Vinogradova, Yu.G. Papulov, I.G. Davydova, and O.V. Obshivkova, Phenomenological calculation of the heat of formation of substituted ethylenes, *Zh. Fiz. Khim.* **71**(11), pp.1992-2002 (1997).
8. M.G. Vinogradova, Yu.G. Papulov, and R. Yu. Papulov, Nonvalent interactions and properties of substituted benzenes: phenomenological study, *Zh. Fiz. Khim.* **72**(4), pp. 604-608 (1998).
9. M.G. Vinogradova, R. Yu. Papulov, Yu. G. Papulov, and Yu. A. Korshunova. Nonvalent interactions of atoms and properties of substituted cyclopropanes, *Russ. J. Phys. Chem.* 74(Suppl.2), pp. S353-S356 (2000).
10. Yu.G. Papulov, V.M. Smolyakov and T.G. Kemenova, in: *Calculation Methods of Investigation in Chemistry* (Tver Univ. Press, Tver, 1990), pp. 23-51.
11. V.M. Tatevskii, *Theory of Physicochemical Properties of Molecules and Substances* (Moscow Univ. Press, Moskva, 1987).
12. Yu.G. Papulov, *Structure of molecules* (Tver Univ. Press, Tver, 1995).
13. Yu.G. Papulov and M.G. Smolyakov, *Physical Properties and Chemical Structure* (Kalinin Univ. Press, Kalinin, 1981).
14. M.G. Vinogradova, Yu.G. Papulov and V.M. Smolyakov, *Quantitative Correlation Structure-Property of Alkanes. Additive Schemes of Calculation* (Tver Univ. Press, Tver, 1998).
15. Yu.G. Papulov and V. M. Smolyakov, in: *Mathematical methods in contemporary chemisty*, edited by S. Kuchanov (Gordon and Breach Publ., New York, 1996), pp. 143-180.
16. K. Fajans, Die energie der atombindungen in diamanten und in aliphatishen kohlenwasserstoffen, *Ber. Deut. Chem. Ges. B* **53**, 643-665 (1920).
17. C.T. Zahn, The significance of chemical bond energies, *J. Chem. Phys.* **2**(10), 671-680 (1934).
18. T.L. Allen, Bond energies and the interactions between next-nearest neighbours. I. Saturated hydrocarbons, diamond, sulfanes, S_8 and organic sulfur compounds, *J. Chem. Phys.* **31**(4), 1039-1049 (1959).
19. Yu.G. Papulov and M.G. Vinogradova, in: *Abstr. IV Intern. Conference on mathematical modelling* (Stankin Press, Moskva, 2000), pp. 88.
20. Yu.G. Papulov, T.I. Chernova, V.M. Smolyakov, and V.N. Polyakov, Use of topological indices for construction of correlation structure-property, *Zh. Fiz. Khim.* **67**(2), pp.203-209 (1993).
21. J.B. Pedley, R.D. Naylor and S.P. Kirby, *Thermochemical data of organic compounds* (Chapman and Hall, London & New York, 1986).

QUANTUM—CHEMICAL MODELS OF THE STRUCTURE AND THE FUNCTIONS OF THE ACTIVE CENTRES OF THE POLYNUCLEAR COMPLEXES

Ludmila Ju. Vasil'eva*

1.INTRODUCTION

The fundamental problem of contemporery is the clusters physics is the formulation of the general theoretical structure conceptions and the clusters stability. The complication of such problem is that the using of the calculating methods to the clusters systems is problematicaly. So it is necessary to use such qualitative method as the method of the correlation diagrams based on the legand–field theory.

The application of the molecular orbitals method(MO) for the many–atomac complex combinations and the generalization of the crystal–field theory notions is called the ligand–theory. The complex is regarded as a single whole. The considerable simplification of Schrodinger equation may be achieved by the using of the ligand–field theory the complex molecular theory. According to the ligand–field theory the complex molecular orbital wave function may be written in the following way[1] :

$$\psi_{MO} = k\psi_0 + f\psi ; \tag{1}$$

where: ψ_0 is the contralion atomic orbital (AO),

$$\psi' = C_1\psi_1 + C_2\psi_2 + \ldots + C_i\psi_i$$

– is the ligands molecular orbital and k, f, C_i are the constants, ψ_i is the ligand atomic orbital.

* Ludmila Ju. Vasil'eva Tver' State University, 170002 Tver, Russia

ψ_{MO} relates to the definite type of symmetry, ψ_0 and ψ' must conform to each other according to the symmetry properties. ψ_{MO} are called the group orbital. The algorithm of the electron structure construction for the mono- and polynuclear biologically active complexes of the transition metals was discussed previously.[2]

The polynuclear metalloenzymes have some reculiarities since the cluster complexes are the active centres connecting with the protein macromolecules. The protein environment provides the necessary conditions for the cluster functioning, stabilizes the cluster structure and reacts on any electron redistributions in the ligand- metal bonds. So the model of the active centre was constructed by only taking info account the nearest environment of the transition metal ions: the ns–, np–, (n–1)d valence atomic orbitals and the rules of group orbital construction based on the symmetry properties of the molecular orbitals.

The Common Idea atbont The Functional Mechanism of the Active Centers of The Transition Metals Biologically Active Complexes.

An analysis of the literature data leads to the following conclusions: (1) There is a well–defined correlation between the structure and functional properties of metalloenzymes. (2) Althoung the functions of metalloenzymes differ, the mechanism of the action of their active sites is the same: accepting and donating electrons.

The functioning of the active centers of the biologically active transition metal complexes is determened by the changes of the charge state of the central ions from the quantum–chemical point of vuw. Such changes are induced by the removal or coming of electron. The electron coming is induced the attraction of the positive groups and the repulsion of the negative groups. From the physical point of view it is the process of the polarization or the process of the formation of the polaron, which destroys then the electron is removed and the system returns to the initial state. It is called the electron–confomational changes keeps the metalloenzym active center in "working" state. At the change of the central ion change states the ionic radius change also, inducing the change of bond lengths and sometimes the displacement of the central ion. Besides the electron removal or coming induces the redistribution of the electron density. All active center or change its symmetry. Owing to the confomational changes the protein ligands keep the active centre symmetry providing the active centre functioning.

2.THE QUANTUM – CHEMICAL MODEL OF THE ACTIVE CENTRE OF NITROGENASE.

The basic function of enzyme nitrogenase is the fixation of the molecular nitrogen is the fixation of the molecular nitrogen. The enzyme active form is maked up by the combined interaction of two components: the ironcontaining protein (Fe Б) and ironmolybdenum or molybdenumsulphur complex. The enzyme eatalytic activity is defined by unification of these proteins in a single whole. The function of Fe Б including Fe_4S_4 – cluster is the reduction of Mo^{3+} in the cluster and the mechanism of its function is analogous to the mechanism of the ferredoxine active centre function.[3] Mo-cluster interacts with the molecular nitrogen directly (fig. 1a).

The enzyme active centre, including Mo^{3+}, surrounds by five S^{2-} – ligands. It may be supposed that three pairs of the p-orbital electrons transter to Mo^{3+}thus that one pair forms σ – bond and two electron pairs form π – bonds. Such supposition was

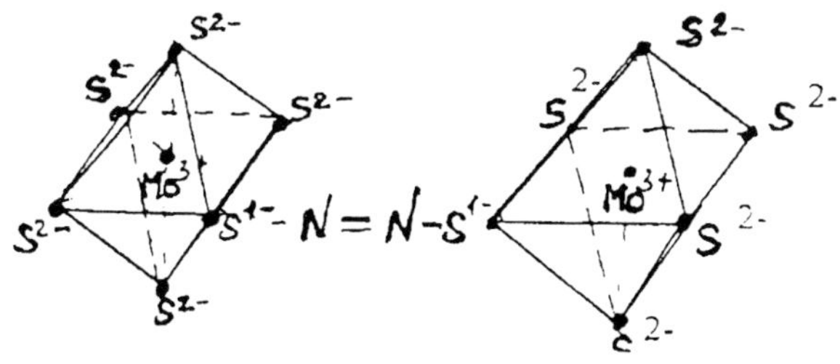

a)

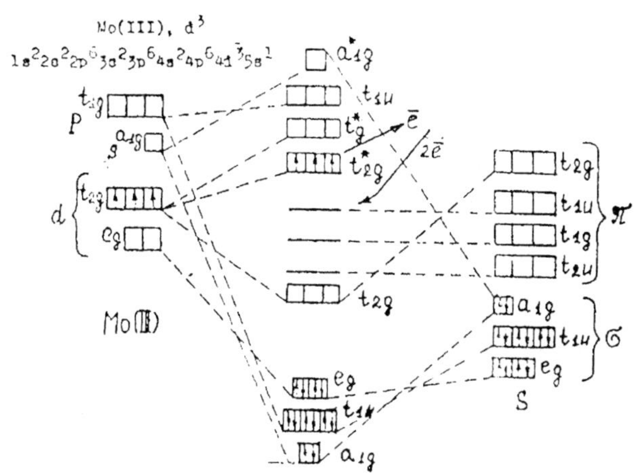

b)

Figure 1. a) Scematically representation of the nitrogenase active centre interacting with the molecular nitrogen (N_2);
b)Correlation cliagram of the molybdenumsulphur complex

made according to the p-orbital symmetry. The scheme of the nitrogenase active centre including molybdenumsulphur complex and its correlation diagram is shown in Fig. 1.The mechanism of the molecular nitrogen fixation is shown in Fig.2. According to the correlation diagram the stability of N_2 is determined by the following way: six valence electrons dispose on six π –orbitals and four electrons are on σ –orbitals, localizing on two nitrogen atoms. These electrons are assumed not capable of the interaction. It is supposed that the electron from the outer restorer goes to Fe–protein. One unpaired electron from sulphur ion of the cysteine residue CS^{1-}) transfers to *Mo*–complex.

It is more probably that proton tarnsfers from the neighboring systems and induces the conformational change of the system. For the keeping of the complex stability it must be took place the redistribution of the electrons from N_2 to Mo^{3+}. Al this moment near nitrogen atoms cere the hydrogen atoms (according to the experimental dates) in – fluencing on the system and as a result of it the bonds S^{1-}=*N*-*N*=S^{1-} are reduced, then two

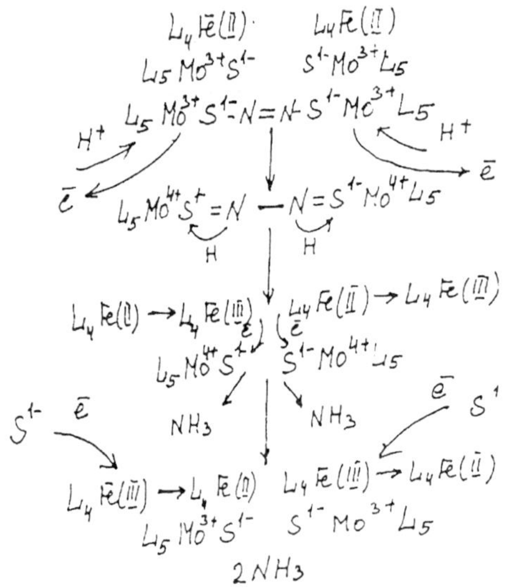

Figure 2. Schematically representation of the mechanism of the molecular nitrogen fixation

NH_3 molecules are formed . the formation of two NH_3 molecules induces the $Fe(II) \rightarrow Fe(III)$ transition in *FeБ*. Owing to the unpaired electron from S^{1-} ion the $Fe(II) \rightarrow Fe(III)$ transition in *FeБ* and $Mo^{4+} \rightarrow Mo^{3+}$ transition in molybdenumsulphur cluster take place on the final stage of the nitrogen fixation process.

3.CONCLUSION

The experimental investigations of the biosystems including nitrogencese with the methods of triplet probes; luminescence and electron paramagnetic resonance (EPR) confirm the discussed scheme and the correctness of the supposition of the stability of the active centre structure at the change of the ion change states.[5,6]

The mechanism of the functioning of the cluster active centre is determined by the electron structure and has many in common with the mononuclear complexes. It may be formulated the following property: for the many–electron systems it is of great importance not only the electron states of the cluster ions but also their bonds with the ligands and within the macromolecule.

The results of the discussed work show that the modelling of the active centre electron configuration must be made on the basis of the ligand–field theory and the protein environment weakly influences on the energetic levels and on the electron redistribution. The protein environment controls the enzyme processes.

REFERENCES

1 J.B. Bersuker, Electronic Structure and Properties of Coordination Compounds (Khimiya, Leningrad, 1976).

2. L.Yu. Vasil'eva, Quantum–chemical models of the Active Centres of Transition Metals Biologically Active Complexes, in: Mathematical Models of Non–hinear Exitations, Transfer, Dynamics and Control in Condensed System and Other Media, edited by L.A. Ivanova, A.E. Arinstein and A.V. Latyshev(Kluver Academic/ Plenum Publishers, New York, 1999), pp.385–398.
3. L.Yu. Vasil'eva, Correlation Between the Structure and Functional Properties of Biologicalli Active Transition Metal Complexes, Russian Jowenal of Phys., Chem., **8**, 1295–1298 (1995).
4. D.S. Chernavskii, N.M. Chernavskaya, "Proteein–Mashine". The Biological Molecular (Janus–K, Moscow, 1999).
5. A.E. Shilov, Intermediate Complexes in Chemical and Biological Nitrogen Fixation, Pure Appl. Chem. **64**, 1409 (1992).
6. L.A. Sirtsova, Formation of The Re–reduced state of The Jronmolybdenum Cluster in Reduced Nitrogenase Active Centre, Oxidative–Restoring Metalenzymes and Their Models (Chernogolovka,).

6. MATHEMATICAL MODELS OF TRANSPORT PROCESSES IN COMPLEX SYSTEMS

ASYMPTOTICS OF TRANSPORT EQUATIONS FOR SPHERICAL GEOMETRY IN L² WITH REFLECTING BOUNDARY CONDITIONS

Degong Song and William Greenberg*

ABSTRACT

The time dependent transport equation in a sphere with reflecting boundary conditions is discussed in the setting of L^2. Some aspects of the spectral properties of the strongly continuous semigroup $T(t)$ generated by the corresponding transport operator A are studied, and it is shown that the spectrum of $T(t)$ outside the disk $\{\lambda : |\lambda| \leq \exp(-\lambda^* t)\}$ (where λ^* is the essential infimum of the total collision frequency $\sigma(r, v)$, or $\lambda^* = \text{ess inf}_{\,r} \lim_{\,v \to 0^+} \sigma(r, v)$) consists of isolated eigenvalues of $T(t)$ with finite algebraic multiplicity, and the accumulation points of $\sigma(T(t)) \bigcap \{\lambda : |\lambda| > \exp(-\lambda^* t)\}$ can only appear on the circle $\{\lambda : |\lambda| = \exp(-\lambda^* t)\}$. Consequently, the asymptotic behavior of the time dependent solution is obtained.

1. INTRODUCTION

Consider the following neutron transport equation[6,13,17]

$$
\text{(I)} \quad \begin{cases}
\dfrac{\partial f(r,v,\mu,t)}{\partial t} = -v\mu \dfrac{\partial f(r,v,\mu,t)}{\partial r} - v\dfrac{1-\mu^2}{r}\dfrac{\partial f(r,v,\mu,t)}{\partial \mu} \\
\qquad -\sigma(r,v) f(r,v,\mu,t) + \dfrac{1}{2}\displaystyle\int_0^{v_M}\int_{-1}^{1} k(r,v,v') f(r,v',\mu',t)\,dv'd\mu', \\
r \in V := [0,R],\ v \in E := (0, v_M],\ \mu \in \Omega := [-1,1],\ t > 0, \\
f(R,v,\mu,t) = \alpha(v,\mu) f(R,v,-\mu,t)\ \forall \mu \in [-1,0),\ v \in E, \\
f(r,v,\mu,0) = f_0(r,v,\mu),
\end{cases}
$$

where the region occupied by the reactor media is a sphere of radius $R > 0$, r is the distance from the center of the sphere, v is the velocity, $0 < v_M < +\infty$, μ is the cosine of the angle the neutron velocity makes with the radius vector, $f(r, v, \mu, t)$ is the neutron distribution at time t, $\sigma(r, v)$ is the total collision frequency, $k(r, v, v')$ is the scattering fission kernel, $\alpha(v, \mu)$ is the boundary reflection coefficient, and $f_0(r, v, \mu)$ is the initial distribution.

*Department of Mathematics and Center for Statistical Mechanics & Mathematical Physics, Virginia Tech, Blacksburg, VA 24061, U.S.A.

Mathematical Modeling: Problems, Methods, Applications
Edited by Uvarova and Latyshev, Kluwer Academic/Plenum Publishers, 2001

Throughout this paper, it is assumed that

(H1) $\sigma(r,v)$ is a real bounded measurable function.

(H2) $\alpha(v,\mu)$ is bounded measurable, and $0 \le \alpha(v,\mu) \le \alpha_0 < 1$, where α_0 is a constant.

(H3) $k(r,v,v')$ is a real measurable function, and there exist positive constants $\delta < 1/2$ and M such that

$$|k(r,v,v')| \le M v^{-\delta}, \tag{1}$$

or

$$|k(r,v,v')| \le M v'^{-\delta}. \tag{2}$$

By the transform $x = r\mu$, $y = r\sqrt{1-\mu^2}$, $\psi(x,y,v,t) = f(r(x,y),\mu(x,y),v,t)$, Eq. (I) can be equivalently written as[13,17]

$$\text{(II)}\quad \begin{cases} \dfrac{\partial\psi(x,y,v,t)}{\partial t} = -v\dfrac{\partial\psi(x,y,v,t)}{\partial x} - \sigma(r,v)\psi(x,y,v,t) \\ \qquad +\dfrac{1}{2r}\displaystyle\int_0^{v_M}\int_{-r}^{r} k(r,v,v')\psi(z,\sqrt{r^2-z^2},v',t)dz, \\ y \in V,\ 0 \le r = \sqrt{x^2+y^2} \le R,\ v \in E,\ t > 0, \\ \psi(-\sqrt{R^2-y^2},y,v,t) = \alpha\left(v,-\sqrt{1-R^{-2}y^2}\right)\psi(\sqrt{R^2-y^2},y,v,t), \\ \psi(x,y,v,0) = \psi_0(x,y,v). \end{cases}$$

Set $D = \{(x,y) : y \ge 0, 0 \le \sqrt{x^2+y^2} \le R\}$, $G = D \times E$, and let $L^2(G)$ represent the Hilbert space composed of all measurable complex functions defined and square integrable over G, with the inner product $(\cdot,\cdot)$ and the norm $\|\cdot\|$ given by

$$(\psi,\varphi) = \frac{1}{2}\int_0^{v_M} dv \int_0^R y dy \int_{-\sqrt{R^2-y^2}}^{\sqrt{R^2-y^2}} \psi(x,y,v)\overline{\varphi(x,y,v)}dx, \qquad \|\psi\| = \sqrt{(\psi,\psi)}.$$

Define operators on $L^2(G)$ as follows:

$$B\psi = -v\frac{\partial\psi}{\partial x} - \sigma(r,v)\psi,$$

$$K\psi = \frac{1}{2r}\int_0^{v_M} dv' \int_{-r}^{r} k(r,v,v')\psi(z,\sqrt{r^2-z^2},v')dz,$$

$$A\psi = B\psi + K\psi,$$

with $D(B) = \{\psi \in L^2(G) : B\psi \in L^2(G),\ \psi(-\sqrt{R^2-y^2},y,v) = \alpha\left(v,-\sqrt{1-R^{-2}y^2}\right)\times \psi(\sqrt{R^2-y^2},y,v)$ for every $y \in V$ and $v \in E\} = D(A)$, $D(K) = L^2(G)$. Then Eq. (II) can be written as

$$\frac{d\psi(t)}{dt} = A\psi(t), \quad \psi(0) = \psi_0.$$

The spectrum of the neutron transport operator was first studied by J. Lehner and G. M. Wing[5] for the one-speed equation. The study of spherical geometry began with Norton's work[13] in 1962 with one-speed non-reentrant boundary conditions. In this case the strongly continuous semigroup $S(t)$ generated by the streaming operator B can be given by an exact expression. Although some work has been done for reentrant boundary conditions,[6,11,17,19] these papers investigate the spectrum of the transport operator A, not the semigroup $T(t)$ generated by A. Since the spectral mapping theorem does not hold for strongly continuous semigroups without any further conditions on the semigroups such as analyticity or norm continuity, the information contained in $\sigma(A)$ can not be directly transferred to $\sigma(T(t))$, and therefore the asymptotic behavior of the time dependent solution cannot be deduced. Because a convenient expression for the streaming semigroup is not available, the semigroup perturbation methods useful for the non-reentry problem have not been entirely successful. Although some conclusions about the asymptotic behavior of the time dependent solution can be obtained by virtue of the inversion of the Laplace transform, it is always required that the initial distribution $\psi_0 \in D(A^2)$. Our goal will be to obtain asymptotic results for the semigroup $T(t)$ without this unphysical condition on the initial distribution.

2. SPECTRAL PROPERTIES OF THE OPERATOR A IN L^2

From the hypothesis (H3), it is evident that K is a bounded operator. Throughout, we shall set $\lambda^* = \text{ess inf}_{(r,v)\in V\times E}\sigma(r,v)$. As in [17] we have the following lemma.

LEMMA 1. *B is a densely defined operator, $\{\lambda : Re\lambda > -\lambda^*\} \subset \rho(B)$, and $\|(\lambda I - B)^{-1}\| \le (Re\lambda + \lambda^*)^{-1}$ for every λ with $Re\lambda > -\lambda^*$. Moreover, $\{\lambda : Re\lambda > \|K\| - \lambda^*\} \subset \rho(A)$. For every λ with $Re\lambda > -\lambda^*$, $\lambda \in P_\sigma(A)$ if and only if $1 \in P_\sigma(K(\lambda I - B)^{-1})$. The set $Pas(A) := \{\lambda : -\lambda^* < Re\lambda \le \|K\| - \lambda^*, \lambda \in \sigma(A)\}$ contains at most countable isolated elements, each of which is an eigenvalue of A with finite algebraic multiplicity.*

LEMMA 2. *For every λ with $Re\lambda > -\lambda^*$, $K(\lambda I - B)^{-1}K$ is an integral operator defined on $L^2(G)$. For every $\psi \in L^2(G)$,*

$$K(\lambda I - B)^{-1}K\psi(x,y,v) = \iiint_G k(\lambda, x, x', y, y', v, v')dx'dy'dv', \tag{3}$$

where

$$k(\lambda, x, x', y, y', v, v') = \frac{y'}{4r\rho}\int_0^{v_M} dv_1 \frac{k(r,v,v_1)k(\rho,v_1,v')}{v_1}\int_{|r-\rho|}^{r+\rho}\frac{1}{s} \\ \times \frac{\exp[-\Sigma_3(\lambda,\sigma,\cdots)] + \alpha(v_1,\sqrt{\cdot})\exp[-\Sigma_4(\lambda,\sigma,\cdots) + \Sigma_3(\lambda,\sigma,\cdots)]}{1-\alpha(v_1,\sqrt{\cdot})\exp[-\Sigma_4(\lambda,\sigma,\cdots)]}ds, \tag{4}$$

with

$$r = \sqrt{x^2+y^2}, \quad \rho = \sqrt{x'^2+y'^2}, \quad \alpha(v_1,\sqrt{\cdot}) = \alpha\left(v_1, \sqrt{r^2 - \Delta_2^2(r,\rho,s)}\right),$$

$$\Sigma_3(\lambda,\sigma,\cdots) = \int_{\Delta_1(r,\rho,s)}^{\Delta_2(r,\rho,s)} \frac{\lambda + \sigma(\sqrt{t^2 + r^2 - \Delta_2^2(r,\rho,s)}, v_1)}{v_1}dt,$$

$$\Sigma_4(\lambda,\sigma,\cdots) = \int_{-\Delta(r,\rho,s)}^{\Delta(r,\rho,s)} \frac{\lambda + \sigma(\sqrt{t^2 + r^2 - \Delta_2^2(r,\rho,s)}, v_1)}{v_1}dt,$$

$$\Delta_1(r,\rho,s) = -\frac{\rho^2-r^2}{2s} - \frac{s}{2}, \quad \Delta_2(r,\rho,s) = -\frac{\rho^2-r^2}{2s} + \frac{s}{2},$$

$$\Delta(r,\rho,s) = \sqrt{R^2 + \frac{1}{4s^2}(\rho^2-r^2)^2 + \frac{s^2}{4} - \frac{1}{2}(\rho^2+r^2)}.$$

Proof. Define operators $J : L^2(D\times E) \longrightarrow L^2(V\times E)$ and $H : L^2(V\times E) \longrightarrow L^2(D\times E)$ by

$$J\psi(r,v) = \frac{1}{2}\int_{-r}^{r} \psi(z, \sqrt{r^2-z^2}, v)dz,\ r = \sqrt{x^2+y^2},$$

$$H\varphi(x,y,v) = \frac{1}{\sqrt{x^2+y^2}}\int_0^{v_M} k(\sqrt{x^2+y^2}, v, v')\varphi(\sqrt{x^2+y^2}, v')dv'.$$

Then $K = HJ$. From [17], p. 14, it is known that

$$J(\lambda I - B)^{-1}H\psi(r,v) = \int_0^{v_M} dv' \int_0^R d\rho h(\lambda, v, v', r, \rho)\psi(\rho, v'),$$

where

$$h(\lambda, v, v', r, \rho) = \frac{k(\rho, v, v')}{2v}\int_{|r-\rho|}^{r+\rho} \frac{1}{s}$$
$$\times \frac{\exp[-\widetilde{\Sigma}_3(\lambda,\sigma,\cdots)] + \alpha(v, \sqrt{\cdot}\,)\exp[-\widetilde{\Sigma}_4(\lambda,\sigma,\cdots) + \widetilde{\Sigma}_3(\lambda,\sigma,\cdots)]}{1-\alpha(v,\sqrt{\cdot}\,)\exp[-\widetilde{\Sigma}_4(\lambda,\sigma,\cdots)]}ds.$$

with

$$\widetilde{\Sigma}_3(\lambda,\sigma,\cdots) = \int_{\Delta_1(r,\rho,s)}^{\Delta_2(r,\rho,s)} \frac{\lambda + \sigma(\sqrt{t^2+r^2-\Delta_2^2(r,\rho,s)}, v)}{v}dt,$$

$$\widetilde{\Sigma}_4(\lambda,\sigma,\cdots) = \int_{-\Delta(r,\rho,s)}^{\Delta(r,\rho,s)} \frac{\lambda + \sigma(\sqrt{t^2+r^2-\Delta_2^2(r,\rho,s)}, v)}{v}dt.$$

Thus,

$$J(\lambda I - B)^{-1}HJ\psi(r,v) = \int_0^{v_M} dv' \int_0^R d\rho h(\lambda, v, v', r, \rho)\cdot\frac{1}{2}\int_{-\rho}^{\rho}\psi(z, \sqrt{\rho^2-z^2}, v')dz,$$

and

$$HJ(\lambda I - B)^{-1}HJ\psi(x,y,v) = \int_0^{v_M} dv' \int_0^R d\rho \int_{-\rho}^{\rho}\psi(z,\sqrt{\rho^2-z^2}, v')dz$$
$$\times \frac{1}{2\sqrt{x^2+y^2}}\int_0^{v_M} k(\sqrt{x^2+y^2}, v, v_1)h(\lambda, v_1, v', r, \rho)dv_1.$$

By virtue of the transform $x' = z,\ \ y' = \sqrt{\rho^2-z^2}$, we get

$$K(\lambda I - B)^{-1}K\psi(x,y,v) \doteq HJ(\lambda I - B)^{-1}HJ\psi(x,y,v)$$
$$= \int_0^{v_M} dv' \iint_D dx'dy'k(\lambda, x, x', y, y', v, v')\psi(x', y', v'),$$

where $k(\lambda, x, x', y, y', v, v')$ is just the expression given by Eq. (4). This completes the proof. Q. E. D.

LEMMA 3. *Let H be an integral operator on $L^2(G)$ with kernel $h(x, x', y, y', v, v')$. Then*

$$\|H\|^2 \le \iiint_G \iiint_G \frac{y}{y'} \mid h(x, x', y, y', v, v') \mid^2 dx'dy'dv'dxdydv.$$

THEOREM 1. *Set $\widetilde{k}(r,v,v') = v^\delta k(r,v,v')$ if Eq. (1) in (H3) is satisfied, or $\widetilde{k}(r,v,v') = v'^\delta k(r,v,v')$ if Eq. (2) in (H3) is satisfied. Let $\lambda = \beta + i\tau$, $\beta \in [\beta_1, \beta_2]$, $\beta_2 > \beta_1 > -\lambda^*$, and assume the following condition is satisfied:*
(H4) $\alpha(v,\mu), \sigma(r,v)$ and $\widetilde{k}(r,v,v')$ are partially differentiable with respect to v, v', and the corresponding partial derivatives $\frac{\partial\alpha}{\partial v}$, $\frac{\partial\sigma}{\partial v}$, $\frac{\partial\widetilde{k}}{\partial v}$ and $\frac{\partial\widetilde{k}}{\partial v'}$ are uniformly bounded.
Then for every $\varepsilon > 0$, there exists a positive constant $\widetilde{\tau}$ independent of $\beta \in [\beta_1, \beta_2]$, such that $\|K(\lambda I - B)^{-1}K\| \leq \varepsilon$ uniformly in $\{\lambda = \beta + i\tau : \beta_1 \leq \beta \leq \beta_2, |\tau| \geq \widetilde{\tau}\}$.

Proof. We will assume $\widetilde{k}(r,v,v') = v^\delta k(r,v,v')$. The argument is similar if $\widetilde{k}(r,v,v') = v'^\delta k(r,v,v')$. Since $0 \leq \alpha(v_1, \sqrt{\cdot}) \leq \alpha_0 < 1$, we have

$$\frac{1}{1 - \alpha(v_1, \sqrt{\cdot}) \exp[-\Sigma_4(\lambda, \sigma, \cdots)]} = \sum_{n=0}^{\infty} \alpha^n(v_1, \sqrt{\cdot}) \exp[-n\Sigma_4(\lambda, \sigma, \cdots)]. \tag{5}$$

For every $n = 0, 1, 2, \cdots$, set

$$g_{n,1}(\lambda, x, x', y, y', v, v') = \frac{y'}{4r\rho} \int_0^{v_M} dv_1 \frac{k(r,v,v_1)k(\rho,v_1,v')}{v_1}$$
$$\times \int_{|r-\rho|}^{r+\rho} \alpha^n(v_1, \sqrt{\cdot}) \exp\left[-n\Sigma_4(\lambda,\sigma,\cdots) - \Sigma_3(\lambda,\sigma,\cdots)\right] s^{-1} ds,$$
$$g_{n,2}(\lambda, x, x', y, y', v, v') = \frac{y'}{4r\rho} \int_0^{v_M} dv_1 \frac{k(r,v,v_1)k(\rho,v_1,v')}{v_1}$$
$$\times \int_{|r-\rho|}^{r+\rho} \alpha^{n+1}(v_1, \sqrt{\cdot}) \exp\left[-(n+1)\Sigma_4(\lambda,\sigma,\cdots) + \Sigma_3(\lambda,\sigma,\cdots)\right] s^{-1} ds,$$

and define operators $G_{n,1}$, $G_{n,2}$ on $L^2(G)$ by

$$G_{n,j}\psi = \iiint_G g_{n,j}(\lambda, x, x', y, y', v, v')\psi(x', y', v')dx'dy'dv', \quad n = 0, 1, \cdots, \ j = 1, 2. \tag{6}$$

Then from Eqs. (3) – (5), we get

$$K(\lambda I - B)^{-1}K = \sum_{n=0}^{\infty} G_{n,1} + \sum_{n=0}^{\infty} G_{n,2}. \tag{7}$$

First, we consider $G_{n,2}$. From Lemma 4 and Eq. (6), we have

$$\|G_{n,2}\|^2 \leq \frac{1}{16} \iint_D dxdy \iint_D dx'dy' \int_E dv \int_E dv' \frac{y}{y'} \cdot \frac{y'^2}{r^2\rho^2} \left| \int_0^{v_M} dv_1 \frac{k(r,v,v_1)k(\rho,v_1,v')}{v_1} \right.$$
$$\left. \times \int_{|r-\rho|}^{r+\rho} \alpha^{n+1}(v_1, \sqrt{\cdot}) \exp\left[-(n+1)\Sigma_4(\lambda,\sigma,\cdots) + \Sigma_3(\lambda,\sigma,\cdots)\right] s^{-1} ds \right|^2.$$

By virtue of the transform $x = z$, $y = \sqrt{r^2 - z^2}$; $x' = z'$, $y' = \sqrt{\rho^2 - z'^2}$, we get

$$\|G_{n,2}\|^2 \leq \frac{1}{16} \int_0^R dr \int_{-r}^{r} \frac{r}{y} dz \int_0^R d\rho \int_{-\rho}^{\rho} \frac{\rho}{y'} dz' \int_E dv \int_E dv' \frac{y}{y'} \cdot \frac{y'^2}{r^2\rho^2}$$

$$\times\left|\int_0^{v_M} dv_1 \frac{k(r,v,v_1)k(\rho,v_1,v')}{v_1}\int_{|r-\rho|}^{r+\rho}\alpha^{n+1}(v_1,\sqrt{\cdot})\right.$$
$$\left.\times\exp\left[-(n+1)\Sigma_4(\lambda,\sigma,\cdots)+\Sigma_3(\lambda,\sigma,\cdots)\right]s^{-1}ds\right|^2$$
$$=\frac{1}{4}\int_E dv\int_E dv'\int_0^R dr\int_0^R d\rho\left|\int_{|r-\rho|}^{r+\rho}F(n,\lambda,r,\rho,\cdots)s^{-1}ds\right|^2, \qquad (8)$$

where

$$F(n,\lambda,r,\rho,\cdots)=\int_0^{v_M}\frac{k(r,v,v_1)k(\rho,v_1,v')}{v_1}$$
$$\times\,\alpha^{n+1}(v_1,\sqrt{\cdot})\exp\left[-(n+1)\Sigma_4(\lambda,\sigma,\cdots)+\Sigma_3(\lambda,\sigma,\cdots)\right]dv_1. \qquad (9)$$

From this we have

$$F(n,\lambda,r,\rho,\cdots)=\int_0^{v_M}\frac{\partial\exp\{i\tau v_1^{-1}[-2(n+1)\Delta(r,\rho,s)+s]\}}{\partial v_1}$$
$$\times\{i\tau[2(n+1)\Delta(r,\rho,s)-s]\}^{-1}v_1^2v^{-\delta}\tilde{k}(r,v,v_1)v_1^{-\delta}\tilde{k}(\rho,v_1,v')$$
$$\times\,\alpha^{n+1}(v_1,\sqrt{\cdot})\exp\left[-(n+1)\Sigma_4(\beta,\sigma,\cdots)+\Sigma_3(\beta,\sigma,\cdots)\right]v_1^{-1}dv_1.$$

By virtue of integration by parts and the inequality $2\Delta(r,\rho,s)\geq|r-\rho|+s$ we can estimate

$$|F(n,\lambda,r,\rho,\cdots)|\leq\tau^{-1}[2(n+1)\Delta(r,\rho,s)-s]^{-1}v^{-\delta}[C_{10}\alpha_0^{n+1}+C_{11}(n+1)\alpha_0^n]$$
$$\leq C_{12}(n+1)\alpha_0^n\tau^{-1}v^{-\delta}[2(n+1)\Delta(r,\rho,s)-s]^{-1}\leq\frac{C_1\alpha_0^n}{|\tau|\cdot|r-\rho|v^\delta}, \qquad (10)$$

where C_1 as well as all the C_i $(i=2,3,\cdots)$ arising in the following are positive constants.

On the other hand, from Eq. (9), Eq. (1) and the expressions for $\Sigma_3(\lambda,\sigma,\cdots)$, $\Sigma_4(\lambda,\sigma,\cdots)$, and $\Delta(r,\rho,s)$, we have

$$|F(n,\lambda,r,\rho,\cdots)|\leq C_{14}\alpha_0^nv^{-\delta}\int_0^{v_M}v_1^{-(1+\delta)}\exp\left\{\frac{[-2(n+1)\Delta(r,\rho,s)+s](\beta+\lambda^*)}{v_1}\right\}dv_1$$
$$\leq C_{14}\alpha_0^nv^{-\delta}\int_{t_0}^{+\infty}\frac{\exp[-(\beta+\lambda^*)|r-\rho|t]}{t^{1-\delta}}dt,$$

where $t_0=v_M^{-1}$. By use of Hölder inequality (set $p=(1-\delta)^{-1}$, $q=\delta^{-1}$), we have

$$|F(n,\lambda,r,\rho,\cdots)|\leq C_{14}\alpha_0^nv^{-\delta}\left\{\int_{t_0}^{+\infty}\frac{\exp[-(\beta+\lambda^*)|r-\rho|t]}{t}dt\right\}^{1-\delta}$$
$$\times\left\{\int_{t_0}^{+\infty}\exp[-(\beta+\lambda^*)|r-\rho|t]\,dt\right\}^{\delta}.$$

Since the function $\frac{1}{\log u}\int_{t_0}^{\infty}\frac{e^{-ut}}{t}dt$ is analytic in $\{u\in C:0\leq \mathrm{Re}u<1\}$, if $|\lambda+\lambda^*|\cdot|r-\rho|\leq 1/2$ (which implies $(\beta+\lambda^*)|r-\rho|\leq1/2$), then it follows

$$|F(n,\lambda,r,\rho,\cdots)|\leq C_{15}\alpha_0^nv^{-\delta}(\beta+\lambda^*)^{-\delta}|r-\rho|^{-\delta}\,|\log[(\beta+\lambda^*)|r-\rho|]\,|^{1-\delta}\leq C_{15}\alpha_0^nv^{-\delta}$$

$$\times (\beta+\lambda^*)^{-\delta}|r-\rho|^{-\delta}\{|\log[|\lambda+\lambda^*|\cdot|r-\rho|]|+|\log|\lambda+\lambda^*||+|\log(\beta+\lambda^*)|+1\}.$$

Thus, we get the following estimation when $|\lambda+\lambda^*|\cdot|r-\rho|\le 1/2$:

$$|F(n,\lambda,r,\rho,\cdots)|\le C_2\alpha_0^n(\mathrm{Re}\lambda+\lambda^*)^{-\delta}|r-\rho|^{-\delta}v^{-\delta}$$
$$\times\{|\log(|\lambda+\lambda^*|\cdot|r-\rho|)|+|\log|\lambda+\lambda^*||+|\log(\mathrm{Re}\lambda+\lambda^*)|+1\}. \tag{11}$$

From Eqs. (10) and (11), we see that $|F(n,\lambda,r,\rho,\cdots)|$ is dominated by a function P independent of s, i.e.,

$$|F(n,\lambda,r,\rho,\cdots)|\le P(n,\lambda,|r-\rho|,v). \tag{12}$$

From Eq. (8), we know

$$\|G_{n,2}\|^2\le\frac{1}{4}\int_E dv\int_E dv'\int_0^R dr\int_0^R d\rho\left[\int_{|r-\rho|}^{r+\rho}s^{-1}P(n,\lambda,|r-\rho|,v)ds\right]^2$$
$$\le\frac{1}{4}\int_E dv\int_E dv'\int_0^R dr\int_0^R d\rho\,[\ln(2R)-\ln|r-\rho|]^2P^2(n,\lambda,|r-\rho|,v)$$
$$\le C_3\int_E dv\int_E dv'\int_0^R dr\int_0^R d\rho\,\left[C_4+\log^2|r-\rho|\right]P^2(n,\lambda,|r-\rho|,v).$$

Consider the transformation $u=r+\rho,\quad w=r-\rho$. The domain $\{(r,\rho):r,\rho\in[0,R]\}$ is then transferred into $S:=\{(u,w):0\le u-w\le 2R,\ 0\le u+w\le 2R\}$ and

$$\|G_{n,2}\|^2\le\frac{C_3}{2}\int_E dv\int_E dv'\iint_S dudw(C_4+\log^2|w|)P^2(n,\lambda,|w|,v)$$
$$\le C_5\int_E dv\int_E dv'\int_0^R(C_4+\log^2 w)P^2(n,\lambda,w,v)dw. \tag{13}$$

Let $W_1=\{w\in[0,R]:|\lambda+\lambda^*|w\le 1/2\}$, $W_2=\{w\in[0,R]:|\lambda+\lambda^*|w>1/2\}$. Since

$$\log^2 w=\log^2(|\lambda+\lambda^*|w\cdot|\lambda+\lambda^*|^{-1})\le 2\log^2(|\lambda+\lambda^*|w)+2\log^2|\lambda+\lambda^*|,$$

it follows from Eqs. (10) – (13) that

$$\|G_{n,2}\|^2\le 2C_5\int_E dv\int_E dv'\int_0^R\left[\log^2(|\lambda+\lambda^*|w)+\log^2|\lambda+\lambda^*|+C_4\right]P^2(n,\lambda,w,v)dw$$
$$\le C_6\alpha_0^{2n}(\beta+\lambda^*)^{-2\delta}\int_E v^{-2\delta}dv\int_E dv'\int_{W_1}\left[\log^2(|\lambda+\lambda^*|w)+\log^2|\lambda+\lambda^*|+C_4\right]$$
$$\times\left[4\log^2(|\lambda+\lambda^*|w)+4\log^2|\lambda+\lambda^*|+4\log^2(\beta+\lambda^*)+4\right]w^{-2\delta}dw$$
$$+C_7\alpha_0^{2n}|\tau|^{-2}\int_E v^{-2\delta}dv\int_E dv'\int_{W_2}\left[\log^2(|\lambda+\lambda^*|w)+\log^2|\lambda+\lambda^*|+C_4\right]w^{-2}dw.$$

Applying the transformation $z(w)=|\lambda+\lambda^*|w$ to the right–hand side of the above equation, we get

$$\|G_{n,2}\|^2\le C_8\alpha_0^{2n}(\beta+\lambda^*)^{-2\delta}\int_0^{1/2}\left[\log^2 z+\log^2|\beta+\lambda^*+i\tau|+C_4\right] \tag{14}$$
$$\times\left[\log^2 z+\log^2|\beta+\lambda^*+i\tau|+\log^2(\beta+\lambda^*)+1\right]|\beta+\lambda^*+i\tau|^{2\delta-1}z^{-2\delta}dz$$
$$+C_9\alpha_0^{2n}|\tau|^{-2}\int_{1/2}^{\infty}\left[\log^2 z+\log^2|\beta+\lambda^*+i\tau|+C_4\right]|\beta+\lambda^*+i\tau|z^{-2}dz.$$

By a similar procedure, we can get an estimate for $\|G_{n,1}\|$. From Eqs. (7) and (14), the proof is complete. Q. E. D.

LEMMA 4. *Suppose $f \in L^2(G)$ and $m - f_1 \le f \le M - f_2$, where $f_1, f_2 \in L^2(G)$, m and M are constants. Then for every positive constant η, there exists a sequence $\{g_n\}$ composed of polynomial functions such that g_n converges to f almost everywhere and $m - \eta - f_1 \le g_n \le M + \eta - f_2$ on G.*

THEOREM 2. *The conclusion given in Theorem 1 still holds if the hypothesis (H4) is not satisfied.*

Proof. Since $\alpha(v,\mu)$, $\sigma(r,v)$ and $\widetilde{k}(r,v,v')$ are bounded measurable, it follows from Lemma 4 that there exist three sequences $\{\alpha_n(v,\mu)\}$, $\{\sigma_n(r,v)\}$ and $\{\widetilde{k}_n(r,v,v')\}$ such that

(a) for every n, $\alpha_n(v,\mu)$, $\sigma_n(r,v)$ and $\widetilde{k}_n(r,v,v')$ are polynomial functions;

(b) $|\alpha_n(v,\mu)| \le \alpha_0 + (1-\alpha_0)/2 < 1$, and $\alpha_n(v,\mu)$ converges to $\alpha(v,\mu)$ almost everywhere in $E \times \Omega$;

(c) $(\lambda^* - \beta_1)/2 \le \sigma_n(r,v) \le \text{ess sup}_{(r,v)}\sigma(r,v) + (\beta_1 + \lambda^*)/2$, and $\sigma_n(r,v)$ converges to $\sigma(r,v)$ almost everywhere in $V \times E$;

(d) $|\widetilde{k}_n(r,v,v')| \le \text{ess sup}_{(r,v,v')}\widetilde{k}(r,v,v') + 1$, and $\widetilde{k}_n(r,v,v')$ converges to $\widetilde{k}(r,v,v')$ almost everywhere in $V \times E \times E$.

Sct

$$k_n(r,v,v') = \begin{cases} v^{-\delta}\widetilde{k}_n(r,v,v'), & \text{if } \widetilde{k}(r,v,v') = v^{\delta}k(r,v,v'), \\ v'^{-\delta}\widetilde{k}_n(r,v,v'), & \text{if } \widetilde{k}(r,v,v') = v'^{\delta}k(r,v,v'), \end{cases}$$

$$\delta k_n(r,v,v') = k(r,v,v') - k_n(r,v,v'),$$

and define operators on $L^2(G)$ as follows:

$$B_n\psi = -v\frac{\partial\psi}{\partial x} - \sigma_n(r,v)\psi,$$

$$K_n\psi = \frac{1}{2r}\int_E dv' \int_{-r}^{r} k_n(r,v,v')\psi(z, \sqrt{r^2 - z^2}, v')dz,$$

$$\delta K_n\psi = \frac{1}{2r}\int_E dv' \int_{-r}^{r} \delta k_n(r,v,v')\psi(z, \sqrt{r^2 - z^2}, v')dz,$$

with $D(B_n) = \{\psi \in L^2(G) : B_n\psi \in L^2(G),\ \psi(-\sqrt{R^2-y^2}, y, v) = \alpha_n(v, -\sqrt{1 - R^{-2}y^2}) \times \psi(\sqrt{R^2-y^2}, y, v)$ for every $y \in V$ and $v \in E\}$ and $D(K_n) = D(\delta K_n) = L^2(G)$. Obviously, $K = K_n + \delta K_n$.

As for Lemma 1, it can be shown that $\{\lambda : \text{Re}\lambda \ge \beta_1\} \subset \rho(B_n)$, and $\|(\lambda I - B_n)^{-1}\| \le 2(\beta_1 + \lambda^*)^{-1}$ for every λ with $\text{Re}\lambda \ge \beta_1$. Further, for every λ with $\text{Re}\lambda > -\lambda^*$, it can be shown that:

(a) $\delta K_n(\lambda I - B_n)^{-1}K$ is an integral operator on $L^2(G)$ with the integral kernel given by

$$h_{n,1}(\lambda, x, x', y, y', v, v') = \frac{y'}{4r\rho}\int_0^{v_M} dv_1 \frac{\delta k_n(r,v,v_1)k(\rho,v_1,v')}{v_1}\int_{|r-\rho|}^{r+\rho}\frac{1}{s}$$
$$\times \frac{\exp[-\Sigma_3(\lambda,\sigma_n,\cdots)] + \alpha_n(v_1,\sqrt{\cdot})\exp[-\Sigma_4(\lambda,\sigma_n,\cdots) + \Sigma_3(\lambda,\sigma_n,\cdots)]}{1 - \alpha_n(v_1,\sqrt{\cdot})\exp[-\Sigma_4(\lambda,\sigma_n,\cdots)]}ds,$$

where

$$\alpha_n(v_1,\sqrt{\cdot}) = \alpha_n\left(v_1, \sqrt{r^2 - \Delta_2^2(r,\rho,s)}\right),$$

$$\Sigma_3(\lambda, \sigma_n, \cdots) = \int_{\Delta_1(r,\rho,s)}^{\Delta_2(r,\rho,s)} \frac{\lambda + \sigma_n(\sqrt{t^2 + r^2 - \Delta_2^2(r,\rho,s)}, v_1)}{v_1} dt,$$

$$\Sigma_4(\lambda, \sigma_n, \cdots) = \int_{-\Delta(r,\rho,s)}^{\Delta(r,\rho,s)} \frac{\lambda + \sigma_n(\sqrt{t^2 + r^2 - \Delta_2^2(r,\rho,s)}, v_1)}{v_1} dt.$$

(b) $K_n(\lambda I - B_n)^{-1}\delta K_n$ is an integral operator on $L^2(G)$ with the kernel

$$h_{n,2}(\lambda, x, x', y, y', v, v') = \frac{y'}{4r\rho} \int_0^{v_M} dv_1 \frac{k_n(r, v, v_1)\delta k_n(\rho, v_1, v')}{v_1} \int_{|r-\rho|}^{r+\rho} \frac{1}{s}$$
$$\times \frac{\exp[-\Sigma_3(\lambda, \sigma_n, \cdots)] + \alpha_n(v_1, \sqrt{\cdot}) \exp[-\Sigma_4(\lambda, \sigma_n, \cdots) + \Sigma_3(\lambda, \sigma_n, \cdots)]}{1 - \alpha_n(v_1, \sqrt{\cdot}) \exp[-\Sigma_4(\lambda, \sigma_n, \cdots)]} ds.$$

(c) $K(\lambda I - B)^{-1}K - K(\lambda I - B_n)^{-1}K$ is an integral operator on $L^2(G)$ with the kernel

$$h_{n,3}(\lambda, x, x', y, y', v, v') = \frac{y'}{4r\rho} \int_0^{v_M} dv_1 \frac{k(r, v, v_1)k(\rho, v_1, v')}{v_1} \int_{|r-\rho|}^{r+\rho} \frac{1}{s}$$
$$\times \left\{ \frac{\exp[-\Sigma_3(\lambda, \sigma, \cdots)] + \alpha(v_1, \sqrt{\cdot}) \exp[-\Sigma_4(\lambda, \sigma, \cdots) + \Sigma_3(\lambda, \sigma, \cdots)]}{1 - \alpha(v_1, \sqrt{\cdot}) \exp[-\Sigma_4(\lambda, \sigma, \cdots)]} \right.$$
$$\left. - \frac{\exp[-\Sigma_3(\lambda, \sigma_n, \cdots)] + \alpha_n(v_1, \sqrt{\cdot}) \exp[-\Sigma_4(\lambda, \sigma_n, \cdots) + \Sigma_3(\lambda, \sigma_n, \cdots)]}{1 - \alpha_n(v_1, \sqrt{\cdot}) \exp[-\Sigma_4(\lambda, \sigma_n, \cdots)]} \right\} ds.$$

From Lemma 3, by virtue of the transform $x = z$, $y = \sqrt{r^2 - z^2}$; $x' = z'$, $y' = \sqrt{\rho^2 - z'^2}$ and Lebesgue's dominated convergence theorem, it is tedious but not difficult to see that $\|\delta K_n(\lambda I - B_n)^{-1}K\|$, $\|K_n(\lambda I - B_n)^{-1}\delta K_n\|$ and $\|K(\lambda I - B)^{-1}K - K(\lambda I - B_n)^{-1}K\|$ converge to 0 uniformly in $\{\lambda : \text{Re}\lambda \geq \beta_1\}$. Thus, for any $\varepsilon > 0$, there exists an integer n_0 such that

$$\|\delta K_{n_0}(\lambda I - B_{n_0})^{-1}K\| < \varepsilon/4, \tag{15}$$

$$\|K_{n_0}(\lambda I - B_{n_0})^{-1}\delta K_{n_0}\| < \varepsilon/4, \tag{16}$$

$$\|K(\lambda I - B)^{-1}K - K(\lambda I - B_{n_0})^{-1}K\| < \varepsilon/4. \tag{17}$$

Since $\alpha_{n_0}(v, \mu)$, $\sigma_{n_0}(r, v)$ and $\widetilde{k}_{n_0}(r, v, v')$ are polynomial functions, it follows from Theorem 1 that there exists a positive constant $\widetilde{\tau}$ independent of $\beta \in [\beta_1, \beta_2]$, such that

$$\|K_{n_0}(\lambda I - B_{n_0})^{-1}K_{n_0}\| < \varepsilon/4 \tag{18}$$

uniformly in $\{\lambda = \beta + i\tau : \beta_1 \leq \beta \leq \beta_2, |\tau| \geq \widetilde{\tau}\}$. From the relation

$$\begin{aligned} K(\lambda I - B)^{-1}K &= [K(\lambda I - B)^{-1}K - K(\lambda I - B_{n_0})^{-1}K] \\ &\quad + (K_{n_0} + \delta K_{n_0})(\lambda I - B_{n_0})^{-1}(K_{n_0} + \delta K_{n_0}) \\ &= [K(\lambda I - B)^{-1}K - K(\lambda I - B_{n_0})^{-1}K] + K_{n_0}(\lambda I - B_{n_0})^{-1}K_{n_0} \\ &\quad + \delta K_{n_0}(\lambda I - B_{n_0})^{-1}K + K_{n_0}(\lambda I - B_{n_0})^{-1}\delta K_{n_0} \end{aligned}$$

and Eqs. (15) – (18), the proof is complete. Q. E. D.

In fact, the results of Theorem 2 can be refined.

THEOREM 3. *Suppose (H1) – (H3) are satisfied and let $\beta_2 > \beta_1 > -\lambda^*$ be two constants. If Eq. (1) in (H3) holds, then for every $\varepsilon > 0$, there exists a positive constant $\widetilde{\tau}$ independent of $\beta \in [\beta_1, \beta_2]$ such that*

$$\|K(\lambda I - B)^{-1}\| \leq \varepsilon \tag{19}$$

uniformly in the domain $D_{\tilde{\tau}} := \{\lambda = \beta + i\tau : \beta_1 \leq \beta \leq \beta_2, |\tau| \geq \tilde{\tau}\}$. *If Eq. (2) in (H3) holds, then*

$$\|(\lambda I - B)^{-1}K\| \leq \varepsilon \tag{20}$$

uniformly in $D_{\tilde{\tau}}$. *If the constant* δ *in Eq. (1) or Eq. (2) is less than 1/4, then*

$$\|K(\lambda I - B)^{-1}\| \leq \varepsilon \text{ and } \|(\lambda I - B)^{-1}K\| \leq \varepsilon \tag{21}$$

uniformly in $D_{\tilde{\tau}}$.

Proof. As in [12], pp. 713–719, it is easy to show

$$\|K(\lambda I - B)^{-1}\| \leq \left[(\text{Re}\lambda + \lambda^*)^{-1}\|K(\lambda I - B)^{-1}K^*\|\right]^{1/2} \tag{22}$$

$$\|(\lambda I - B)^{-1}K\| \leq \left[(\text{Re}\lambda + \lambda^*)^{-1}\|K^*(\lambda I - B)^{-1}K\|\right]^{1/2} \tag{23}$$

for every λ with $\text{Re}\lambda > -\lambda^*$, where K^* is the adjoint operator of K given by

$$K^*\psi = \frac{1}{2r}\int_0^{v_M} dv' \int_{-r}^{r} k(r, v', v)\psi(z, \sqrt{r^2 - z^2}, v')dz.$$

If Eq. (1) in (H3) is satisfied, then by a procedure similar to that of Theorem 1 and Theorem 2, it can be shown that for every $\tilde{\varepsilon} > 0$, there exists $\tilde{\tau} > 0$ independent of β such that $\|K(\lambda I - B)^{-1}K^*\| < \tilde{\varepsilon}$ uniformly in $D_{\tilde{\tau}} := \{\lambda = \beta + i\tau : \beta_1 \leq \beta \leq \beta_2, |\tau| \geq \tilde{\tau}\}$. This together with Eq. (22) implies Eq. (19). If Eq. (2) in (H3) is satisfied, then we consider $K^*(\lambda I - B)^{-1}K$, and it can be shown that for every $\tilde{\varepsilon} > 0$, there exists $\tilde{\tau} > 0$ independent of β such that $\|K^*(\lambda I - B)^{-1}K\| < \tilde{\varepsilon}$ uniformly in $D_{\tilde{\tau}}$. This together with Eq. (23) implies Eq. (20). If the constant δ in Eq. (1) or Eq. (2) is less than 1/4, then it can be shown that both of these inequalities hold in this case, and so do Eqs. (19) and (20). This completes the proof. Q. E. D.

By virtue of Weierstrass's accumulation principle, we get the main result of this section from Lemma 2 and Theorem 2.

THEOREM 4. $\sigma(A) \bigcap \{\lambda : \beta_1 \leq Re\lambda \leq \beta_2\}$ *contains at most finitely many elements, each of which is an eigenvalue of A with finite algebraic multiplicity. The accumulation points of Pas(A) can only appear on the line* $Re\lambda = -\lambda^*$.

3. SPECTRAL PROPERTIES OF $T(t)$ AND ASYMPTOTIC BEHAVIOR OF THE TIME DEPENDENT SOLUTION

From [17] it is known that both B and A generate strongly continuous semigroups in $L^2(G)$, which are denoted by $S(t)$ and $T(t)$ respectively. This section will be devoted to discussing some aspects of the spectral properties of $T(t)$, and these spectral properties are closely linked to the asymptotic behavior of the solution $\psi(t)$ of Eq. (I). We have[3,10]:

LEMMA 5. *Let* $T(t)$ *be the strongly continuous semigroup generated by the operator A in a Hilbert space H, with* $\omega_0(A)$ *the growth bound of* $T(t)$, *and* $s_0(A)$ *the spectral bound of A, i.e.,* $\omega_0(A) = \lim_{t\to+\infty} t^{-1}\log\|T(t)\|$, $s_0(A) = \sup\{Re\lambda : \lambda \in \sigma(A)\}$. *Then*

$\omega_0(A) = s_0(A)$ if and only if for every $\varepsilon > 0$, there exists a positive constant M_ε such that $\|(\lambda I - A)^{-1}\| \le M_\varepsilon$ uniformly in $\{\lambda : Re\lambda \ge s_0(A) + \varepsilon\}$.

From Theorem 4, the eigenvalues of A lying in the half–plane $\mathrm{Re}\lambda > -\lambda^*$ can be ordered in such a way that the real part decreases. Suppose $\lambda_1, \lambda_2, \cdots, \lambda_m, \lambda_{m+1}, \cdots$ are eigenvalues of A, $\mathrm{Re}\lambda_1 \ge \mathrm{Re}\lambda_2 \ge \cdots \ge \mathrm{Re}\lambda_m > \mathrm{Re}\lambda_{m+1} \ge \cdots > -\lambda^*$, and $\{\lambda : \mathrm{Re}\lambda > -\lambda^*\} \setminus \{\lambda_n : n = 1, 2, \cdots\} \subset \rho(A)$. For every integer m satisfying $\mathrm{Re}\lambda_m > \mathrm{Re}\lambda_{m+1}$, let $\sigma_1 := \{\lambda_1, \lambda_2, \cdots, \lambda_m\}$, $\sigma_2 := \sigma(A) \setminus \sigma_1$. Since σ_1 is a compact set, it follows from [9], p. 70, that there exists a unique spectral composition $L^2(G) = H_1 \oplus H_2$ such that $T_i(t)$, the part of $T(t)$ in H_i $(i = 1, 2)$, is a strongly continuous semigroup. Furthermore, the spectral set of A_i (where A_i is the generator of $T_i(t)$) is equal to σ_i, i.e., $\sigma(A_i) = \sigma_i$, $i = 1, 2$, and A_1 is a bounded operator on H_1. Denoting by P the projection operator of σ_1 corresponding to A, then

$$T_1(t) = T(t)P, \quad T_2(t) = T(t)(I - P),$$

$$(\lambda I - A_1)^{-1} = (\lambda I - A)^{-1}P, \quad (\lambda I - A_2)^{-1} = (\lambda I - A)^{-1}(I - P),$$

$$\sigma(T(t)) = \sigma(T_1(t)) \bigcup \sigma(T_2(t)). \tag{24}$$

Since A_1 is a bounded operator on H_1, we have $\sigma(T_1(t)) = \{\exp(\lambda_n t) : n = 1, 2, \cdots, m\}$.

LEMMA 6. *For every $\varepsilon > 0$, there exists a positive constant $M_{m,\varepsilon}$ such that $\|(\lambda I - A_2)^{-1}\| \le M_{m,\varepsilon}$ uniformly in $\{\lambda : Re\lambda \ge Re\lambda_{m+1} + \varepsilon\}$.*

Proof. From $\sigma(A_2) = \sigma_2$ and the definition of σ_2, it is seen that the spectral bound of A_2 is $\mathrm{Re}\lambda_{m+1}$. Thus, $(\lambda I - A_2)^{-1}$ is a bounded operator for every λ with $\mathrm{Re}\lambda > \mathrm{Re}\lambda_{m+1}$. For every $\varepsilon > 0$, let $\beta_1 = \mathrm{Re}\lambda_{m+1} + \varepsilon$, $\beta_2 = \|K\| - \lambda^* + 1$, $K_\lambda := (\lambda I - B)^{-1}K$. From Theorem 2, there exists $\tilde{\tau} > 0$ independent of $\beta \in [\beta_1, \beta_2]$ such that $\|K_\lambda^2\| < 1/2$ uniformly in the domain $D_1 := \{\lambda = \beta + i\tau : \beta_1 \le \beta \le \beta_2, |\tau| \ge \tilde{\tau}\}$. Obviously, $(I - K_\lambda^2)^{-1}$ exists in D_1, and

$$(I - K_\lambda^2)^{-1} = \sum_{n=0}^{\infty} K_\lambda^{2n}, \qquad \|(I - K_\lambda^2)^{-1}\| \le \sum_{n=0}^{\infty} \|K_\lambda^2\|^n < 2. \tag{25}$$

From the relation $(I + K_\lambda)(I - K_\lambda^2)^{-1} = (I - K_\lambda^2)^{-1}(I + K_\lambda)$ (which can be verified from Eq. (28) and $(I - K_\lambda)(I + K_\lambda) = I - K_\lambda^2$, we have $(I - K_\lambda)^{-1} = (I - K_\lambda^2)^{-1}(I + K_\lambda)$. Therefore,

$$(\lambda I - A)^{-1} = (\lambda I - B - K)^{-1} = [(\lambda I - B)(I - K_\lambda)]^{-1} = (I - K_\lambda)^{-1}(\lambda I - B)^{-1}. \tag{26}$$

From Eqs. (25) and (26), it follows that $\|(\lambda I - A_2)^{-1}\| = \|(\lambda I - A)^{-1}(I - P)\|$ is uniformly bounded in D_1.

Set $D_2 := \{\lambda = \beta + i\tau : \beta_1 \le \beta \le \beta_2, |\tau| \le \tilde{\tau}\}$. Then $D_2 \subset \rho(A_2)$ and $\|(\lambda I - A_2)^{-1}\|$ is continuous with respect to $\lambda \in D_2$. Noting that D_2 is a compact set, we know that $\|(\lambda I - A_2)^{-1}\|$ is uniformly bounded in D_2. Therefore, $\|(\lambda I - A_2)^{-1}\|$ is uniformly bounded in $D_1 \cup D_2 = \{\lambda : \beta_1 \le \mathrm{Re}\lambda \le \beta_2\}$. It is easy to know $\|(\lambda I - A)^{-1}\| \le (\mathrm{Re}\lambda + \lambda^* - \|K\|)^{-1} \le 1$ for every $\lambda \in \{\lambda : \mathrm{Re}\lambda \ge \beta_2\}$, and so is $\|(\lambda I - A_2)^{-1}\|$. This completes the proof. Q. E. D.

From Lemma 5 and Lemma 6, it is known that the growth bound of $T_2(t)$ is $\text{Re}\lambda_{m+1}$, and thereby the spectral radius of $T_2(t)$ is $\text{Re}\lambda_{m+1}$. Since m can be selected arbitrarily, we get the following conclusion.

THEOREM 5. *The spectrum of $T(t)$ outside the disk $\{\lambda : |\lambda| \leq \exp(-\lambda^* t)\}$ consists of isolated eigenvalues of $T(t)$ with finite algebraic multiplicity, and the accumulation points of the set $\sigma(T(t)) \bigcap \{\lambda : |\lambda| > \exp(-\lambda^* t)\}$ can only appear on the circle $\{\lambda : |\lambda| = \exp(-\lambda^* t)\}$.*

COROLLARY 1. *Suppose $\lambda_1, \lambda_2, \cdots, \lambda_m, \lambda_{m+1}, \cdots$ are eigenvalues of A, $\text{Re}\lambda_1 \geq \text{Re}\lambda_2 \geq \cdots \geq \text{Re}\lambda_m > \text{Re}\lambda_{m+1} \geq \cdots > -\lambda^*$, and $\{\lambda : \text{Re}\lambda > -\lambda^*\} \setminus \{\lambda_n : n = 1, 2, \cdots\} \subset \rho(A)$. Furthermore, let $P_n(t)$ be the projection operator associated with λ_n to A, i.e.,*

$$P_n(t) = \sum_{j=1}^{k_n} \frac{t^{j-1} B_j^{(\lambda_n)}}{(j-1)!},$$

where k_n is the algebraic multiplicity of λ_n,

$$B_1^{(\lambda_n)} = \frac{1}{2\pi i} \oint_{C_{\lambda_n}} (\lambda I - A)^{-1} d\lambda, \quad C_{\lambda_n} = \{\lambda : |\lambda - \lambda_n| = r_{\lambda_n}\},$$

$$B_{j+1}^{(\lambda_n)} = (A - \lambda_n I) B_j^{(\lambda_n)}, \quad j = 1, 2, \cdots, k_n - 1,$$

r_{λ_n} is sufficiently small such that $\{\lambda : |\lambda - \lambda_n| \leq r_{\lambda_n}\} \bigcap \sigma(A) = \{\lambda_n\}$. Then for every $b \in (\text{Re}\lambda_{m+1}, \text{Re}\lambda_m)$ and $\psi_0 \in L^2(G)$,

$$\left\| T(t)\psi_0 - \sum_{n=0}^{m} \exp(\lambda_n t) P_n(t) \psi_0 \right\| \leq \exp(bt) \|\psi_0\|.$$

With additional assumptions,[17] the dominant eigenvalue β_0 of A exists and $\beta_0 > -\lambda^*$. In this case, we have

COROLLARY 2. *Suppose $\beta_0 > -\lambda^*$ is the dominant eigenvalue of A. Then for every $b < \beta_0$ and $\psi_0 \in L^2(G)$,*

$$\|T(t)\psi_0 - \exp(\beta_0 t)(\psi_0, \psi_{\beta_0}^*)(\psi_{\beta_0}, \psi_{\beta_0}^*)^{-1} \psi_{\beta_0}\| \leq \exp(bt) \|\psi_0\|.$$

As $t \to +\infty$, the solution $\psi(t)$ of Eq. (I) tends to 0, $+\infty$, or $(\psi_0, \psi_{\beta_0}^*)(\psi_{\beta_0}, \psi_{\beta_0}^*)^{-1} \psi_{\beta_0}$, depending on $\beta_0 < 0$, $\beta_0 > 0$ and $\beta_0 = 0$ respectively.

Finally some closing remarks. All the conclusions obtained in the previous sections still hold if the velocity domain $E = (0, v_M]$ is replaced by $E = [v_m, v_M]$, where $0 < v_m < v_M < +\infty$.

Suppose there exists a constant $c \geq 0$ such that $\sigma(r, v) \geq \lambda_0^* - cv$, where[7] $\lambda_0^* = \text{ess inf}_{r \in V} \lim_{v \to 0^+} \sigma(r, v)$. Then all the conclusions except Theorem 3 and Lemma 1 still hold if λ^* (the infimum of $\sigma(r, v)$ on $V \times E$) is replaced by λ_0^*. This result can be obtained by making some amendments in the proof of Theorem 1 and Theorem 2.

With some additional assumptions, it is known that the spectral bound of B is $-\lambda^*$ (cf. [17]). Since $S(t)$ (the strongly continuous semigroup generated by B) is positive, it follows[2,9] that the growth bound of $S(t)$ is also $-\lambda^*$, and the spectral disk of $S(t)$ is $\{\lambda : |\lambda| \leq \exp(-\lambda^* t)\}$. Thus, the conclusions obtained in [1],[4],[8],[14],[15],[16],[18] are extended to transport equations with reentry boundary conditions.

REFERENCES

1. G. Greiner, Spectral properties and asymptotic behavior of the linear transport equation, *Math. Z.* **185**, 167–177 (1984).
2. G. Greiner and R. Nagel, On stability of strongly continuous semigroups of positive operators on $L^2(\mu)$, *Ann. Scuola Norm. Sup. Pisa* **10**, 257–262 (1983).
3. Falun Huang, Characteristic conditions for exponential stability of linear dynamical systems in Hilbert spaces, *Ann. Diff. Eqs.* **1:1**, 43–56 (1985).
4. K. Jörgens, An asymptotic expansion in the theory of neutron transport, *Comm. Pure Appl. Math.* **11**, 219–242 (1958).
5. J. Lehner and G. M. Wing, Solution of the linearized Boltzmann transport equation for slab geometry, *Duke Math. J.* **23**, 125–142 (1956).
6. Peng Lei and Mingzhu Yang, On the spectrum of the transport operator in a homogeneous sphere with partly reflection boundary conditions, *Kexue Tongbao* **31:24**, 1867–1871 (1986) (in Chinese).
7. E. W. Larsen and P. F. Zweifel, On the spectrum of the linear transport operator, *J. Math. Phys.* **15**, 1987–1997 (1974).
8. M. Mokhtar–Kharroubi, Time asymptotic behavior and compactness in transport theory, *Eur. J. Mech. B: Fluids* **11**, 39–68 (1992).
9. R. Nagel ed., *One–Parameter Semigroups of Positive Operators*, (Lect. Notes in Math. 1184, Springer–Verlag, Berlin, 1986).
10. J. Pruss, On the spectrum of C_0 semigroups, *Transl. Amer. Math. Soc.* **284:2**, 847–857 (1984).
11. D. C. Sahni, N. S. Garis and N. G. Sjostrand, Spectrum of one-speed neutron transport operator with reflective boundary conditions in slab geometry, *Transport Theory and Statistical Physics* **24**, 629–656 (1995).
12. Degong Song, Miansen Wang and Guangtian Zhu, A note on the spectrum of the transport operator with continuous energy, *Transp. Theory Stat. Phys.* **22:5**, 709–721 (1993).
13. R. van Norton, On the real spectrum of a mono–energetic neutron transport operator, *Comm. Pure Appl. Math.* **15**, 149–158 (1962).
14. I. Vidav, Spectra of perturbed semigroups with applications to transport theory, *J. Math. Anal. Appl.* **30**, 264–279 (1970) .
15. J. Voigt, A perturbation theorem for the essential spectral radius of strongly continuous semigroups, *Mh. Math.* **90**, 153–161 (1980).
16. J. Voigt, Spectral properties of the neutron transport equation, *J. Math. Anal. Appl.* **106**, 140–153 (1985).
17. Haiyan Wang, Mingzhu Yang and Wenwei Wang, On the spectrum of a spherical symmetric transport operator with generalized boundary conditions, *J. Southeast Univ.* **19:5**, 7–17 (1989) (in Chinese).
18. L. W. Weis, A generalization of the Vidav–Jörgens perturbation theorem for semigroups and its application to transport theory, *J. Math. Anal. Appl.* **129**, 6–23 (1988).
19. Xianwen Zhang and BenzhongLiang, On the spectrum of a one-velocity transport operator with Maxwell boundary condition, *J. Math. Anal. Appl.* **202:3**, 920–939 (1996).

TRAVELING HEAT WAVES IN HIGH TEMPERATURE MEDIUM

E.A.Larionov, E.I. Levanov, and P.P.Volosevich*

1.INTRODUCTION

The group methods, leading to constructing of the self-similar solutions, exponential solutions, traveling wave solutions and etc., are the necessary tool in general program of mathematical modelling and numerical experiments. They allow to obtain wide information about characteristic features of the process being studied and main regularities of its development. Combining numerical calculations and self-similar methods allow to study the process in question in detail.

The traveling wave solutions are popular among other invariant-group types solutions.

In many works the method of traveling waves is used for analysis the so-called structure of the front of shock waves defined by various dissipative processes. In base of mathematical theory the structure of the front of shocking compression lie the assumption that this structure is quasistationar. In present work analogous[1,2] is assumed that traveling wave arise in the medium due to influence of substantialy non-stationary sources of energy: the temperature or the heat. Due to such approach more general form of traveling wave is considered. The properties of such traveling wave are closed to properties of the processes described by self-similar solutions of equations conductivity and gas dynamics[2]. The analysis show that there are various types of traveling waves[3,4]: classical (ordinary) – the searching functions are represented in form $F(x,t) = f(x-Dt)$, $D = \text{const}$, classical waves with exponent $(\Psi(x,t) = e^{nt}\Psi(x - Dt))$.

In present work are demonstrated the examples of the description of heat transfer processes with aid various form of traveling waves in high-temperature medium with the heat conductivity and nonlinear sources of various nature.

---

* E.A.Larionov, E.I. Levanov, P.P.Volosevich, Institute for Mathematical Modelling of RAS, Moscow, Miusskaya sq., 4-A, 125047.

Mathematical Modeling: Problems, Methods, Applications
Edited by Uvarova and Latyshev, Kluwer Academic/Plenum Publishers, 2001

Let $\rho \equiv \rho_0$ - constant density of medium, x - spatial coordinate, $m = \rho_0 x$ – mass coordinate, t – the time, T – the temperature, W – the density of heat flow due to heat conductivity. We count that internal energy is linear function of temperature ($\varepsilon = C_v T$) and coefficient of heat conductivity $K(T)$ and intensity of the sources $G = G(T)$ are power functions of T.

In the case planar symmetry the heat conduction can be described by the equation system of the form:

$$W = -K_0 T^{a_0} \frac{\partial T}{\partial m}, \qquad C_v \frac{\partial T}{\partial t} = -\frac{\partial W}{\partial m} - G_0 T^{b_0} \tag{1.1}$$

where C_v, K_0 - are positive constants, $G_0 < 0$ for the sources and $G_0 > 0$ for the looses for energy. We count that $a_0 \geq 1$, $a_0 + b_0 > 1$.

We suppose that due to the heat regime

$$T(0,t) = T_*(t) \tag{1.2}$$

the traveling waves can propagate. In this case for $t = 0$ and any $t > 0$ before the front of wave we have

$$T \equiv 0 \tag{1.3}$$

It is known that for $a_0 > 1$ and a condition (1.3) heat waves spread in medium with finite velocity[2,3].

2.CLASSICAL TRAVELING WAVES

Well known traveling waves described by functions $F(m,t) = \hat{F}(D_f t - m)$, where $D_f = \text{const}$, are called classical traveling waves. Corresponding equations we write in dimensionless form.

We introduce the substitution of variables

$$S = t/t_0 - m/D, \qquad T(m,t) = T_0 f(s), \qquad W(m,t) = W_0 w(s), \tag{1.4}$$

where t_0, D, T_0, W_0 are dimensional constants, s – is dimensionless independent variable, $f = f(s)$ dimensionless function of temperature, $W = W(s)$ – is the density of heat flow.

Let a coordinate of the front of traveling wave is defined by formula

$$m = m_f = (D/t_0)t, \tag{1.5}$$

that is in variables (1.4) $s = 0$. The velocity of the spread of wave is constant:

$$\frac{dm_f}{dt} = D_f = \text{const} .$$

With respect to the formulas

$$T_0 = (t_0^{-1} D^2 C_v K_0^{-1})^{1/a_0}, \quad W_0 = T_0 t_0^{-1} D C_v , \tag{1.6}$$

we obtain the system of ordinary differential equations

$$\frac{df}{ds} = \frac{dw}{ds} - \hat{G}_0 f^{b_0}, \qquad w = f^{a_0} \frac{df}{ds}, \tag{1.7}$$

where $\hat{G}_0 = G_0 t_0 C_v^{-1} (t_0 D^{-2} C_v^{-1} K_0)^{(1-b_0)/a_0}$ - , is dimensionless constant. In the domain $s \leq 0$ there is trivial solution of the system (1.7):

$$f(s) \equiv 0, \qquad w(s) \equiv 0. \tag{1.8}$$

In variables (1.4) the value $m = 0$ correspond to the coordinate

$$s = s_0 = t / t_0 . \tag{1.9}$$

The heat regime of type (1.2), is expressed by formula:

$$T_*(t) = \left(t_0^{-1} D^2 C_v K_0^{-1}\right)^{1/a_0} f(s_*), \tag{1.10}$$

where the function $f = f(s_*)$ is defined directly from the solution of equation system (1.7) under the boundary conditions $f(0)=0$, $w(0)=0$.

For solution analysis of the problem (1.7), (1.8) the dimensionless temperature f is chosen as independent variable and the system (1.7) is described in the form

$$\frac{dw}{df} = 1 + \hat{G}_0 (f^{a_0 + b_0} / w), \tag{1.11}$$

$$\frac{ds}{df} = f^{a_0} / w . \tag{1.12}$$

For $\hat{G}_0 = 0$ in the domain $s \geq 0$ we obtain known analytic solution of the problem in question (see, for example[2]):

$$f = (a_0)^{1/a_0} s^{1/a_0}, \qquad w = f . \tag{1.13}$$

For $\hat{G}_0 \neq 0$ the asymptotic solution of system (1.11), (1.12) in neighborhood of the traveling wave front under assumption ($s = 0, f = 0, w = 0$) $a_0 + b_0 > 1$ with exactitude to main members coincide with (1.13). However, in further, the change of functions $s = s(f)$ and $w = w(f)$ lead to the dependence the character of considered solution from the sign of constant $\hat{G}_0$.

In the case $\hat{G}_0 > 0$ (bulk looses of the energy), $b_0 \leq a_0 + 1$, considered solution of the traveling wave types exist when $s \leq \infty$ and therefore for any $t>0$. For $b_0 > a_0 + 1$ the solution exist only finite time $0 < t \leq s_k t_0 = C_0$.

Qualitative character of the function distribution $f = f(s)$ in the case $\hat{G}_0 \geq 0$, $a_0 > 1$ for different relationships between parameters a_0 and b_0 is presented in fig.1.

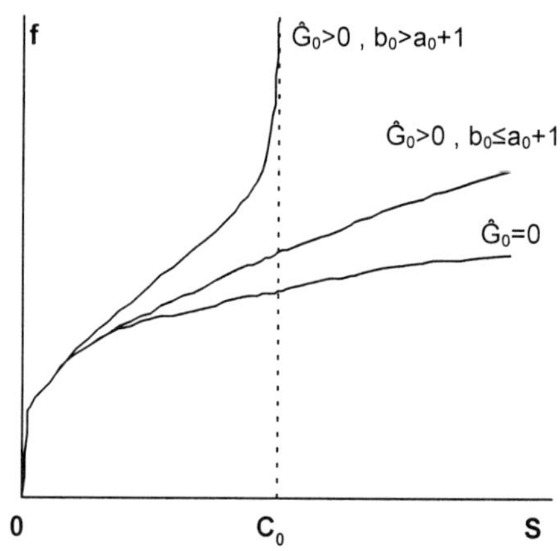

Figure 1. Function distribution $f = f(s)$ for classical traveling waves in the case $\hat{G}_0 = 0$

Let $\hat{G}_0 < 0$ (the bulk source of energy exist). The analysis shows that considered solution has physical sense only on finite interval $s \leq s_k$, and consequently for $t \leq s_k t_0$.

Qualitatively character of the functions distribution $w = w(f)$ and $f = f(s)$ in the case classical traveling heat wave and when $\hat{G}_0 < 0$, $a_0 > 1$ is shown in fig. 2. In fig. 2a) by dotted line is represented the isocline of zeros of Eq. (1.11).

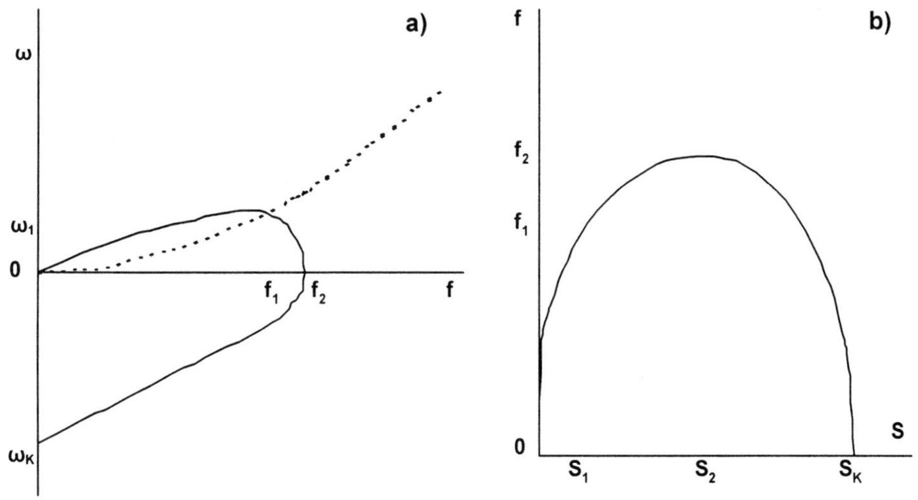

Figure 2. Distribution of heat flow $w = w(f)$ (fig. 2a) and temperature $f = f(s)$ (fig. 2b) for classical traveling waves in the case $\hat{G}_0 < 0$

3.LOGARITHMIC TRAVELING WAVE

We introduce the change of variables

$$S = \ln\frac{t}{t_0} - \frac{m}{D}, \qquad T(m,t) = T_0\left(\frac{t}{t_0}\right)^{n_0} f(s), \quad W(m,t) = W_0\left(\frac{t}{t_0}\right)^{n_1} w(s), \qquad (1.14)$$

where the constants T_0 and W_0 are defined by formulas (1.6). Used (1.14) we get that equation system (1.1) is reduced to the corresponding system of ordinary differential equations with respect to s if there are follow conditions:

$$n_0 = -1/a_0, \qquad n_1 = n_0 - 1, \qquad b_0 = a_0 + 1. \qquad (1.15)$$

Indicated system has the form

$$\frac{df}{ds} = \frac{dw}{ds} - \hat{G}_0 f^{a_0+1} + \frac{1}{a_0} f, \qquad w = f^{a_0}\frac{df}{ds}, \qquad (1.16)$$

where the dimensionless constant $\hat{G}_0(b_0 = a_0 + 1)$ is defined by expression $\hat{G}_0 = -G_0 D^2 K_0^{-1}$. Considered traveling wave we call logarithmic traveling heat wave. The coordinate of its front we define by value $s = 0$, i.e.

$$m = m_f = D\ln(t/t_0)\,, \qquad t \le t_0\,. \tag{1.17}$$

Perturbed medium in variables (1.14) there is in the domain $s > 0$. When $s \le 0$ the trivial solution (1.8) exist. The velocity of spreading of logarithmic traveling wave is not constant $D_f = \dfrac{dm_f}{dt} = D/t$. In variables (1.17) the formula (1.2) has the form

$$T_*(t) = T_0\left(\frac{t}{t_0}\right)^{-1/a_0} f(s_*)\,, \tag{1.18}$$

where $s_* = \ln(t/t_0), f(s_*)$ is defined from the solution equation system (1.16) under the boundary conditions $f(0) = w(0) = 0$.

Analogous preceding we take f for the independent variable. Corresponding equation system is reduced to form

$$\frac{dw}{df} = 1 - \frac{1}{a_0}\frac{f^{a_0+1}(1-\hat{G}_0\, af^{a_0})}{w}\,, \qquad \frac{ds}{df} = \frac{f^{a_0}}{w} \tag{1.19}$$

In neighborhood $s = 0$ the solution of system (1.19) with exactitude to main members is defined by formulas

$$w = f + \ldots, \qquad s = f^{a_0}/a_0 + \ldots \tag{1.20}$$

The analysis of the system (1.19) in the domain $f > 0$, $s > 0$ shows that in case $\hat{G}_0 \le 0$ qualitative character of the function distribution $w = w(f)$ and $s = s(f)$, described logarithmic traveling wave is analogous to the case $\hat{G}_0 < 0$ for classical traveling wave (see fig.2). The solution of logarithmic traveling wave type as well with presence ($\hat{G}_0 < 0$) as with absence ($\hat{G}_0 = 0$) of the bulk sources of energy exist only at finite time.

Let at present there are the bulk sinks of the energy, i.e. $\hat{G}_0 > 0$. From (1.22) it follows that in this case the behavior of the functions $w = w(f)$ and $s = s(f)$ can be different on dependence of the parameder $\hat{G}_0$. The set of considered solutions in the case $a_2 = 2$ is given in fig. 3.

In this figure by continuos lines are represented the integral curves of first equation of the system (1.22) in plane (f,w). For different values of parameter $\hat{G}_0$: I. $\hat{G}_0 = 1.0$, II. $\hat{G}_0 = 0.3$, III. $\hat{G}_0 = 0.15$, IV. $\hat{G}_0 = 0.075$, V. $\hat{G}_0 = 0.05$.

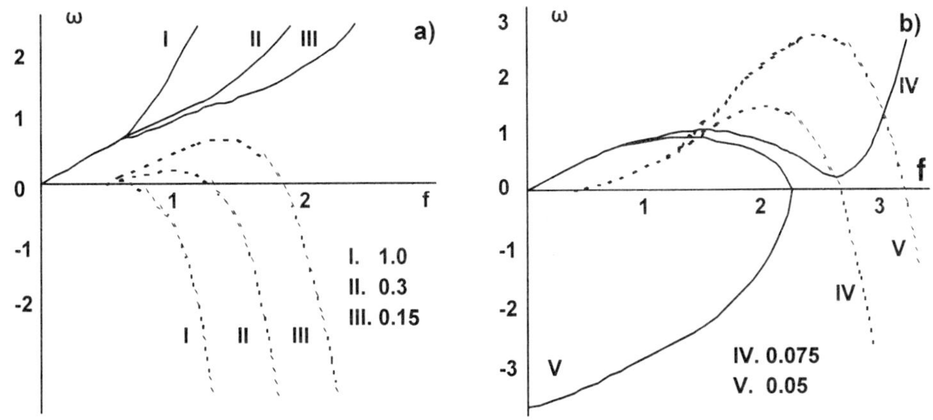

Figure 3. Function distribution $w = w(f)$ for logarithmic traveling waves at various $\hat{G}_0 < 0$.

By dotted lines is shown corresponding curves of the form

$$w = f^{a_0+1}\left(1 - \hat{G}_0\, a_0 f^{a_0}\right) / a_0 \tag{1.21}$$

The analysis shows that for comparatively values of the parameter $\hat{G}_0$ we have two cases. In first case the integral curves $w = w(f)$ do not intersect the curves (1.21) and they are monotones curves (see fig. 3a). In second case the curves $w = w(f)$ intersect lines (1.21) and they are nomonotones curves (see fig. 3b). At last with decreasing $\hat{G}_0$ the picture arise analogous to represented in fig.2a) (see curves V in fig.3b)). In this case logarithmic traveling wave exist only when $s \leq s_k < \infty$ and therefore at finite time.

REFERENCES

1. A.A.Samarskii, S.P.Kurdiumov, and P.P.Volosevich,. Traveling waves in medium with nonlinear heat conductivity, *Zhurn. Vychisl. Matem. and Matem. Fiz.*, **5**, № 2, 1965, p.199-217.
2. P.P.Volosevich and E.I.Levanov, *Self-similar solutions of gasodynamic and heat transport problems* (MIPT PUBLISHER, Moscow, 1997)- 240 pages.
3. P.P.Volosevich, N.A.Darin, E.I.Levanov, and N.M.Skhirtladze, *A problem on piston in gas with sources and arains (self-similar solutions).*(Tbilisi, the University, Publishing House, 1986)-239 pages.
4. P.P.Volosevich, I.I.Galiguzova, and E.I.Levanov, Group-invariant traveling waves type of solution of the gasdynamic equations. K. Isd.UNIPRESS, 1998, - p.13-14.

SMOOTH LYAPUNOV MANIFOLDS AND CORRECT MATHEMATICAL SIMULATION OF NONLINEAR SINGULAR PROBLEMS IN MATHEMATICAL PHYSICS

Nadezhda B. Konyukhova and Alexander I. Sukov*

1. INTRODUCTION

Several problems in various fields of mathematical physics occur lead to the systems of ordinary differential equations (ODEs) having singularities or being defined on an infinite interval. For different classes of linear and nonlinear ODEs, there are many publications dealing with correct statement of singular boundary value problems (BVPs) and their reduction to the equivalent regular ones (see, e.g., the reviews[1, 2]). This paper deals with some results of [3-5] obtained in the indicated direction for autonomous systems (ASs) of nonlinear ODEs and their application to singular BVPs arising from hydrodynamics.

We consider AS of nonlinear ODEs which has in a phase space a (pseudo)hyperbolic equilibrium point. We study a time-invariant stable initial manifold (SIM) of solutions which exists in the neighbourhood of this point, i.e., a stable separatrix "surface" of a saddle. A smooth time-invariant SIM is represented in terms of a solution to the Lyapunov type problem for associated stationary system of quasilinear partial differential equations (PDEs) with a degeneracy. The application of the results to singular BVPs for AS of nonlinear ODEs on an infinite interval is given: a limiting boundary condition of the type where the solution tends to a stationary value is transferred from infinity by the requirement that the values of solutions belong to the SIM.

We give examples to correct statement and reduction on a finite interval of some singular BVPs arising from the incompressible fluid mechanics. In particular, we eliminate inaccuracies admitted in the literature by certain authors. We propose a stable method of an optimal shooting[6] beginning with smooth SIM and represent the numerical results.

* Nadezhda B. Konyukhova, Computing Center of Russian Academy of Sciences, Moscow, 117967, Russia. Alexander I. Sukov, Moscow State University of Technology "STANKIN", Moscow, 101472, Russia.

2. SMOOTH SIMS AND SINGULAR BVPS FOR ASS OF NONLINEAR ODES

Notation. $\mathbf{K} \in \{\mathbf{R}, \mathbf{C}\}$, $|\cdot|$ *is a norm in* $\mathbf{K}^n$ *(the associated matrix norm is denoted by the same symbol);* $\mathbf{L}(\mathbf{K}^n)[\mathbf{L}(\mathbf{K}^n, \mathbf{K}^m)]$ *is the linear space of* $n \times n$*-matrices [*$m \times n$*-matrices],* $\lambda(A)$ *is an eigenvalue (EV) of the matrix* A, $\partial p / \partial x = (\partial p_i / \partial x_j)_{i=1,\ldots,m; j=1,\ldots,n}$ *is the Jacobi matrix,* $\partial p / \partial x \in \mathbf{L}(\mathbf{K}^n, \mathbf{K}^m)$; $\Omega_x^n(a) = \{x : x \in \mathbf{K}^n, |x| < a\}$, $\overline{\Omega}_x^n(a) = \{x : x \in \mathbf{K}^n, |x| \le a\}$, $a > 0$, $\overline{\Omega}_{x,\rho}(a) = \overline{\Omega}_x^n(a) \times \overline{\Omega}_\rho^m(a)$; $C_{l,a_0} = C_l(\overline{\Omega}_{x,\rho}(a_0))$ *is the class of vector-functions* $\varphi(x,\rho)$, $\varphi : \overline{\Omega}_x^n(a_0) \times \overline{\Omega}_\rho^m(a_0) \to \mathbf{K}^l$, *which are continuous with respect to the set of variables* $\{x, \rho\}$ *in* $\overline{\Omega}_{x,\rho}(a_0)$, $0 < a_0$ *is fixed;* $C_{l,a_0}^k = C_l^k(\overline{\Omega}_{x,\rho}(a_0))$, $k \ge 1$, *is the class of* $\varphi(x,\rho)$, $\varphi : \Omega_{x,p}(a_0) \to \mathrm{K}^l$, *which are continuous with respect to the set of variables* $\{x, \rho\}$ *in the* $\Omega_{x,p}(a_0)$ *together with their partial derivatives with respect to* $x_1, \ldots, x_n, \rho_1, \ldots, \rho_m$ *up to order* k; $H_{l,a_0} = H_l(\overline{\Omega}_{x,\rho}(a_0))$ *is the class of* $\varphi(x,\rho)$, $\varphi : \overline{\Omega}_{x,\rho}(a_0) \to \mathbf{K}^l$, *holomorfic with respect to* $\{x, \rho\}$ *in some neighbourhood of the region* $\overline{\Omega}_{x,\rho}(a_0)$; $[x]_\nu$ *is an absolutely convergent power series in coordinates* x, $x \in \mathbf{K}^n$, *starting with terms of degree not lower than* ν, $0 \le \nu$ *is integer.*

2.1 Boundary Conditions at Infinity and Their Transfer to a Finite Point

Consider AS of N nonlinear ODEs

$$z' = Q(z), \quad -\infty < \tau < \infty, \tag{1}$$

where $z \in \mathbf{K}^N$, $Q \in G_N(G_N)$, G_N is the region in $\mathbf{K}^N$ or all the space. Let z_s be a stationary point of (1), $z_s : \{z_s \in G_N, Q(z_s) = 0\}$, and in the neighbourhood of the point z_s let $Q(z)$ be represented in the form

$$Q(x) = A(z - z_s) + P(z), \; P(z) = o(z - z_s), \; |z - z_s| \to 0. \tag{2}$$

Here A is a constant matrix, $A \in \mathbf{L}(\mathbf{K}^N)$; $P(z) \in G_N(G_N)$ and satisfies the Lipschitz condition with constant $L_P(G_N)$ which can be chosen arbitrarily small by means of determination of G_N as a small neighbourhood of the point z_s. The stationary points of (1) are called also the critical points, the equilibrium ones or the equilibriums.

Definition 1. *The equilibrium* $z = z_S$ *of the system* (1) *is called a saddle-node of* (μ, n)*-type for* $\mu \geq 0$ $[(\mu, N-n)$*-type for* $\mu \leq 0]$ *where* n *is integer,* $0 \leq n \leq N$, *if, for these* μ, n, *the matrix* A *in* (2) *has* n *the EVs satisfying condition* $\operatorname{Re}\lambda(A) < -\mu$, *and* $N-n$ *the EVs satisfying condition* $\operatorname{Re}\lambda(A) > -\mu$. *If* $n(N-n) \neq 0$ *and* $\mu = 0$ $(\mu \neq 0)$ *then* z_S *is called a hyperbolic (pseudohyperbolic) saddle or equilibrium.*

Let in (1) the point $z = z_S$ be a saddle-node of (μ, n)-type for $\mu \geq 0$ and $0 \leq n \leq N$. Set a limiting condition at infinity

$$\lim_{\tau\to\infty} \exp(\mu\tau)[z(\tau) - z_S] = 0 \tag{3}$$

and consider the problem (1), (3) as a singular Cauchy problem. This problem has an n-parameter family of solutions which values form in the neighbourhood of the point $z = z_S$ of the phase space a τ-invariant n-dimensional SIM $\mathbf{M}_+^{(n)}$ [4, 7]. Then Eq. (3) is equivalent to the requirement

$$z(\tau) \in \mathbf{M}_+^{(n)}, \quad \tau \geq \tau_\infty, \tag{4}$$

for large enough τ_∞. We say that (4) is a result of the boundary condition transfer from infinity to a finite point. To realize such transfer, we must assign (exactly or approximately) the equation of SIM $\mathbf{M}_+^{(n)}$.

Definition 2. *The limiting condition* (3) *is called a correct boundary condition at infinity for system* (1) *if the stationary point* $z = z_S$ *is a saddle-node of* (μ, n)*-type,* $0 \leq n \leq N$.

Remark 1. *Obviously, the case* $n = N$ *(stable node) implies that all solutions of* (1) *sufficiently close to* z_S *satisfy condition* (3), *and the case* $n = 0$ *that only the solution* $z(\tau) \equiv z_S$ *satisfies* (3). *In the general case when formulating singular BVPs for* (1), *it has to be taken into account that the number of nonlinear relations at a finite point which are equivalent in aggregate to the limiting condition* (3) *is* $N-n$, *and the relations themselves are given by the equation of the SIM. If not, the following errors, encountered in the formulation and solution of physical problems in the literature, can arise: the boundary condition at infinity might be inaccurate (the limits of the individual components rather than the whole solution being indicated), the problem might be set incorrectly as regards the total number of boundary conditions at both ends of an infinite or semi-infinite interval, and the condition of type* (3) *at a finite point might be approximated incorrectly.*

In addition, let in (1) $z = \tilde{z}_S$ be a saddle-node of $(\nu, N-n)-$ type for $\nu \leq 0$, and let the condition

$$\lim_{\tau\to-\infty} \exp(\nu\tau)[z(\tau) - \tilde{z}_S] = 0 \tag{5}$$

be given. Then the problem (1), (3), (5) is set correctly with respect to the total number of boundary conditions in the neighbourhoods of the points $\tau = -\infty$ and. $\tau = +\infty$. The singular BVPs of such type arise, e.g., in the nonlinear field theory and for reaction diffusion equations in biological and chemical models where their solutions are called the kink-, wall- or front-type solutions if $z \neq \tilde{z}_s$ and the splash-, bell- or billow-type solutions otherwise.

Now suppose that $Q(z) \in C_N^1(G_N)$, $z_s : z_s \in G_N$, $Q(z_s) = 0$, $(\partial Q/\partial z)(z_s) \neq 0$, and the limiting condition at infinity has the form

$$\lim_{\tau \to +\infty} \frac{\varphi_0[z(\tau) - z_s]}{|z(\tau) - z_s|} = 0, \tag{6}$$

where $\varphi_0 \in \mathbf{L}(\mathbf{K}^N, \mathbf{K}^{N-n})$ and has a rank $N-n$, $0 \le n \le N$ (the solution tends in the direction to get into given linear subspace). It will be assumed by the definition that $z(\tau) \equiv z_s$ satisfies (6).

Definition 3. *The condition* (6) *is called an admissible boundary condition at infinity for system* (1) *if, in the neighbourhood* $z = z_s$, *there is a smooth* n *- dimensional SIM* $\mathbf{M}_+^{(n)}$, *generated by values of solutions of* (1) *sufficiently close to* z_s, *and the equation*

$$\mathbf{K}_0^n : \varphi_0(z - z_s) = 0, \; z \in \mathbf{K}^N, \tag{7}$$

assigns an n *- dimensional tangent subspace to* $\mathbf{M}_+^{(n)}$ *at the point* $z = z_s$. [*Thus if* $\mathbf{M}_+^{(n)}$ *is given by the equation* $\varphi(z) = 0$, $\varphi \in \mathbf{K}^{N-n}$, $\varphi(z_s) = 0$, *then* $\varphi_0 = P(\partial\varphi/\partial z)(z_s)$ *where* P *is an arbitrary nonsingular matrix,* $P \in \mathbf{L}(\mathbf{K}^{N-n})$, *e.g.,* $P = E_{N-n}$].

The class of admissible φ_0, by analogy with the linear case[2], can be obtained as follows.

Proposition[4, 5]. *Let* $\mathbf{H}$ *be the space of rows of length* N. *We will choose any* $\mu \ge 0$ *such that* $A = (\partial Q/\partial z)(z_s)$ *does not have EVs on the straight line* $\operatorname{Re}\lambda = -\mu$. *Let* $\mathbf{X}$ *be the subspace of* $\mathbf{H}$ *generated by all left root vectors of* A, *for which* $\operatorname{Re}\lambda(A) > -\mu$. *As* φ_0 *we take any basis in* $\mathbf{X}$. *Then the condition* (6) *will be admissible and equivalent to* (3). [*Thus condition* (6) *is transferred to a finite point with the help of the equation for* $\mathbf{M}_+^{(n)}$ *which in linear approach is given by* (7)].

2.2. Smooth SIMs and Associated Lyapunov Singular Problems for PDEs

In the Cartesian product space $\mathbf{K}^n \times \mathbf{K}^m = \mathbf{K}^N$, let us consider AS of N nonlinear ODEs

$$\frac{dx}{d\tau} = f(x, y), \quad \frac{dy}{d\tau} = g(x, y), \quad x \in G_n, \; y \in G_m, \; -\infty < \tau < \infty, \tag{8}$$

and a nonlinear relation

$$y = \rho(x),\ x \in \widetilde{G}_n \subseteq G_n, \tag{9}$$

where $f \in G_n(G_N)$, $g \in G_m(G_N)$, $G_N = G_n \times G_m$, $\rho(x) \in C_m(\widetilde{G}_n)$.

Definition 4. *We say that the relation* (9) *assigns an* n *- dimensional* τ *- invariant SIM* $\mathbf{M}_{+}^{(n)}$ ($\mathbf{M}_{-}^{(n)}$) *in an* N *- dimensional phase space of the system* (8) *if each solution which values satisfy* (9) *at the moment* $\tau = 0$ *remains the same property for all* $\tau \geq 0$ ($\tau \leq 0$). *The SIM* (9) *is called smooth if* $\rho \in C_m^k(\widetilde{G}_n)$, $k \geq 1$.

The following not comlicated theorems are valid.

Theorem 1[5]. *The relation* (9) *is the equation of a smooth SIM for* (8) *if and only if* $\rho(x)$ *is a solution of the equation*

$$\frac{\partial \rho}{\partial x} f(x,\rho) = g(x,\rho). \tag{10}$$

Theorem 2[5]. *Let* $\{x, y\} = \{0,0\}$ *be an equilibrium of* (8), *i.e.,* $\{0,0\} \in G_N$, $f(0,0) = 0$, $g(0,0) = 0$. *This stationary solution of* (8) *belongs to the smooth SIM* (9) *if and only if* $\rho(x)$ *is a solution of the equation* (10) *satisfying condition* $\rho(0) = 0$.

The Lyapunov type singular problem has the form

$$\frac{\partial \rho}{\partial x}[\Lambda_1 x + F_1(x,\rho)] = \Lambda_2 \rho + F_2(x,\rho),\ \ x \in \Omega_x^n(c),\ \rho(0) = 0. \tag{11}$$

Here $x \in \mathbf{K}^n$, $\rho \in \mathbf{K}^m$, $\partial \rho / \partial x$ is the Jacobi matrix, $\partial \rho / \partial x \in \mathbf{L}(\mathbf{K}^n, \mathbf{K}^m)$, $\Lambda_1 \in \mathbf{L}(\mathbf{K}^n)$, $\Lambda_2 \in \mathbf{L}(\mathbf{K}^m)$, $F_1 : \overline{\Omega}_{x,\rho}(a_0) \to \mathbf{K}^n$, $F_2 : \overline{\Omega}_{x,\rho}(a_0) \to \mathbf{K}^m$.

About the next Lyapunov theorems see in[3, 8] in detail.

Theorem (Lyapunov's first theorem). *Let the real parts of all EVs of* Λ_1 *be non-zero and of the same sign, and suppose that there are no relations of the form*

$$\sum_{j=1}^{n} l_j \lambda_j(\Lambda_1) = \lambda_s(\Lambda_2),\ \ s = 1,\ldots,m, \tag{12}$$

for integer $l_j \geq 0$, $j = 1,\ldots,n$, $\sum_{j=1}^{n} l_j > 0$; *let* $F_1 \in H_{n,a_0}$, $F_2 \in H_{m,a_0}$ *and let for* $(x,\rho) \in \overline{\Omega}_{x,\rho}(a_0)$ *the relations*

$$F_1(x,\rho) = [x;\rho]_2, \tag{13}$$

$$F_2(0,\rho) = [\rho]_2,\ \ F_2(x,0) = [x]_\nu \tag{14}$$

be valid where $\nu \geq 1$. *Then* $\exists c_0$, $0 < c_0 < a_0$, *such that* $\forall c$: $0 < c < c_0$, *in the class* $H_m(\Omega_x^n(c))$ *a unique solution* $\hat{\rho}(x)$ *of the problem* (11) *exists, and*

$$\hat{\rho}(x) = [x]_\nu, \quad x \in \Omega_x^n(c). \tag{15}$$

Theorem (Lyapunov's second theorem). *In the hypotheses of the preceding theorem, let* (13) *be replaced by the conditions*

$$F_1(x,0) = [x]_2, \; F_1(0,\rho) = [\rho]_1, \; F_2(x,\rho) = [x,\rho]_2, \; (x,\rho) \in \overline{\Omega}_{x,\rho}(a_0),$$

and let for (12) *the relation* $\sum_{j=1}^n l_j > 0$ *be replaced by the inequality* $\sum_{j=1}^n l_j > 1$. *Then* $\exists c_0$, $0 < c_0 \leq a_0$, *such that* $\forall c$: $0 < c \leq c_0$, *problem* (11) *has a unique solution* $\hat{\rho}(x)$ *in the class* $H_m(\Omega_x^n(c))$ *satisfying the additional requirement* $(\partial\rho / \partial x)(0) = 0$, *and if* (14) *holds for* $\nu \geq 2$ *then* (15) *is valid.*

The theorems given below are the generalizations of Lyapunov's theorems to the systems of PDEs with nonholomorphic given functions. They concern, in particular, to the smooth SIM Eq. (9) for the system (8) where $f(x,y) = \Lambda_1 x + F_1(x,y)$, $g(x,y) = \Lambda_2 y + F_2(x,y)$ and the equilibrium $\{x,y\} = \{0,0\}$ is a saddle-node of $(\mu,n)-$ type, $\mu \geq 0$, $0 < n < N$.

Condition 1. $\exists \mu \geq 0$ *such that* $\operatorname{Re}\lambda(\Lambda_1) < -\mu < \operatorname{Re}\lambda(\Lambda_2)$ *for all EVs* $\Lambda_{1,2}$.

Condition 2. $F_1 \in C^1_{n,a_0}$, $F_2 \in C^1_{m,a_0}$ *and satisfy the relations*

$$F_j(0,0) = 0, \; \frac{\partial F_j}{\partial x}(0,0) = 0, \; \frac{\partial F_j}{\partial \rho}(0,0) = 0, \; j = 1,2. \tag{16}$$

Let **Conditions 1**, **2** hold. Denote

$$J_1 = \int_0^\infty |\exp((\Lambda_1 + \mu E_n)\tau)| \, d\tau, \; J_2 = \int_0^\infty |\exp(-(\Lambda_2 + \mu E_m)\tau)| \, d\tau, \; J = \max_{j=1,2} J_j,$$

$$K = \sup_{\tau \geq 0} |\exp((\Lambda_1 + \mu E_n)\tau)|,$$

by virtue of **Condition 1**, we have $0 < J < \infty$, $1 \leq K < \infty$. We fix a number $q \in (0,1)$ and choose $\delta \in (0, a_0)$ from the condition

$$\max_{j=1,2} \sup_{(x,\rho) \in \overline{\Omega}_{x,\rho(\delta)}} \left\{ \max\left[\left| \frac{\partial F_j}{\partial x}(x,\rho) \right| ; \left| \frac{\partial F_j}{\partial \rho}(x,\rho) \right| \right] \right\} \leq q/(2J)$$

that is possible by virtue of **Condition 2**.

Theorem 3[3, 17]. *Let* **Conditions 1** *and* **2** *be realised and for fixed* $q \in (0,1)$ *let* δ *be defined as indicated above. Then the following statements hold:* $\forall c : 0 < c \leq (1-q)\delta / K$, *problem* (11) *has a unique solution* $\hat{\rho}(x)$ *in the class* $C_m^1\left(\Omega_x^n(c)\right)$, *moreover* $(\partial\rho / \partial x)(0) = 0$, *and* $|\hat{\rho}(x)| \leq Kq|x|/(1-q)$, $x \in \Omega_x^n(c)$; *if* $F_1 \in C_{n,a_0}^k$, $F_2 \in C_{m,a_0}^k$, $k \geq 1$, *then* $\hat{\rho} \in C_m^k\left(\Omega_x^n(c)\right)$; *if* $F_1 \in H_{n,a_0}$, $F_2 \in H_{m,a_0}$, *then* $\hat{\rho} \in H_m\left(\Omega_x^n(c)\right)$, *and, in addition, if* (14) *is valid for* $\nu \geq 2$ *then* (15) *holds.*

Corollary. *Let one of the conditions in* (16) *be replaced by the more general one* $(\partial F_2 / \partial x)(0,0) = D \neq 0$, $D \in \mathbf{L}(\mathbf{K}^n, \mathbf{K}^m)$, *and otherwise let the conditions of* **Theorem 3** *hold. Let the* $(m \times n)$*-matrix* α *be a solution of the equation* $\Lambda_2\alpha - \alpha\Lambda_1 = -D$, *i.e.,* $\alpha = -\int_0^\infty \exp(-\Lambda_2\tau) D \exp(\Lambda_1\tau) d\tau$, *and let* $F_j(x,\rho)$, $j = 1,2$, *be defined in the region* $\overline{\Omega}_x^n(a_0) \times \{\rho : |\rho| \leq (1+|\alpha|)a_0\}$. *Then the replacement* $\rho = \alpha x + \widetilde{\rho}$ *will reduce* (11) *to the analogous problem for* $\widetilde{\rho}$ *with functions* $\widetilde{F}_1(x,\widetilde{\rho}) = F_1(x, \alpha x + \widetilde{\rho})$, $\widetilde{F}_2(x,\widetilde{\rho}) = F_2(x, \alpha x + \widetilde{\rho}) - \alpha F_1(x, \alpha x + \widetilde{\rho}) - Dx$, *for which the conditions of* **Theorem 3** *now hold.*

Remark 2. *In*[9] *the following singular Cauchy problem for ODEs have been examined (as an auxiliary problem)*

$$rh' + bh = rf(r,h) + h^2 g(r,h), \quad h(0) = 0, \tag{17}$$

where $f(r,h)$, $g(r,h)$ *are real analytic near* $(0,0)$. *Then, for* $b > -1$, *there exists a unique analytic solution of the problem* (17) *(see* **Theorem 3.1** *in*[9] *with two different proofs , one of them was suggested by L. Nirenberg). But this statement at once follows from the* **first Lyapunov's theorem**. *Moreover, according to* **Theorem 3** *and* **Corollary**, *it is a unique solution in the class* $C_1^1[-r_0, r_0]$, $r_0 > 0$.

Remark 3. *It is easy to construct an example in which problem* (11) *has more than one solution if the* **Condition 2** *is not fulfilled. Consider the scalar equation* $\rho'(\lambda_1 x + mp) = \lambda_2\rho$, *where* $m \neq 0$. *It has at least two solutions vanishing for* $x = 0$, *namely,* $\rho_1 \equiv 0$ *and* $\rho_2 = (\lambda_2 - \lambda_1)x/m$. *For the first solution* $\rho_1 \equiv 0$, *and only for it, the hypotheses of* **Lyapunov's second theorem** *and the next* **Theorem 4** *are satisfied.*

Condition 3. $F_1 \in C_{n,a_0}^1$, $F_2 \in C_{m,a_0}^1$, $0 < a_0 \leq 1$, *and satisfy the relations* $F_j(0,0) = 0$, $j = 1,2$, $(\partial F_1 / \partial x)(0,0) = 0$, $(\partial F_2 / \partial\rho)(0,0) = 0$, $(\partial F_1 / \partial\rho)(0,0) = M \neq 0$, $M \in \mathbf{L}(\mathbf{K}^m, \mathbf{K}^n)$, $\left|(\partial F_2 / \partial x)(x,\rho)\right| \leq C(|x|^\gamma + |\rho|^\gamma)$, $(x,\rho) \in \overline{\Omega}_{x,\rho}(a_0)$, *where* $0 < C = const$, $0 < \gamma = const$.

Let **Conditions 1, 3** be fulfilled. Let us take arbitrary q and Δ, $0 < q < 1$, $0 < \Delta < \gamma$, and-choose $\delta \in (0, a_0)$ such that

$$\sup_{(x,\rho)\in\overline{\Omega}_{x,\rho(\delta)}} \left\{ \max\left[2\left|\frac{\partial F_1}{\partial x}(x,\rho)\right|; 2\left|\frac{\partial F_1}{\partial \rho}(x,\rho) - M\right|; \left|\frac{\partial F_2}{\partial \rho}(x,\rho)\right| \right] \right\} \le q/(2J),$$

$$|M|\,\delta^{\Delta} \le q/(2J_1), \quad 2C\delta^{\gamma-\Delta} \le q/(2J_2),$$

here $\delta = \delta(\Delta) \to 0$ as $\Delta \to 0$ or $\Delta \to \gamma - 0$.

Theorem 4[3]. *Let* **Conditions 1** *and* **3** *be fulfilled and for fixed* $q \in (0,1)$ *and* $\Delta \in (0,\gamma)$ *let* δ *be defined as indicated above. Then the following statements hold:* $\forall c : 0 < c \le (1-q)\delta / K$, *problem* (11) *has a unique solution* $\hat{\rho}(x) \in C_m^1\left(\Omega_x^n(c)\right)$ *satisfying the additional requirement* $\rho(x) = O(|x|^{1+\Delta})$, $|x| \to 0$; *if* $F_1 \in C_{n,a_0}^k$, $F_2 \in C_{m,a_0}^k$, $k \ge 1$, *then* $\hat{\rho} \in C_m^k\left(\Omega_x^n(c)\right)$; *if* $F_1 \in H_{n,a_0}$, $F_2 \in H_{m,a_0}$, *then* $\hat{\rho} \in H_m\left(\Omega_x^n(c)\right)$, *and, in addition, if* (14) *is valid for* $\nu \ge 2$ *then* (15) *holds.*

Remark 4. *Let us demonstrate on the example that the omission of some assumptions of both* **Theorems 3** *and* **4** *implies nonuniqueness of solution to the problem* (11) *and nonexistence of solution satisfying condition:* $\exists \Delta > 0$ *such that*

$$\rho(x) = O\left(|x|^{1+\Delta}\right), \quad |x| \to 0. \tag{18}$$

Consider the equation

$$\rho'(\lambda x + m\rho) = -\lambda\rho + ax^k, \tag{19}$$

where $m \ne 0$, $\lambda \ne 0$, $k \ge 1$, $a \ne 0$ *for* $k > 1$ *and* $a > -\lambda^2/m^2$ *for* $k = 1$. *This equation is equivalent to* $(\lambda x\rho + m\rho^2/2)' = ax^k$ *whence* $\lambda x\rho + m\rho^2/2 = ax^{k+1}/(k+1) + const$, *and there exist two solution satisfying condition* $\rho(0) = 0$,

$$\rho_{\pm} = -\lambda x/m \pm \sqrt{\lambda^2 x^2/m^2 + 2a\,x^{k+1}/[m(k+1)]}.$$

Moreover, if $k > 1$ *then a unique solution* $\rho = \rho_+(x)$ *satisfies additional requirement* (18) *while if* $k = 1$ *then condition* (18) *is not fulfilled for both solutions.*

Remark 5. *The singular Cauchy problem for ODEs*

$$\frac{dy}{dt}[\lambda_1 t + f(t,y)] = \Lambda_2 y + F_2(t,y), \quad |t| \le t_0, \quad t_0 > 0, \quad y(0) = 0, \tag{20}$$

where λ_1 *and* f *are scalar quantities, is a special case of problem* (11). *In particular, problem* (20) *with* $f(t,y)\equiv 0$, $\lambda_1=-1$, *which is of interest in itself, was studied in detail under quite general assumptions, in particular, for functional-differential and operator equations. As an example, we have the equation* (19) *with* $m=0$ (*it is a bifurcation value of* m) *and a unique solution of* (19) *such that* $\rho(0)=0$ *is given by formula* $\rho = ax^k/[\lambda(k+1)]$. *For* $0<k<1$, *this function is also a solution of the problem although it is not differentiable in the point* $x=0$ *so that an existence and uniqueness of solution, which are in accordance, e.g., with* **Theorem 2** *in*[10] , *do not follow from the theorems of the present paper.*

3. CORRECT MATHEMATICAL SIMULATION AND NUMERICAL SOLUTION OF SOME PROBLEMS FROM HYDRODYNAMICS

We consider some examples of correct statement and reduction on a finite interval of singular BVPs arising from incompressible fluid mechanics[11-13, 18] and propose a stable method to solve them numerically. We use **Proposition** to obtain the smooth SIM equation in linear approach (7) and **Theorems 1-3** for more exact approximations.

3.1. Example 1: A Plane-Parallel Flow in the Mixture Layers

This problem is formulated for automodelling solutions of the boundary layer equation for the stream function with the zero gradient of a pressure and has the form[11] :

$$F''' + FF'' - [(m-1)/m](F')^2 = 0,\ -\infty<\tau<\infty, \tag{21}$$

$$\lim_{\tau\to-\infty} F'(\tau)=0, \tag{22}$$

$$F(0)=0; \tag{23}$$

$$\lim_{\tau\to\infty} F(\tau)/\tau^m = b,\ 0<b \text{ is fixed,} \tag{24}$$

where $m>0$ is a given parameter. Eq. (21) has in a phase space $\mathbf{R}^3$ an infinite set of stationary points $(F,F',F'')=(-a,0,0)$, where a is an arbitrary real number. For $\tau<0$ and any fixed $a>0$, such point is a pseudohyperbolic saddle (or equilibrium) with one-dimensional stable separatrix [in other words, this point is a saddle-node of $(-\varepsilon,1)$-type $\forall\varepsilon:\ 0<\varepsilon<a$]. Then the condition (22) must be reformulated in more exact form:

$$\lim_{\tau\to-\infty} \{\exp(-\varepsilon\tau)[F(\tau)+a,\ F'(\tau),\ F''(\tau)]\} = \{0,0,0\},$$

where $0<a$ is unknown number, $0<\varepsilon<a$. Using the equation of analytic SIM $\mathbf{M}_-^{(1)}$, we obtain, instead of (22), two conditions at a finite point $\tau=-T$, $T>>1$, approximately in the form

$$F(-T)+a-F''(-T)/a^2-[3/(4a^5 m)][F''(-T)]^2 \approx 0, \tag{25}$$

$$F'(-T)-F''(-T)/a-[1/(2a^4 m)][F''(-T)]^2 \approx 0. \tag{26}$$

Thus we must solve BVP (21), (23), (25), (26) on the segment [-T,0] and choose the parameter $a > 0$ to satisfy condition (24) by solving the Cauchy problem from $\tau = 0$ rightwards to large enough τ (according to[11], it is possible to satisfy (24) for $m > 1/2$). To solve BVP (21), (23), (25), (26) for fixed $a > 0$, we propose a stable shooting method beginning with smooth SIM (25), (26) and solving rightwards the auxiliary Cauchy problems with one unknown parameter, e.g., $F''(-T)$ to satisfy condition (23). Some results of numerical experiments see on the Fig. 1.

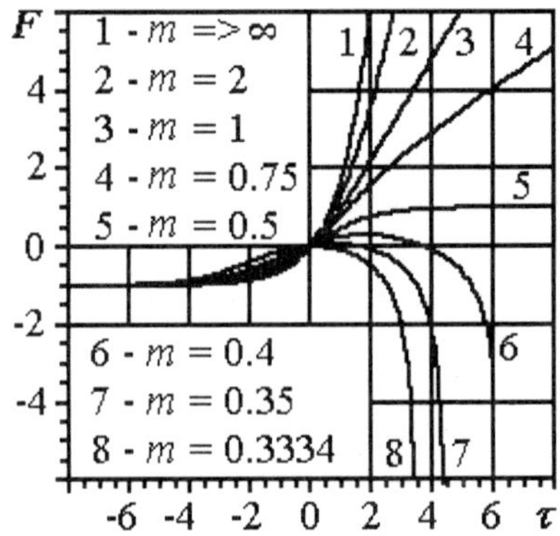

Figure 1. A plane-parallel flow in the mixture layers: $a = 1$; 1 - $F(\tau) = \exp(\tau) - 1$; 5 - $F(\tau) = \tanh(\tau/2)$.

3.2. Example 2: A Flow Near a Rotating Disk of an Infinite Radius

This problem is stated from the Navier-Stokes and continuity equations and has the form[12, 18]:

$$F''' - FF'' + (F')^2/2 - 2G^2 = 0, \tag{27}$$

$$G'' - FG' + F'G = 0, \quad 0 \le \tau < \infty, \tag{28}$$

$$F(0) = F'(0) = 0, G(0) = 1, \tag{29}$$

$$\lim_{t \to \infty} F'(\tau) = \lim_{\tau \to \infty} G(\tau) = 0. \tag{30}$$

The system (27), (28) has in a phase space $\mathbf{R}^5$ an infinite set of stationary points $(F, G, F', G', F'') = (-a, 0, 0, 0, 0)$ where a is an arbitrary real number. For any fixed $a > 0$, such point is a pseudohyperbolic equilibrium with analytic two-dimensional SIM $\mathbf{M}_+^{(2)}$ [it is a saddle-node of $(\varepsilon, 2)$-type $\forall \varepsilon: 0 < \varepsilon < a$]. Thus the condition (30) must be reformulated as

$$\lim_{\tau \to +\infty} \exp(\varepsilon\tau)[F(\tau) + a, G(\tau), F'(\tau), G'(\tau), F''(\tau)] = \{0,0,0,0,0\},$$

where $0 < a$ is unknown number, $\varepsilon: 0 < \varepsilon < a$. Using the equation of smooth SIM $\mathbf{M}_+^{(2)}$ in the linear approach, we obtain, instead of (30), three conditions at a finite point $\tau = T$, $T >> 1$, approximately in the form

$$F(T) + a - F''(T)/a^2 \approx 0, \; G(T) + G'(T)/a \approx 0, \tag{31}$$

$$F'(T) + F''(T)/a \approx 0\,. \tag{32}$$

Thus it is incorrectly to pose two conditions at the point $\tau = T$ approximately in the form $F'(T) \approx 0$, $G(T) \approx 0$. We must put three conditions (31), (32) with unknown parameter $a > 0$. To solve numerically the approximate regular BVP (27), (28), (29), (31), (32) with a parameter $a > 0$ we use a stable method of an optimal shooting beginning with smooth SIM (31), (32) and solving leftwards the Cauchy problems with three unknown parameters $a, F(T), G(T)$. More in detail, we introduce a goal function $D^*(a, F(T), G(T)) = F^2(0) + [G(0)-1]^2 + [F'(0)]^2$ and use the optimization methods[19] to minimize this function so that to satisfy conditions (29). The results of numerical experiments see on the Fig. 2.

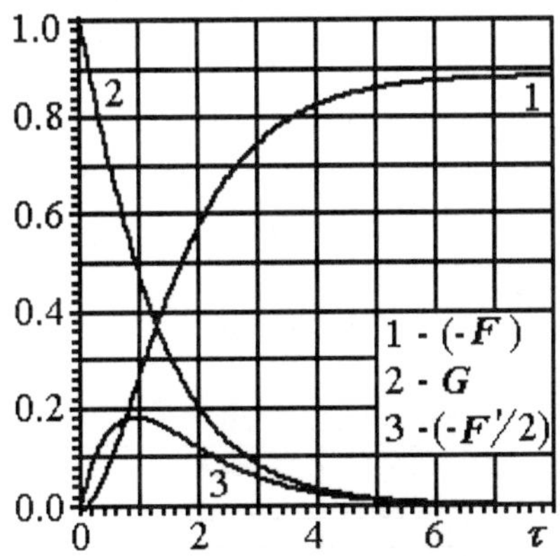

Figure 2. A flow near a rotating disk of an infinite radius: $a = 0.8854\ldots$.

3.3. Example 3: A Flow Near an Immovable Infinite Base due to a Fluid Rotation Far from the Wall (Without and in the Presence of a Magnetic Field).

This problem is obtained from the Navier-Stokes and continuity equations (or from the magnetohydrodynamics equations) and has the form[12, 13]:

$$F''' + [(3-\chi)/2]FF'' + \chi(F')^2 - sF' + G^2 - 1 = 0\,, \tag{33}$$

$$G'' + [(3-\chi)/2]FG' + (\chi - 1)F'G - s(G-1) = 0\,,\quad 0 \le \tau < \infty\,, \tag{34}$$

$$F(0) = F'(0) = G(0) = 0\,, \tag{35}$$

$$\lim_{\tau\to\infty} F'(\tau) = 0\,,\quad \lim_{\tau\to\infty} G(\tau) = 1\,, \tag{36}$$

where χ and s are the parameters, $|\chi| \le 1$, $s \ge 0$ ($s \ne 0$ or $\chi \ne -1$ in the presence of a magnetic field). The system (33), (34) has in a phase space $\mathbf{R}^5$ an infinite set of stationary points $(F, F', F'', G, G') = (-a, 0, 0, 1, 0)$, where a is an arbitrary real number. For any fixed a, such point is a pseudohyperbolic equilibrium with analytic two-dimensional SIM $\mathbf{M}_+^{(2)}$. Using the equation of smooth SIM $\mathbf{M}_+^{(2)}$, we obtain, instead of (36), three conditions at a finite point $\tau = T$,

$T >> 1$. For example, in the case $\chi = 1$ these conditions, in the linear approach, are the following (other cases see in detail in[5, 6]):

$$(s^2/2)[F(T)+a]+(as/2)F'(T)-(s/2)F''(T)+a[G(T)-1]-G'(T)\approx 0, \quad (37)$$

$$[\lambda(a)-a][G(T)-1]+G'(T)\approx 0, \quad (38)$$

$$[(a-2\lambda(a))(\lambda(a)-a)/2]F'(T)+[(a-2\lambda(a))/2]F''(T)+G(T)-1\approx 0, \quad (39)$$

where $\lambda(a) = a/2+\sqrt{a^2/4+s}$.

The Holt problem (33)-(36) is considered by some authors as a difficult for solving (see[14-16]). But these papers and[13] contain the inaccuracies: it is unsuccessfully to pose two boundary conditions at the point $\tau = T >> 1$ in the form $F'(T)=0$, $G(T)=1$ and solve an approximate nonsingular BVP from zero to the right by unstable methods because of a movement in the direction to a saddle point. We must put three conditions (37)-(39) with unknown parameter a. To solve BVP (33)-(35), (37)-(39) numerically we introduce a goal function $G^*(a, F(T), G(T)) = F^2(0)+G^2(0)+[F'(0)]^2$, defined by shooting method from SIM (37)-(39) leftwards and use the optimization methods[19] to minimize this function so that to satisfy conditions (35). The results of numerical experiments see on the Fig. 3.

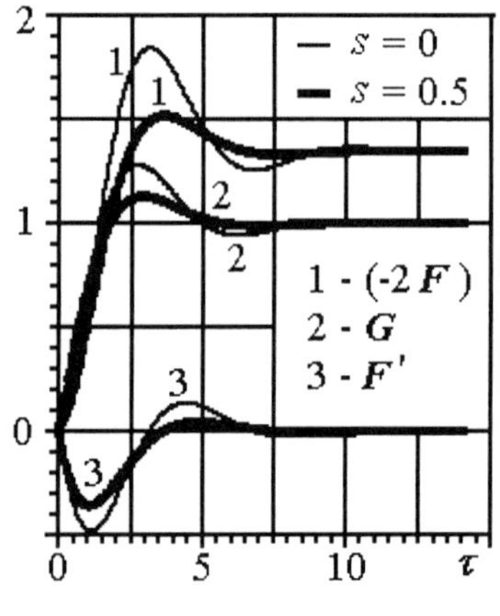

Figure 3. A flow near an immovable infinite base due to a fluid rotation far from the wall: $\chi = -1$, $a = 0.6721\ldots$.

4. CONCLUSIONS

We represent an approach to correct statement of limiting boundary conditions at infinity for ASs of nonlinear ODEs as a convergence for solution to get into a (pseudo)hyperbolic equilibrium. We describe the method of boundary condition transfer from infinity to a finite point by the construction of smooth SIM in the neighbourhood of the equilibrium point of a phase space. The examples of correct statement and reduction on a finite interval for singular BVPs arising from hydrodynamics are given. A stable method to solve the approximate regular BVPs and the numerical results are represented.

5. ACKNOWLEDGEMENTS

The authors were partly supported in this work by the Russian Foundation for Basic Research (RFBR projects No. 99-01-00331 and No. 00-01-00674) and by the project NATO PST/CLG.976878.

REFERENCES

1. A.A. Abramov, N.B. Konyukhova, and K. Balla, *Comput. Math. Banah Center Pubs.*, v.**13**, 319-351 (1984) (in Russian).
2. A.A. Abramov, N.B. Konyukhova, *Sov J. Numer. Anal. Math. Modelling,* v.**1**(4), 245-265 (1986).
3. N.B. Konyukhova, *Diff. Eq.*, v.**30**(8), 1284-1294 (1994).
4. N.B. Konyukhova, *Comp. Maths Math. Phys.*, v.**34**(10), 1179-1195 (1994).
5. N.B. Konyukhova, *Soobsch. po Prikl. Mat. VC RAN* (VC RAN, Moscow, 1996) (in Russian).
6. N.B. Konyukhova., A.I. Sukov, in: *The basic physical-mathematical problems and the technical-technological system simulation* (STANKIN, Moscow, 2001), 117-121 (in Russian).
7. E.A. Coddington and N. Levinson, *Theory of Ordinary Differential Equations* (McGraw-Hill, New York, 1955).
8. A.M. Lyapunov, *General Motion Stability Problem* (GITTL, Moscow, 1950) (in Russian).
9. P. Guan and Y.Y. Li, *Commun. Pure and Appl. Math.*, v.**L,** 789-811 (1997).
10. N.B. Konyukhova, *USSR Comput. Maths Math. Phys.*, v.**23**(3), 72-82 (1983).
11. V.N. Diesperov, *Soobsch. po Prikl. Mat. VC AN SSSR* (VC AN SSSR, Moscow, 1986) (in Russian).
12. H. Schlichting, *Grenzschicht-Theorie* (Karlsruhe, 1951).
13. J.H. Holt, *Communs. Assoc. Comput. Machinery*, v.**7**(6), 366-373 (1964).
14. S.M. Roberts and J.S. Shipman, *J. Optimizat. Theory and Appl.*, v.**7**(4), 301-318 (1971).
15. U. Ascher, J. Christiansen, and R.D. Russell, *ASM Trans. Math. Softw.*, v.**7**(2), 209-222 (1981).
16. M. Macconi and A. Pascuali, *J. Optimizat. Theory and Appl.*, v.**19**(3), 367-379 (1976).
17. N.B. Konyukhova, *Diff. Eq.*, v.**28**(9), 1283-1293 (1992).
18. L.D. Landau and E.M. Lifshits, *Theoretical Physics*, v.**VI**: *Hydrodynamics* (Nauka, Moscow, 1986) (in Russian).
19. N.S. Bakhvalov, N.P. Zhidkov, and G.M. Kobel'kov, *Numerical methods* (Nauka, Moscow, 1987) (in Russian).

COMPUTATIONAL METHODS FOR THE ESTIMATION OF THE AEROSOL SIZE DISTRIBUTIONS

A.Voutilainen*, V. Kolehmainen, F. Stratmann+, and J.P. Kaipio

1. INTRODUCTION

Aerosol particles play an important role in many physical and chemical processes in the atmosphere[1]. Physical and chemical behaviour of aerosol particles is strongly dependent on particle size and thus the size cannot be ignored in the evaluation and theoretical prediction of the effects caused by airborne particles. Since the particle diameter d_p can range from few nanometers to about 100 micrometers, a size distribution function is used to describe how certain property, e.g. number, surface area or mass, of particles per unit gas volume is distributed on different particle sizes. The determination of the size distribution function is a very important fundamental task in aerosol research. However, the size distribution cannot be measured directly but it has to be reconstructed on the basis of indirect observations using computational methods. From the mathematical point of view the determination of the size distribution function is an ill-posed problem since the problem does not have a unique solution. The purpose of this chapter is to describe the problem and give a brief review on some computational methods proposed for the reconstruction of particle size distributions.

2. DISCRIPTION OF THE PROBLEM

In general, the relation between the size distribution $f(x)$ and an observation y_i is

$$y_i = \int_{x_l}^{x_u} k_i(x) f(x) dx + \varepsilon_i = \mu_i + \varepsilon_i, \quad i = 1,\dots,n, \tag{1}$$

where $k_i(x)$ is a kernel function of the system, μ_i is an error-free observation, ε_i is the measurement error, x is the size-related parameter with x_l and x_u being the lower and

* A.Voutilainen, V. Kolehmainen, , and J.P. Kaipio, University of Kuipio, Departament of Applied Physics, P.O.Box 1627, FIN-70211 Kuipio, Finland.
+ F. Stratmann, Institute for Tropospheric Research, Permosersty.15, D-04318 Leipzig, Germany.

upper limits of the size range of interest, and n is the number of observations[2]. A discrete approximation for Eq.(1) can be written as

$$y_i = \sum_{j=1}^{m} w_j k_j(x_j) f(x_j) + \varepsilon_i \quad i = 1,\dots,n, \tag{2}$$

where $x_1,\dots,x_m$ are grid points and the coefficients w_j are the weights of a numerical quadrature. The number of grid points must be large enough to keep the numerical integration accurate. The set of equations (2) can be written in the matrix form as

$$y = Hf + \varepsilon , \tag{3}$$

where $y, \varepsilon \in R^n$ and the elements of the matrix $H \in R^{n\times m}$ are $H_{ij} = w_j k_i(x_j)$.

Now the problem is to find an estimate for the (discrete) size distribution function f on the basis of the observations y. Since the number of observations is usually less than the number of unknown parameters, the problem is underdetermined. Thus, if the rows of the matrix are the linearly independent, the matrix equation has infinitely many exact solutions and the problem is sad to be ill-posed. The ill-posedness is a characteristic feature for inverse problems, that arise in various applications in which inferences on certain quantity have to be made based on indirect observations. In order to find a feasible estimate for the size distribution function some additional information about the solution must be incorporated to the problem. This can be understood such that the original ill-posed problem is replaced by an approximative well-posed problem the solution of which can be assumed to be close to the actual size distribution function.

3. RECONSTRUCTION METHODS

3.1 Least Square Method

Various reconstruction methods have been proposed to overcome the ill-posedness in the estimation of aerosol size distributions. Least square methods are widely used in parameter estimation problem but in this case the (weighted) least square solution

$$\hat{f}_{1,s} = arg\, \min_{f} \| L_1 (Hf - y) \|^2 \tag{4}$$

is not unique. A common approach to stabilise the problem is to assume that the size distribution is of some functional form $f = f(\psi)$, and determine the minimizer of the norm of the (weighted) residual with respect to the parameter vector ψ[3,4]. A commonly used parametric model is the log-normal function, which has been found to describe the particle size distributions relatively well. Usually it is necessary to describe the size distribution as a combination of several log-normal functions. The use of the log-normal model leads to a non-linear least square problem and the solution can be determined using iterative metods.[5] An option to simplify the problem is to express the size distribution function as a linear combination of certain properly selected basis function, i.e. $f = B\phi$, where the column vectors of the matrix B contain the basis functions and the parameter vector ϕ is to be determined.[6] If the basis functions are selected appropriately, the least squares is unique. A drawback of both approaches is that if the actual size distribution cannot be described with the functional or with the basis functions selected, the estimate may be severely inaccurate.

3.2. Tikhonov Regularization Methods

If no assumptions on the shape of the size distribution are made, the underdetermined least squares problem (4) must be modified in order to find a unique solution. Tikhonov type regularization methods are widely used techniques in solving discrete ill-posed problems arising also in size distribution measurements.[7,8,9] The basic idea of these approaches is not to determine the minimizer of the norm of the (weighted) residual but accept a small deviation from the minimum in order to obtain a stable solution with acceptable and realistic properties. The Tikhonov regularized solution in general form is

$$\hat{f}_\lambda = \arg\min_f \left\{ \left\| L_1(Hf - y) \right\|^2 + \lambda \Phi(f) \right\}, \tag{5}$$

where $\Phi(f)$ is a regularizing functional and $\lambda > 0$ is a regularization parameter. The prior information on the solution is incorporated to the problem with $\Phi(f)$ and λ, and the quality and accuracy of the estimate depends on these terms. Since particle size distribution functions are generally assumed to be smooth functions, the regularizing functional is often selected to be of the form $\Phi(f) = \|D_2 f\|^2$, where the matrix D_2 is the second - order difference matrix. Since the size distribution functions are non-negative by definition, the minimization problem (5) is usually solved subject to the constraint $f \geq 0$.

3.3. Iterative Methods

Probably the best-known iterative size distribution reconstruction method is the Tvomey's method [10] in which the size distribution is represented in the non-parametric form. Starting from an initial guess $f^{(0)}$, the next estimate is obtained by multiplying the initial guess with coefficients that area selected such that new estimate has better agreement with the data. The recursion is of the form

$$f_j^{(t+1)} = \left(1 + \left(a_i^{(t)} - 1\right)k_{ij}\right) f_j^{(t)},$$

where $k_{ij} = k_i(x_j)$ and the coefficient $a_i^{(t)}$ is the ratio of the measured and computed observations , i.e. $a_i^{(t)} = y_i / \left(H_i f^{(t)}\right)$ with H_i, being the ith cow of the matrix H. The iteration yields a non-negative estimate that may be highly oscillative. In addition, the solution depends on the initial guess. The Twomey's method is an improved modification of the Chahine's method[11] that was originally proposed for the estimation of vertical temperature profiles of the atmosphere and later adopted for the estimation of particle size distributions.

3.4 Statistical Methods

The observations obtained from any measurement system are always more or less noisy. In the methods described above the statistical nature of the observations is not explicitly considered. In general, the observations can be regarded as samples from certain probability density function. In statistical approaches the information on the noise statistics is included to the problem formulation and the parameters to be estimated are treated as random variables.

In the statistical method proposed by Ramachandran and Kandlikar,[12] the size distribution was assumed to be a bi-modal log-normal function and the observations were assumed to be Gaussian with zero mean. Let the vector ψ contain the unknown parameters and the size distribution function is denoted as $f(\psi;x)$. First, a total of T parameter vectors are drawn from certain appropriate uniform probability density. The solution for the inverse problem is obtained as the weighted average of the sampled parameters

$$\hat{\psi} = \sum_{t=1}^{T} r^{(t)}\psi^{(t)} \Big/ \sum_{t=1}^{T} r^{(t)}. \tag{7}$$

Given the observations y, the weights are obtained as the values of the probability density function of the observation error at points $\psi^{(t)}$, i.e.

$$r^{(t)} = (2\pi)^{-n/2}|\Gamma_\varepsilon|^{-1/2}\exp\left\{-\tfrac{1}{2}(y-\mu^{(t)})^{\mathrm{T}}\Gamma_\varepsilon^{-1}(y-\mu^{(t)})\right\}, \tag{8}$$

where $\mu^{(t)} = \int_{x_1}^{x_u} K(x) f(\psi^{(t)};x)\,\mathrm{d}x$ with $K(x)$ being defined as $K(x) = (k_1(x), k_2(x), \ldots, k_n(x))^{\mathrm{T}}$ and Γ_ε is the covariance of the observations errors

Maher and Laird[13] considered the reconstruction of the size distribution from Poisson distributed observations that are mutually independent. Under these assumptions the most probable non-negative solution was determined using the Expectation-Maximization (EM) algorithm. The solution was obtained recursively as

$$f_j^{(t+1)} = \frac{f_j^{(t)}}{\sum_{i=1}^{n} H_{ij}} \frac{\sum_{i=1}^{n} y_i H_{ij} f_j^{(t)}}{\sum_{j=1}^{m} H_{ij} f_j^{(t)}}. \tag{9}$$

The iteration converges to a non-negative solution and if $\mathrm{rank}(H) = m$, the solution is unique. Otherwise, the solution depends on the initial value of the iteration.

In the extreme value estimation[14] the objective is not to find a single solution that is in some sense optimal but the aim is to determine a set of acceptable solutions and compute ,certain confidence intervals. The observation errors are assumed to be independent Gaussian with zero-mean and the set of acceptable solutions is defined by means of the chi-square sum $Q(f)=\sum_{i=1}^{n}(H_i f - y_i)^2/\sigma_i^2$, where σ_i is the standard deviation of the ith measurement error. The set of acceptable solutions is defined as

$$S=\{f \in \mathbf{R}^m \mid Q(f) \leq Q_{\text{opt}} + \Delta Q,\ f \geq 0\}, \tag{10}$$

where $Q_{\text{opt}} = \min_f Q(f)$ and ΔQ is a confidence parameter. Next, the minimum $f_{i,\min}$ and maximum $f_{i,\max}$ values of each component $f_i, i=1,\dots,m$ are determined such that $f \in S$. These values are used as the solution of the problem and the actual size distribution lies between the maximum and minimum values at 90 % probability when the confidence parameter is set to the recommended value $\Delta Q = 3$. If the matrix H is ill-conditioned or the problem is under-determined, the intervals $\lfloor f_{i,\min}, f_{i,\max} \rfloor$ are very wide and therefore not very informative. In such a case the size distribution can be represented as a linear combination of feasible basis functions $b_l, l=1,\dots,p,$ i.e. $f = B\phi$. The set S is determined with respect to the new parameters ϕ and the maximum and minimum values of each component f_j are determined within this set.

A number of other methods have been proposed for solving the inverse problem associated with the aerosol measurement data processing but most of them are not very well-known or widely used. In the following sections three novel reconstruction methods proposed by the authors are briefly reviewed.[15,16,17] The methods are evaluated using simulated differential mobility particle sizer (DMPS) data, but, in principle, the methods are applicable also in other measurement systems. In all simulations the measurement systems consisted on a short Vienna type DMA and a TSI 3025A CPC.[18,19]

4. STATISTICAL INVERSION METHOD

In statistical approaches both the observations y and unknown parameters f are treated as random variables with some joint probability density $p(f,y)$. Given the observations, the conditional probability density of the parameters is constructed using feasible probability models and the Bayes' theorem. The conditional probability density is of the form

$$p(f \mid y) = p(y \mid f)p(f)/p(y) \propto p(y \mid f)p(f), \tag{11}$$

where $p(y \mid f)$ is called the likelihood density and $p(f)$ is the prior density. If the data is obtained from the DMPS measurement system, it is appropriate to assume that the observations are samples from certain Poisson distributions. Thus, assuming that the observations are statistically independent, the likelihood density is obtained as

$$p(y \mid f) = \prod_{i=1}^{n} (y_i!)^{-1} (H_i f) \exp(-H_i f) = \left(\prod_{i=1}^{n} (y_i!)^{-1} \right) \exp\left(y^{\mathrm{T}} \ln(Hf) - \mathbf{1}^{\mathrm{T}} Hf\right), \tag{12}$$

where $H_i f$ is the expectation of y_i and $\mathbf{1} = (1,1,\ldots,1)^{\mathrm{T}}$. Since the particle size distributions are assumed to be smooth functions, we select the prior density to be of the standard form

$$p(f) \propto \exp\left(-\tfrac{1}{2} \lambda f^T D_2^T D_2 f\right) p_+(f), \tag{13}$$

where $\lambda > 0$ is a smoothing parameter and $p_+(f) = 1$ if $f \geq 0$. Otherwise $p_+(f) = 0$. Thus, the non-normalized conditional density of f given the observations y is

$$p(f \mid y) \propto \exp\left(y^{\mathrm{T}} \ln(Hf) - \mathbf{1}^{\mathrm{T}} Hf - \tfrac{1}{2} \lambda f^{\mathrm{T}} D_2^{\mathrm{T}} D_2 f\right) p_+(f) \tag{14}$$

and it is called the posterior density. The posterior density characterizes the solution and it can be understood as the solution of the inverse problem. However, if the dimension of the problem is large, the posterior density may not be very illustrative. Thus, in order to describe the solution in a more convenient way, different point estimates and confidence intervals are computed from the posterior density. The most commonly used point estimates are the maximum a posteriori (MAP) estimate and the conditional mean (CM). The MAP estimate is the maximizer of the posterior density and it can be understood as the most probable solution given the observations and the prior model $p(f)$. In this case the MAP estimate is obtained as the non-negative maximizer of the exponent term, i.e.

$$\hat{f}_{\mathrm{MAP}} = \arg\min_{f \geq 0} \left(y^T \ln(Hf) - \mathbf{1}^{\mathrm{T}} Hf - \tfrac{1}{2} \lambda f^{\mathrm{T}} D_2^{\mathrm{T}} D_2 f\right). \tag{15}$$

The constrained optimisation problem can be solved using e.g. the interior point algorithm.[20] The CM estimate is defined as

$$\hat{f}_{\mathrm{CM}} = \int_{\mathbf{R}^m} f\, p(f \mid y)\, df. \tag{16}$$

In high-dimensional problems analytic evaluation of the integral (16) is impossible in practice and it can be evaluated numerically using Markov chain Monte Carlo (MCMC) methods.[21] The key principle of MCMC methods is to draw (dependent) samples from the posterior density and approximate the mean as the sample mean.

A simulation study was carried out in which artificial DMPS data were generated corresponding to a situation in which the observed particle counts were low. First, the MAP estimate and the Tikhonov regularized solution, see Eq. (5), were determined using parameter $\lambda = 2 \times 10^{-3}$. The estimates are shown in Fig. 1. The weight matrix L_1 was chosen as $L_1^T L_1 = \Gamma_\varepsilon^{-1}$ and Γ_ε was approximated as $\Gamma_\varepsilon = \mathrm{diag}(y_1 + 1, y_2 + 1, \ldots, y_n + 1)$. It can be shown that Tikhonov regularization with the standard smoothness side constraint and the non-negativity constraint yields the MAP estimate corresponding to Gaussian observation errors and the prior density (13). As can be seen, the estimates differ from each other especially in small particle sizes, which is due to different statistical assumptions and it can be shown that the MAP estimate is on average closer to the actual size distribution than the Tikhonov solution.[15] As the counts become higher the difference between the MAP estimate and the Tikhonov regularized solution decreases. The CM estimate and the 90% confidence intervals obtained with an MCMC algorithm are shown in Fig. 2. The tail in the lower diameter end is due to the non-negativity constraint since the actual size distribution is very close to the boundary of the feasible region ($f \geq 0$) and at the same time the variances of individual components of f are large.

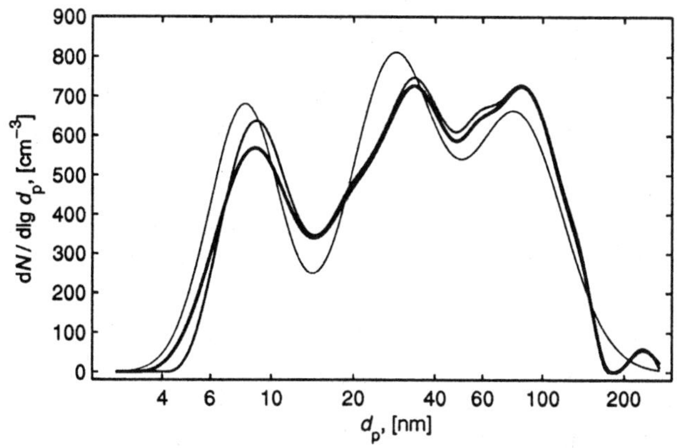

Figure 1. The actual (simulated) size distribution (thin line), the MAP estimate (medium line) and the Tikhonov regularized solution (thick line).

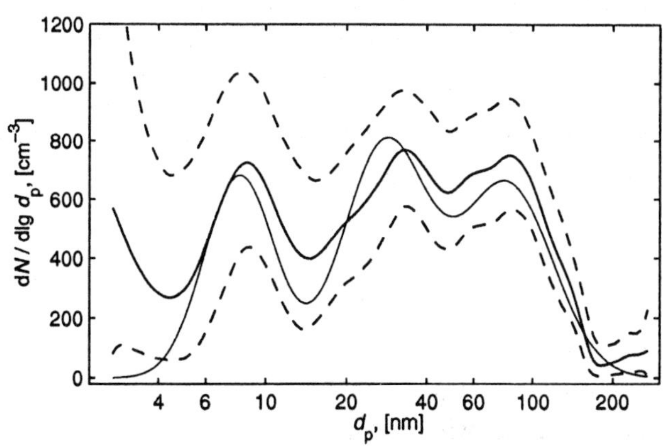

Figure 2. The actual (simulated) size distribution (thin line), the CM estimate (medium line) and the 90% confidence intervals (dashed lines).

5. NON-HOMOGENEOUS REGULARIZATION METHOD

The Tikhonov regularization with the standard smoothness side constraint usually yields reliable estimates in the reconstruction of atmospheric size distributions from DMPS data. However, in certain situations the use of the standard smoothness side constraint does not necessarily guarantee an accurate estimate. The second-order difference matrix has an equally strong smoothing effect on the whole size range, which means that by increasing the value of the regularization parameter λ the estimate becomes smoother on the whole size range and vice versa. However, aerosol size distributions are not always homogeneously smooth, i.e. the absolute value of the second derivative (or the second difference in the discrete case) of the size distribution function may have very large values on certain size ranges while it may be very small in other size intervals. Then the prior information included to the problem is infeasible since the side constraint has large value at the point corresponding to the actual size distribution and the estimate may not be acceptable. Situation of this kind is encountered e.g. when the size distribution to be estimated has very narrow peaks compared to the whole size range that is measured.

The problem associated with the use of the standard smoothness side constraint was briefly discussed by Wolfenbarger and Seinfeld.[7] They also proposed an approach to overcome the problem. They used standard side constraints that were modified by using special weighting functions and the modified smoothing constraint would be of the form $\Phi(f) = \| WD_2 f \|^2$, where W is a diagonal weight matrix and the diagonal elements represent a discrete weight function. Unfortunately, the selection of a proper weight matrix may be very difficult and no algorithm was given for these purposes. However, if certain prior requirements are set for the weight function, estimates for the weight function and for the size distribution function can be obtained simultaneously. Since the differences between the elements of W may be of several orders of magnitude, the weight matrix is expressed as $W = \mathrm{diag}(10^{\beta_1}, 10^{\beta_2}, \ldots, 10^{\beta_{m-2}}) \doteq 10^{\beta}$. First, it is assumed that the logarithm of the weight function, i.e. the (discrete) function β is smooth. Secondly, we assume that the weight function is close to 1, i.e. the elements of β are assumed to be close to zero. Prior assumptions about the logarithm of the weight function can be incorporated to the problem using standard constraints $\Phi_1(f) = \| D_{2,\beta} \beta \|^2$ and $\Phi_2(f) = \| \beta \|^2$, where $D_{2,\beta}$ is a second-order difference matrix with proper dimensions. Thus, the new functional to be minimized is

$$\Xi(f,\beta) = \left\| L_1(Hf - y) \right\|^2 + \lambda \left\| 10^{\beta} D_2 f \right\|^2 + \gamma_1 \left\| D_{2,\beta} \beta \right\|^2 + \gamma_2 \left\| \beta \right\|^2 \tag{17}$$

and it is minimized with respect to both f and β. Since the size distribution is non-negative, the solution is sought subject to $f \geq 0$. The relative weights of the terms in the right-hand side are controlled with regularization parameters λ, γ_1, and γ_2. The minimization of functional (17) is an inequality constrained non-linear least squares problem that can be solved with inequality constrained Gauss-Newton recursion or with the interior or exterior point algorithms.[5,20]

We simulated Tandem-DMA (two DMAs in series) measurements in which the size distribution to be reconstructed had a narrow peak compared to the whole size range. The regularization parameters were $\lambda = 3\times10^{-1}$, $\gamma_1 = 900$ and $\gamma_2 = 1.0$. The estimate for the size distribution and the logarithm of the weight function are shown in Fig. 3. As it can be seen the estimate is very close to the actual size distribution and the norm of the estimation error is $\| f_{\text{actual}} - \hat{f}_{\Xi} \| = 304.4$. For comparison, the minimum estimation error that can be obtained with homogeneous regularization technique, i.e. using the Tikhonov regularization with the standard smoothness constraint, is $\| f_{\text{actual}} - \hat{f}_{\lambda} \| = 530.3$. The Tikhonov regularized solution possesses features of both under- and over-regularization, i.e. the estimate is oscillative on the smooth parts of the actual size distribution while on the region of the peak the solution is too smooth.

6. DYNAMICAL INVERSION METHOD

In all reconstruction methods the actual size distribution is assumed to be time-invariant during each measurement cycle. However, in general the size distribution functions are time-varying quantities due to various non-stationary particle sources and since many physical and chemical processes have effects on particle size and concentration. Thus, if the duration of the measurement cycle is long and if temporal changes in the size distribution are rapid, the estimate obtained with conventional methods may be inaccurate. The time-evolution of the size distribution can be taken into account by using the state space representation of the measurement system. The time-evolution of the size distribution can be described with the difference equation

$$f_{t+1} = F_t f_t + \omega_t , \qquad (18)$$

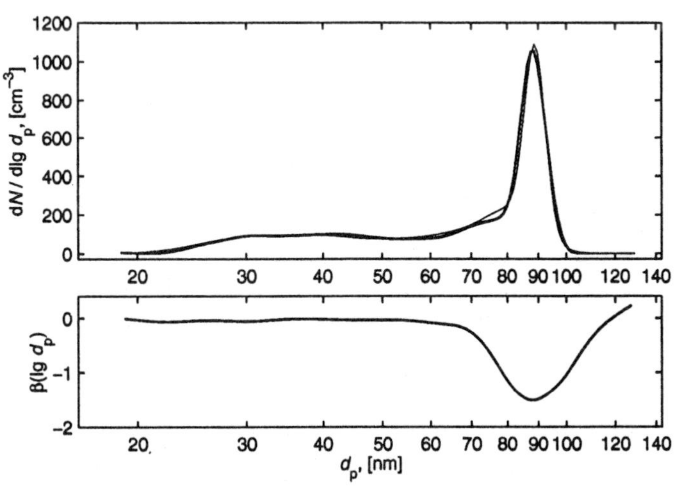

Figure 3. The actual particle size distribution (thin line) and the estimate obtained with the non-homogeneous regularization method (thick line). The estimated weight function β is shown in the lower figure.

where F_t is a state transition matrix and ω_t is a zero-mean Gaussian state noise process with covariance $\Gamma_{\omega,t}$. The subscript t refers to a discrete time index. In the DMPS system observations are obtained one by one and the relation between the unknown size distribution and an observation can be approximated as

$$y_t = H_t f_t + \varepsilon_t, \tag{19}$$

where H_t is a row vector representing the integral operator, see Eq. (3), and ε_t is a zero-mean Gaussian observation error. Equations (18) and (19) are called the state space representation of the system. In on-line data processing the objective is to determine an estimate for the size distribution f_t based on observations $y_1, y_2, \ldots, y_t$. An optimal estimate in the minimum mean squares sense can be determined with the Kalman filter algorithm.[22] The Kalman filter equations can be written as

$$\begin{aligned} G_t &= \Gamma_{t|t-1} H_t^{\mathrm{T}} (H_t \Gamma_{t|t-1} H_t^{\mathrm{T}} + \Gamma_{\varepsilon,t})^{-1} \\ f_{t|t} &= f_{t|t-1} + G_t (y_t - H_t f_{t|t-1}) \\ \Gamma_{t|t} &= \Gamma_{t|t-1} - G_t H_t \Gamma_{t|t-1} \\ f_{t+1|t} &= F_t f_{t|t} \\ \Gamma_{t+1|t} &= F_t \Gamma_{t|t} F_t^T + \Gamma_{\omega,t} \end{aligned} \tag{20}$$

where $f_{t|t}$ and $f_{t|t-1}$ are the estimates computed based on the sets $\{y_1, \ldots, y_t\}$ and $\{y_1, \ldots, y_{t-1}\}$, respectively, and matrices $\Gamma_{t|t}$ and $\Gamma_{t|t-1}$ are their covariances. The estimates can also be computed using all the observations $\{y_1, \ldots, y_M\}$ and then the estimates for each state are obtained by performing a backward run with equations

$$\begin{aligned} A_{t-1} &= \Gamma_{t-1|t-1} F_{t-1}^{\mathrm{T}} \Gamma_{t|t-1}^{-1} \\ f_{t-1|M} &= f_{t-1|t-1} + A_{t-1} (f_{t|M} - f_{t|t-1}) \\ \Gamma_{t-1|M} &= \Gamma_{t-1|t-1} + A_{t-1} (\Gamma_{t|M} - \Gamma_{t|t-1}) A_{t-1}^{\mathrm{T}} \end{aligned} \tag{21}$$

This is called the fixed-interval smoother algorithm.[23]

We constructed a time-varying size distribution, see Fig. 4, and generated realistic DMPS measurement data. The data were generated such that 10 measurement cycles were performed and each cycle consisted on 24 channels. The duration of each cycle was 20 min = 1200 s. The estimates were determined with Kalman filter and fixed-interval smoother algorithms. The estimates were computed using the random-walk evolution model, i.e. $F_t \equiv I$. The state noise covariance was $\Gamma_{\omega,t} \equiv 3 \times 10^4 I$ and $\Gamma_{\varepsilon,t} = y_t + 1$. Since the problem is severely underdetermined, spatial regularization was applied in order to obtain smooth estimates.[24] In the spatial regularization the smoothing parameter was $\lambda = 1 \times 10^{-4}$ and the covariance of artificial observations was set to $\Gamma = 10 I$. The non-negativity constraint was also used. For comparison, the Tikhonov regularized solutions were also determined using smoothing parameter $\lambda = 4 \times 10^{-2}$. The norms of the

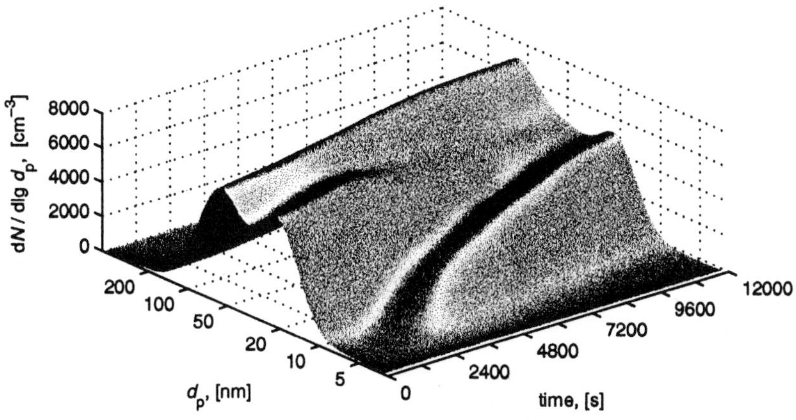

Figure 4. The time-evolution of the true size distribution function.

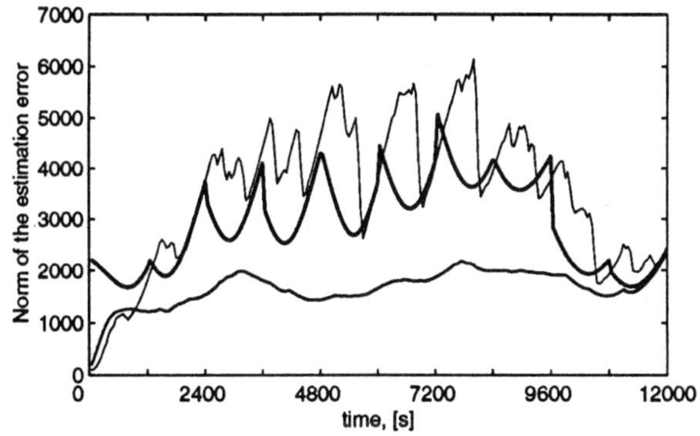

Figure 5. The norms of the estimation errors of the estimates obtained with the Kalman filter (narrow line), fixed-interval smoother (medium line) and the Tikhonov regularization (thick line).

estimation errors are shown in Fig. 5. The fixed-interval smoother yields most accurate results and in this case the Tikhonov regularized solutions are more accurate than the estimates obtained with the Kalman filter despite the time resolution is poorer. The accuracy of the estimates can be improved by using more realistic evolution models.

7. CONCLUSIONS

Several computational methods have been proposed for solving the ill-posed problem arising in the aerosol size distribution measurements and some techniques were reviewed in this chapter. In addition, three novel reconstruction methods proposed by the authors were briefly described. Since the problem is usually severely underdetermined, the solution is not unique. In order to stabilize the problem some further prior information on the solution must be incorporated to the problem. Usually it is assumed that the actual size distribution can be expressed in certain parametric form or the solution is presumed to be a smooth function. However, in certain measurement situations, information e.g. on the time-evolution or non-homogeneous smoothness properties should be exploited to improve the accuracy of the estimates. Even though only aerosol size distribution reconstruction methods were considered here, the methods can, in principle, be used for solving similar problems arising in other applications.

REFERENCES

1. J.H. Seinfeld and S.N. Pandis, *Atmospheric Physics and Chemistry. From Air Pollution to Climate Change* (John Wiley & Sons, 1998).
2. M. Kandlikar and G. Ramachandran, Inverse methods for analysing aerosol spectrometer measurements: a critical review, *J. Aerosol Sci.* **30**, 413-437 (1999).
3. T.G. Dzubay and H. Hasan, Fitting multimodal lognormal size distributions to cascade impactor data, *Aerosol Sci. Tech.* **13**, 144-150 (1990).
4. O.G. Raabe, A general method for fitting size distributions to multicomponent aerosol data using weighted least-squares, *Env. Sci. Tech.* **12**, 1162-1167 (1978).
5. Å. Björck, *Numerical Methods for Least Squares Problems* (SIAM, 1996).
6. U. Amato, D. Di Bello, F. Esposito, C. Serio, G. Pavese, and F. Romano, Intercomparing the Twomey method with a multimodal lognormal approach to retrieve the aerosol size distribution, *J. Geophys. Res. D* **101**, 19,267-19,275 (1996).
7. J.K. Wolfenbarger and J.H. Seinfeld, Inversion of aerosol size distribution data, *J. Aerosol Sci.* **21**, 227-247 (1990).
8. V.S. Bashurova, K.P. Koutzenogil, A.Y. Pusep, and N.V. Shokhirev, Determination of atmospheric aerosol size distribution functions from screen diffusion battery data: mathematical aspects, *J. Aerosol Sci.* **22**, 373-388 (1991).
9. U. Amato, M.R. Carfora, V. Cuomo, and C. Serio, Objective algorithms for the aerosol problem, *Appl. Opt.* **34**, 5442-5452 (1995).
10. S. Twomey, Comparison of constrained linear inversion and an iterative nonlinear algorithm applied to the indirect estimation of particle size distributions, *J. Comput. Phys.* **18**, 188-200 (1975).
11. M.T. Chahine, Determination of the temperature profile in an atmosphere form its outgoing radiance, *J. Opt. Soc. Am.* **58**, 1634-1637 (1968).
12. G. Ramachandran and M. Kandlikar, Bayesian analysis for inversion of aerosol size distribution data, *J. Aerosol Sci.* **27**, 1099-1112 (1996).
13. E.F. Maher and N.M. Laird, EM algorithm reconstruction of particle size distributions from diffusion battery data, *J. Aerosol Sci.* **16**, 557-570 (1985).
14. P. Paatero, *The Extreme Value Estimation Deconvolution Method with Applications in Aerosol Research,* Technical Report No. HU-P-250, University of Helsinki, Department of Physics (1990).
15. A.Voutilainen, V. Kolehmainen, and J.P. Kaipio, Statistical inversion of aerosol size measurement data, *Inv. Probl. Eng.* (2001), in press.
16. A. Voutilainen, F. Stratmann, and J.P. Kaipio, A non-homogeneous regularization method for the estimation of narrow aerosol size distributions, *J. Aerosol Sci.* **31**, 1433-1445 (2000).
17. A. Voutilainen and J.P. Kaipio, Estimation of non-stationary aerosol size distributions using the state-space approach, *J. Aerosol Sci.* (2001), in press.
18. W. Winklmayr, G.P. Reischl, A.O. Lindner, and A. Berner, A new electromobility spectrometer for the measurement of aerosol size distributions in the size range from 1 to 1000 nm, *J. Aerosol Sci.* **22**, 289-296 (1991).
19. TSI Inc. (St. Paul, MN, USA, January 10, 2001); http://www.tsi.com
20. A.V. Fiacco and G.P. McCormick, *Nonlinear Programming: Sequential Unconstrained Minimization Techniques* (SIAM, 1990).
21. W.R. Gilks, S. Rickhardson, and D.J. Spiegelhalter, editors, *Markov Chain Monte Carlo in Practice* (Chapman & Hall, 1996).
22. C.K. Chui and G. Chen, *Kalman Filtering* (Springer-Verlag, 1987).
23. B.D.O Anderson and J.B. Moore, *Optimal Filtering* (Prentice-Hall, 1979).
24. J. Kaipio and E. Somersalo, Nonstationary inverse problems and state estimation, *J. Inv. Ill-Posed Problems* **7**, 273-282 (1999).

TWO DISPERSE PARTICLES IN THE FIELD OF THE ELECTROMAGNETIC RADIATION

Irina V. Krivenko, Aleksei V. Klinger and Ludmila A. Uvarova*

1. INTRODUCTION

In this work we investigated the heat transfer in the system of two disperse particles. It was received the solution of the stationary Maxwell's equatons for the system of two spherical absorbent particles within the framework of Mie theory. On this basis it was found the expressions for the heat sources and the solution of the heat transfer equation for these sources.

2. SOLUTIONS OF MAXWELL'S EQUATIONS AND THE HEAT PROBLEM IN THE SYSTEM OF TWO PARTICLES

The problem of control of the disperse system with the aid of the electromagnetic radiation (the laser) is one of the actual problems of the physics of disperse systems. The heat source, coused by the electromagnetic radiation, may be represented as the electromagnetic energy, absorbed with particle per unit of volume in the unit of time. As a result, there is the heating of the particles and the evaporation or the heat explosion, in dependence on various conditions.

If take into cosideration the system of only two disperse particles in the external field, it is possible to receive the exact analytical solutionsof the equations, describing the electromagnetic interacting and the heat transfer such systems in number of cases.

The choice of the electromagnetic interaction model of the two spherical absorbent particles system in the electromagnetic radiation field depends on the

Irina.V.Krivenko, Aleksei.V.Klinger, Tver Technical State University, Tver, 170026, Af. Nikitin emb., Russia
Ludmila A. Uvarova, Moscow State University of Technology "STANKIN", Moscow, 103055, Vadkovskii st., 3a, Rassia

Mathematical Modeling: Problems, Methods, Applications
Edited by Uvarova and Latyshev, Kluwer Academic/Plenum Publishers, 2001

parameters of the system of the particles and the external radiation. If the wavelength of the initiative plane wave is a lot over than the size of the spheres, it is possible to consider that the electromagnetic field is the steady within the particle volume. So it can be connect the bispherical coordinate system with the system of spheres and solve the Laplace equation in this coordinates[1,2]. The exact solution of the problem of the scattering and absorption of the electromagnetic radiation on single particle has been received by J. Mie in 1908. This work became the basis of large number of publications on this problem, for example[3-6]. Subsequently the Mie`s theory and using of the modern mathematical methods allowed to consider the problem of scatteringand absorption of the electromagnetic radiation by an ensemble of spherical particles [7-13]

In this work it is considered a disperse system model, including two spherical particles with the different optical parameters, which are placed in the field of the plane electromagnetic wave with wavelength λ.

Two spherical particles with radii R_1 ani R_2 are placed in a homogeneous medium at the distance O_1 O_2 (Fig.1) . Z axis is along the line joining centers of the particles. The wave vector **k** of initiative plane wave forms the θ angle with the positive direction of the z axis. The Decart's coordinate system x'y'z' is connected with the center O_1 of the first particle. The z' axis is along **k** vector, the x' axis is along the electric vector **E**, the y' axis is along the magnetic vector **H**. The Decart`s coordinate systems x_1 y_1 z and x_2 y_2 z are connected with the centers of the spheres (the paires of the axes x_1 and x_2, y_1 and y_2 are parallel). The direction of the x_1 axis is choosed thus, that vector **k** was lay at the plane x_1O_1z, and y_1-axis was along so, that coordinate system x_1y_1z was right. The spherical coordinate systems $r_1\theta_1\varphi$, $r_2\theta_2\varphi$, $r`\theta`\varphi`$ are connected with the same Decart`s systems. The straight line of intersection of planes x_1O_1z and $x'O_1y'$ is indicated through L' at the Fig.1. The electric vector **E** forms the angle β with the plane $\varphi = 0$ (β is the angle between the lines L' and x').

Supposing the time dependence of **E** and **H** vectors according to ω is described by the factor $e^{-i\omega t}$, can be written in the following way for the amplitudes of **E** and **H** vectors:

$$\nabla^2 \mathbf{E} + k^{(j)2} \mathbf{E} = 0, \tag{1}$$

$$\nabla^2 \mathbf{H} + k^{(j)2} \mathbf{H} = 0, \tag{2}$$

where $k^{(j)}$ is the wave number in the j medium. The superscript j has meaning 0, 1, 2 (0 refers to surrounding medium, 1 and 2 to the first and the second particles accordingly).

The influence of the neighbouring particle may be taken into account of the following way: the external with regard to the particle electromagnetic field $\mathbf{E}_j$ can be represented with the sum of two components: the incident plane wave radiation field **E** and the field, scattered from the neighbouring particle $\mathbf{E}_k^S$

$$\mathbf{E}_j = \mathbf{E} + \mathbf{E}_k^S \tag{3}$$

The subskripts j and k takes the meaning: j=1,2; j=k.

Using the Mie theory[14], it is possible to express the electric **E** and magnetic **H** vectors in terms of the Debye potentials. In the spherical coordinate systems, connected with the center of every particle, the Debye potentials of an incident radiation will be represented with the following way

$$U^{i} = \frac{1}{k^{(0)}} \sum_{l=1}^{\infty} i^{l-1} \frac{2l+1}{l(l+1)} j_l(k^{(0)} r_1) P_l^{1}(\cos\theta')\cos\phi' \;, \tag{4}$$

the Debye potentials, corresponding to the scattered by sphere with number j wave

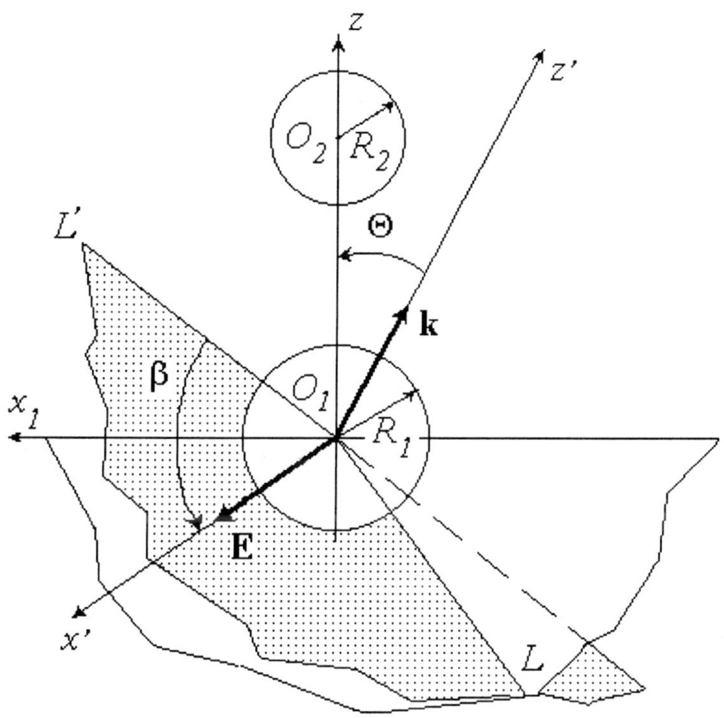

Figure 1. The coordinate systems, connected with two particles

$$U_j^{s} = \frac{1}{k^{(0)}} \sum_{n=o}^{\infty} \sum_{m=-n}^{n} B_{nm}^{(j)} h_n(k^{(0)} r_j) P_n^{m}(\cos\theta_j) e^{im\phi} , \tag{5}$$

and the Debye potentials inside of the sphere j

$$U_j^w = \frac{1}{k^{(j)}} \sum_{n=o}^{\infty} \sum_{m=-n}^{n} D_{nm}^{(j)} j_n(k^{j0)} r_j) P_n^m(\cos\theta_j) e^{im\phi} . \tag{6}$$

There superscript w refers to the incident field, s - to the scattered field, w - to the absorbed radiation; B and D - independent coefficients, that can be determined from the boundary conditions.

The similar expressions may be written for the magnetic Debye potentials V. The boundary conditions at the surface of the sphere j can be expressed in terms of scalar potentials thus

$$\frac{\partial}{\partial r_j}\left\{r_j(U^j + U_k^s + U_j^s)\right\} = \frac{\partial}{\partial r_j}\left\{r_j U_j^w\right\} \Big|_{r_j} = R_j, \tag{7}$$

$$k_j^{(0)}\left\{r_j(U^j + U_k^s + U_j^s)\right\} = k_j^{(0)} r_j U_j^w \Big|_{r_j} = R_j. \tag{8}$$

The boundary conditioms can be written for magnetic Debye potentials by analogous way. In the expressions (7), (8) the Debye potentials of the incident radiation are expressed in the coordinate system x'y'z' and Debye potentials of the particle k scattering radiation - in the coordinate system $x_k y_k z$ (k = 1,2). For the solution of this problem it is nessesary to represent the Debye potentials of the particle k scattering radiation as the own spherical functions of j - sphere expansion $P_n^m(\cos\theta_j)e^{im\phi}$. That transformation is carried out with using of the methods of groop theory. Three angles of the turn $\pi/2 - \beta$, θ, $3\pi/2$ are determined by the rotation g relatively point O_1, that transfers the coordinate system x'y'z' to the system x_1y_1z. So, for Legendre polynomials, including in Eq. (4), it can be received the next expression (for j=1).

$$P_n^1(\cos\theta')\cos k\phi' = \sum_{l=-n}^{n} L_{1l}^n P_n^l(\cos\theta_1)e^{il\phi} \tag{9}$$

The coefficients L^n_{1l} are given by

$$L_{1l}^n = \frac{1}{2}\left(\frac{(n+1)!(n-l)!}{(n-1)!(n+l)!}\right)^{1/2} i^{1-m}\left(e^{-i\beta}P_{1,l}^n(\cos\theta) + \right. \tag{10}$$

$$\left. +e^{i\beta}P_{1,-l}(\cos\theta)\right)$$

The function P^l_{mn} is determined[15]. Taking into account, that it can be matched the coordinate systems x1y1z and x2y2z with the parallel carrying along the z axis[16], it can be possible to find the Debye potentials of the scattering on the k - th particle radiation field in the form of the infinite sums of j - th particle coordinates functions. Further, determing the independent coefficients $B^{(j)}_{lm}$, $D^{(j)}_{lm}$ [17] and using the connection of the Debye potentials with **E** and **H** vectors, it may be find the values of field vectors in the any point of the spherical particle.

Note, that in the case, when it is necessary to take into account the weak dependence of the dielectric permittivity of the particle substance on the field, it may be done on the basis of the received solution by using of the successive approximation method.

The density value of the heat sources in the any point inside of the particle j is determined from following equation[18-19]

$$q_j = 4pn_j m_j |\mathbf{E}|^2 \, I / n_0 \lambda \tag{11}$$

where n is the index of refraction, m is the absorption coefficient, λ is the electromagnetic initiative radiation wavelength, I is the intensity of the initiative radiation. So, the value of q_j can be determined with using of the early received solution of the electromagnetics problem.

The temperature distribution in the system can be determined from the solution of the quasylinear Laplace equation (for continuous medium) and the Poisson equation (inside the particles) with the boundary conditions of the fourth kind:

$$\nabla(k_j \nabla T_e) = 0 \tag{12}$$

$$\nabla(k_j \nabla T_j) = -q_j \tag{13}$$

$$k_e \left.\frac{\partial T_e}{\partial r_j}\right|_{r_j=R_j} = k_j \left.\frac{\partial T_j}{\partial r_j}\right|_{r_j=R_j} \tag{14}$$

$$\left.T_e\right|_{r_j\to\infty} = T_\infty \tag{15}$$

$$\left. T_e - K_{T_j} l \frac{\partial T_e}{\partial r_j}\right|_{r_j=R_j} = \left. Tj \right|_{r_j = R_j} \tag{16}$$

Where k_e is the coefficient of the heat conductivity in the surrounding media, k_j is the coefficient of the heat conductivity inside of the particle j,

$k_j = k_{j0} f_j(T)$, K_{Tj} is the temperature jamp, l is the average free length in the surroundings. In the case of the same functional dependences of heat conductivity coefficients on the temperature under the condition $l/R_j << 1$, it is possible to transform the Eqs. (12) - (16) into the linear form for the determination of φ by using of the Kirchhoff's substitution[20] $d\varphi = f(T)dT$.

If the dependence of the heat conductivity coefficients on temperature is not taking into account, the Eqs. (12) - (16) will be linear and it will be possible to find solution in bispherical coordinate system[21].

It was received the expressions for the temperature inside of particles:

$$T_j = \frac{\sqrt{2}R_j^2 q_j}{3k_j}\sqrt{ch\xi - \cos\eta}\sum_{n=0}^{\infty} e^{-(n+1/2)\xi} \times$$

$$\times\left[2sh^2\xi_j(n+\frac{1}{2})(cth|\xi_j| - cth|\xi|) + \frac{k_j}{k_e} + \frac{q_k}{q_j}\left(\frac{R_k}{R_j}\right)^2 e^{(2n+1)\xi_k}\right] \times \tag{17}$$

$$\times P_n(\cos\eta) + T_{e\infty}.$$

Where η, ξ are the bispherical coordinates. The surface of the sphere is the coordinate surface in the bispherical coordinate system. The equations of the surfaces of the first and second spheres in the bispherical coordinate system can be written accordingly in the following form: ξ_1 = const, ξ_2 = const. There are taking place the next correlations:

$$ch\xi_j = \frac{R^2 + R_j^2 - R_k^2}{2RR_j}, \quad R_1 = \frac{a}{sh\xi_1}, \quad R_2 = -\frac{a}{sh\xi_2},$$

where a is the polar distance. Thus, having received the real distributions of the heat sources, caused by the electromagnetic radiation influenced on the particle system, it can be possible to define the real heat fields in the same system. If the difraction parameters of particles are less than one, then the heat sources distributions will be near to homogeneous at the certain directions of the incident wave (it will be discussed below). It is possible to average the heat sources density with regard to the particle volume. Then it will be possible to use the Eq. (17) for the calculation of the temperature in the any point inside of the particle, where q_j will be the average value of the heat sources density.

3. THE RESULTS OF CALCULATIONS

The calculations of the heat sources density distribution on the cross sections of the particles, the average values of the heat sources density and the temperature in the concrete aerosol systems were carried out on the basis of the received solutions. The diagrams of the distribution of the lectric vector amplitude square on the cross section of the water aerosol particle in the second particle presence are shown in Fig. 2. The wavelength of the incident radiation is equal to 10.63 mcm. The radii of the particles are equal to 1 mcm, 3 mcm. The complex refraction parameters of every particle are equal to 1.173 + 0.0823 i. Let's designate d through the value $R/(R_1 + R_2)$. It was considered the different kinds of the arrangements of the line joining centers of the spheres with regard to the vectors **k**, **E**. The diagrams, characterizing the $|E|^2$ distribution into the cross section of the first particle at the spherical coordinate $\theta_1 = 90^0$ at the different values of spherical coordinates φ, r_1 and the Eulerian angles ϑ, β and d = 5, are shown in the Fig. 2 (a, b, c). With the decrease of r_1, the energy distribution into the particle cross section becames more symmetrical, and exactly for the cases (a) and (b) the diagrams are symmetrical "figure - of - eight" and for the case (c) it is a circumference. So one can see from Fig. 2 that at the angles $\theta = 90^0$, $\beta = 0^0$ the distribution of $|\mathbf{E}|^2$ into the cross section of the particle is most even, therefore exactly for this case for the heat problem solution it is possible to take into consideration the average value of the heat sources density as the heat source.

It was considered the systems of the two equal spherical carbon particles with radii equal to 0.06 mcm in the helium - neon laser field with the wavelength equals to 0.6328 mcm. The complex carbon refractive parameter equals to 1.96 + 0.82 i. In the Fig. 3 there are represented the diagrams of $|\mathbf{E}|^2$ distribution into the cross cross section $\theta_1 = 90^0$ of the first particle at $\theta = 90^0$, $\beta = 0^0$, d =5. The diagram has "the teeth" near the surface of the particle. With the decreasing of r_1 "the teeth" smoothes out - the energy distributes even on the cross section of the particle.

The diagrams of the electric vector amplitude square distribution inside the carbon particle with radius 0.06 mcm for the various values of the radius of the second particle are shown in Fig. 4. There are no "the teeth" on the diagrams at the small radii of the first particle (smaller than 0.03 mcm).

It was received the diagrams with "the teeth" for the water aerosol particles too. "The teeth" were observed for the particles with radii more, than 2 mcm for initiative radiation wavelength equals to 10.6 mcm.

The calculations of the average values of the heat sources densities at the volume in the system of two spherical water drops with radii 1 mcm, and with the distance between the centers 20 mcm were carried out on the basis of the solution of the electrodynamics problem. The particles were placed in the air. The wavelength of the laser radiation, induced the heat transfer, was equal to 10.63 mcm.

It was denoted over q* the value $q^* = q/I$, where q is the heat source, I is the laser intensity.There were carried out the calculations for the various positions of the two particles system with regard to the vectors **E**, **H** and **k**, characterizing by the angles θ and β. The results of the calculations are represented in the Tab. 1.

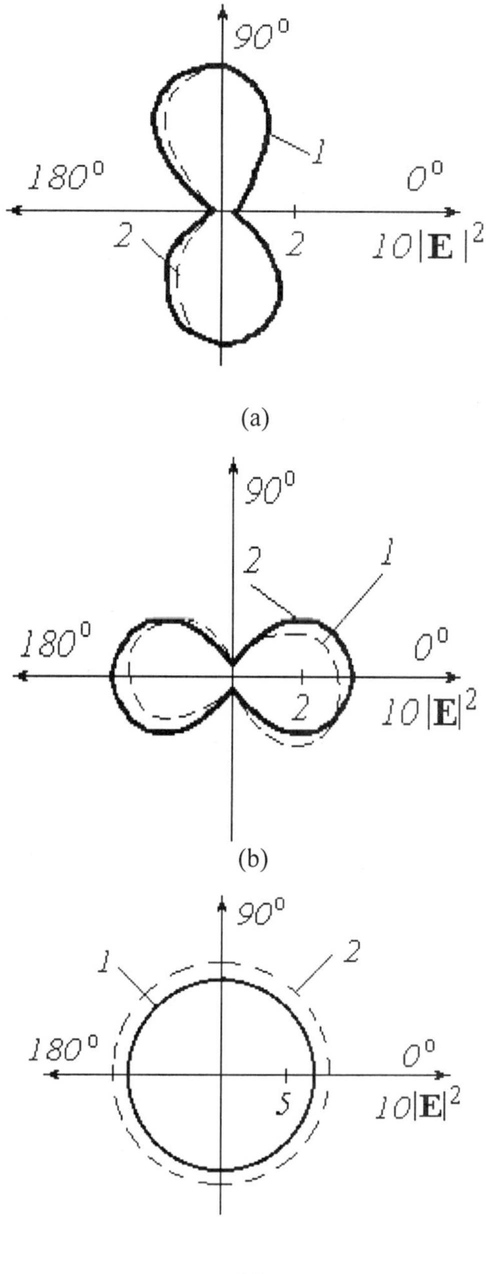

Figure 2. The diagrams, characterizing the dependence of $|\mathbf{E}|^2$ on the angle φ inside the particle of the water aerosol at d = 5 and $r_l = 0.9R_l$ (1) and $r_l = 0.5R_l$ (2); $\theta = \beta = 90^0$ (a); $\theta = 90^0$, $\beta = 5^0$ (b); $\theta = 90^0$, $\beta = 0^0$ (c).

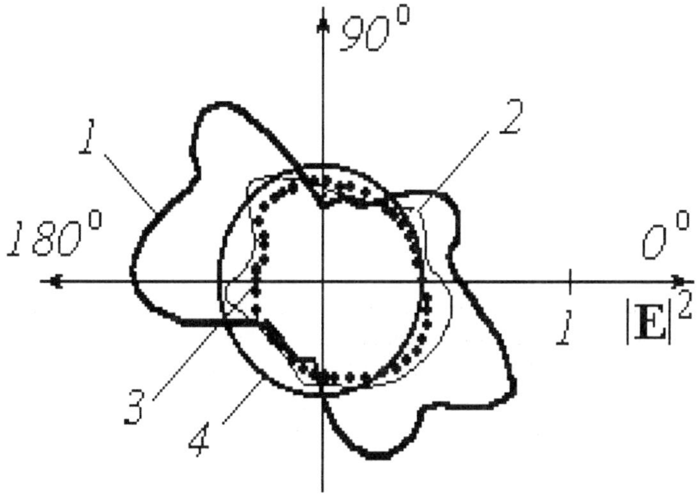

Figure 3. The diagrams, characterizing the dependence of the electric vector amplitude square $|\mathbf{E}|^2$ on the angle φ inside the carbon particle ($\theta_1 = 90^0$). The curves (1), (2), (3), (4) are build for the following r_1 values: $0.9R_1$, $0.8R_1$, $0.7R_1$, $0.1r_1$ respectively

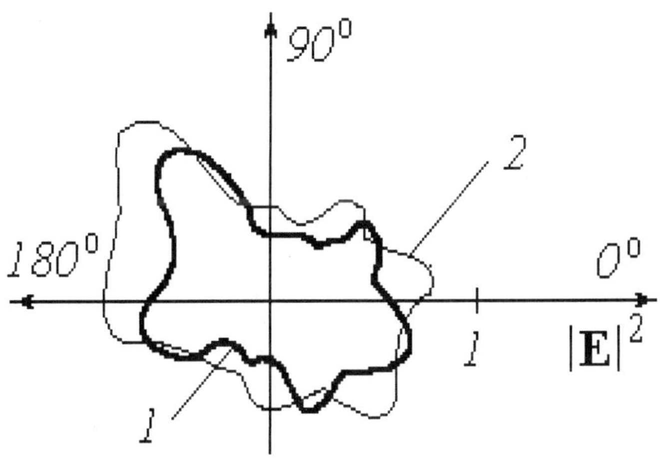

Figure 4. The diagrames, characterizing the dependence of the electric vector amplitude square $|\mathbf{E}|^2$ on the angle φ inside carbon particle ($\theta_1 = 90^0$) at the various values of the radius of the second particle. The curve (1) corresponds to the value $R_2 = 0.06$ mcm, the curve (2) - $R_2 = 0.01$ mcm. The distance between the centers of the particles equals to 3.6 mcm

Table 1. The dependence q* on the angles θ and β

θ	β	$q^{*} \bullet 10^{-4}$, м$^{-1}$
90^0	0^0	8.657
90^0	90^0	5.652
45^0	45^0	6.336

According to the presented calculations, the average value of the heat sources density significantly depends on the position of two particles system with regard to the wave vector **k** and the electric fild vector, induced heat transfer. The calculations of the temperature inside first particle for the same particle system by using Eq. (17) for the angles $\theta = 90^0$, $\beta = 0^0$ were carried out. For it there were used the results, given in the Tab.1. In the Tab. 2 there were listed the values $\Delta T = T_1 - T_e$, where T_1 is the temperature in the

Table 2. The distribution of the temperature into the cross section of the water aerosol particles at the various intensities of the laser radiation

I, MW/m^2	θ_1	ΔT, K	ΔT', K
10	0^0	12.98	0.97
	90^0	12.01	
	180^0	11.74	
	270^0	12.01	
40	0^0	51.92	4.96
	90^0	48.05	
	180^0	46.96	
	270^0	48.05	
60	0^0	77.88	5.80
	90^0	72.08	
	180^0	70.45	
	270^0	72.08	
70	0^0	90.86	6.77
	90^0	84.09	
	180^0	82.12	
	270^0	84.09	

choosed point M on the surface of the particle, T_e is the temperature in surroundings. According to the Tab. 2, ΔT depends on the angle θ_1 (the coordinate in spherical coordinate system, connected with the center of the first particle), and this dependence is increasing with the rise of the intensity of the laser radiation, inducing the heat transfer.

Table 3. The values of temperature in the point M

R_2, mcm	T,K	T_s, K	T, K	T_s, K
	$I = 10^6$ W/m^2		$I = 10^7$ W/m^2	
1	274.26	274.20	285.63	284.99
1.1	274.29	274.21	285.93	285.09
1.2	274.33	274.22	286.30	285.20
1.3	274.37	274.23	286.74	285.34
1.4	274.43	274.25	287.25	285.49

There is indicated through $\Delta T'$ the value $\Delta T' = \Delta T_0 - \Delta T_{90}$, where $\Delta T_0 - \Delta T_{90}$ are the differences of the temperatures in the points, corresponding to the angles $\theta_1 = 0^0$ and $\theta_1 = 90^0$. The value $\Delta T'$ is increasing with the rise of the laser intensity, so then the temperature distribution into the particle cross section becames more nonuniform. The temperature of surroundings is 273 K. In the Tab. 3 there were listed the results of the calculation of the temperature in the point M on the surface of the first particle (the coordinate of the point is in the bispherical coordinate system $\eta = 180^0$). It was supposed, that $T_e = 273$ K. The values of the laser intensities were varied. It was indicated through T_s the value of the temperature in the same point, calculated for the average heat sources density, but without the consideration of the heat influence of the second particle. The radius of the first particle equals to 1 mcm, the second - R_2. The distance between the centers of the particles equals to 20 mcm. From the Tab. 3, in particular, follows, that the influence of the neighbouring particle becames noticeably with increasing of the laser radiation intensity.

4. CONCLUSION

The carried out calculating experiments shown, that depending on the direction of the radiation spreading with regard to the line joining the particles centers, the physical - chemical properties and the size of the particles a) the influence of the second particle on the distribution of the absorptive energy inside the first particle may be considerable in the "large" ($R/(R_1 + R_2) >> 1$) distances between particles also; b) the diagrams, characterizing the distribution of the absorptive energy density into the particle cross section, have various kind: the circle, "figure - of - eight", "the figure with teeth". "The teeth" take place on the diagrams, when the size of the particle is more of some critical value, which depends on the concrete conditions.

The carried out calculations for vrious aerodisperce systems shown, that at the angles $\theta = 90^0$, $\beta = 0^0$ the value of $|\mathbf{E}|^2$, and, consequently, the heat source densities, considerably depend only on radial coordinate r inside the particle and are more close to homogeneous.

In the case when the wave vector **k** is directed perpendiculary to the line joining the particle centers, the average of the heat sources with regard to the volume of the particle is well - founded.

The executive computations shown, that with the rise of the size of the second particle and with increasing of the laser radiation intensity, the temperature difference on the particle surface and in surroundings is increasing, and the distribution of the temperatures into the particle cross section becames more inhomogeneous.

The discussed in this work model of the interaction of two particle system with the external laser radiation makes it possible to predict the distribution of the absorptive energy and the temperature inside particles, by varying of various parameters of the system of the particles and initiative radiation.

5. ACKNOWLEDGEMENTS

The authors were partly supported in this work by the Russian Foundation for Basic Research (RFBR No. 00-01-00674) and by the project of Federal program "Integration" №A 0106 – 2.1

REFERENCES

1.P.K. Aravind, A. Nitzan. Interaction between electromagnetic resonances, Surf. Sci., v.110, 189-204 (1981).
2. R. Ruppin. Optical absorption of two spheres, J. Of Phys. Soc. of Japan, v.58, 1446-1451 (1989)
3. P.A. Wyatt. Scattering of electromagnetic plane waves from inhomogeneous spherically symmetric objects. Phys. Rew, v.127, 1837-1843 (1962).
4. A.Bott, W.Zdunkowski. Electromagnetic energy within dielectric spheres, J. Opt. Soc.Am. A., v.4, 1361-1365 (1987).
5. A.P. Prishivalko, Optical and heat fields inside radiation scattering particles (Nauka i technica, Minsk, 1983) (in Russian).
6. N.N. Belov. Distribution of radiation field inside spherical particle, Reports of RAS, v.292, 1360 –1363 (1987) (in Russian).
7.E.A. Ivanov. Scattering of electromagnetic plane wave by an ensemble of spheres, Dif. Uravnen. , v.2, 1285-1297 (1969) (in Russian).
8. K.A. Fuller, J.W. Kattawar. Consumate solution to the problem of classical electromagnetic scattering by an ensemble of spheres. 1.: Linear chains, Opt. Letters, v.13, 90-92 (1988).
9. K.A. Fuller, J.W. Kattawar. Consumate solution to the problem of classical electromagnetic scattering by an ensemble of spheres. 2.: Clusters of arbitrary configuration, Opt. Letters, v.13, 1063-1065 (1988).
10. B.U. Felderhof, R.B. Jones. Addition theorems for spherical wave solutions of the vector Helmholtz equation, J. Math. Phys., v.28, 836-839 (1987).
11. Yu.A. Yeremin, N.V. Orlov, V.I. Rozenberg. Analysis of wave scattering by an ensemble of particles, Optica i spectroskopiya, v.75, 115 –122 (1993) (in Russian).
12. I.V.Krivenko. Calculation of interaction of electromagnetic radiation with spherical absorbent particles, *Mathematical methods in chemistry,* 137-144 (Tver State University , 1994) (in Russian).
13. N. I. Gamayunov, I.V. Krivenko, L.A. Uvarova, Yu.Z. Bondarev. Peculiarities of electromagnetic radiation transport in aerosol systems "particle – continuum medium

– particle", *Proceedings of International Aerosol Symposium IAS-2,* v.**1**, 27 (Moscow, 1995).
14. M. Born and E. Wolf, Principles of Optics (Pergamon Press, New York., 1959).
15. N.Ya. Vilenkin, 1956, The special functions and theory of the representation of groups, Science, Moscow, 588 p.
16. V.I. Rozenberg,, Scattering and relaxation of the electromagnetic radiation by atmospheric particles (Gidrometizdat, Leningrad, 1972).
17. N.I. Gamayunov, I.V. Krivenko, L.A. Uvarova, Yu.Z. Bondarev. The peculiarities of the spreading of electromagnetic radiation and initiated by it heat transfer in the system aerosol particles - surrounding medium, Journal of Physical Chemistry, v. 71(12), 2270 – 2274 (1997).
18. A.P. Prishivalko, L.G.Astaf'eva. Distribution of energy inside radiation scattering particles, *Preprint (Institute of Physics, Minsk, 1974) (in Russian).*
19. N.N.Belov. Structure of optical field inside spheres and heat explosion of particles, *Fizika goreniya i vzryva*, v.**4**, 44-48 (1987) (in Russian).
20. A.V. Lykov, *Theory of heat transfer* (Vysshaya Shkola, Moscow, 1967) in Russian.).
21. E.R. Shchukin, N.V. Malay. The photophoretical and thermodiffusiophoretical motion of aerosol particles, *Engineering - Physical Journal*, v.. **54** (4) , 628 – 634 (1988) (in Russian).

TRANSPORT PROCESSES IN AERODISPERSE SYSTEMS: TRANSITIONAL GROWTH OF NON-SPHERICAL PARTICLES AND MOBILITY OF IONS

Alexey B.Nadykto[1]

1.INTRODUCTION

This paper consists of three different problems related to the transport processes in the aerodisperse systems. Transitional growth rates for non-spherical particles, derived by the modified boundary sphere method (see *e.g.* Fuchs (1964), Seinfeld and Pandis (1998)), are presented in the second section. The third section discusses a transitional condensational growth of a multicomponent non-spherical particle and the fourth section considers a mobility of ions.

2.TRANSITIONAL CONDENSATIONAL GROWTH OF A NON-SPHERICAL PARTICLE IN BINARY GAS MIXTURE

Ice particles, growing from humid air, which contain supersaturated water vapour, play an important role in mixed and ice clouds' formation in the atmosphere. Diffusion of the condensable vapour to the particle surface, governing (together with the surface physics) the growth process, limits the rate at which the particle can growth. Particles in the atmosphere can grow in continuum, transition or free molecular regimes, depending on their size and ambient conditions. When the particle size is sufficiently large compared to the mean free path of the diffusing vapour molecules, the particle condenses or evaporates in the continuum regime. In this case the mass flux to/from the particle surface can be derived by solving the equations of continuity for mass and heat flows. Continuum transport equations are no longer valid, when the particle size becomes comparable with mean free path. Vapour transport in this regime, which is called transitional, is governed

[1]Department of Applied Physics, University of Kuopio, P.O. Box 1627, FIN-70211 Kuopio, Finland

rigorously by the Boltzmann equation. Since solving the Boltzmann equation (see Sitarsky and Nowakovsky (1979), Loyalka (1983)) is very complex mathematically, there does not exist general solution to this equation as well as a theory, accurately predicting the condensation rates of particles of simplest spherical shape, which is valid for full range of particle sizes for arbitrary masses of the diffusing species and surrounding gas (see Williams and Loyalka (1993), Seinfeld and Pandis (1998)). Particles, which are small compared to the mean free path, condense or evaporate in free molecular regime. Simple kinetic theory of the molecular transport can be applied in this case (Hirshfelder *et al.* (1954)).

In cloud physics, the continuum regime theory, suggested by Pruppacher and Klett (1978), is used to predict the growth rates of snow and ice particles at atmospheric conditions. Fuchs and Sutugin (1970), Dahneke (1983) and Loyalka (1983) shown theoretically and their predictions were confirmed by laboratory experiments (see *e.g.* Lie and Davis (1995)), that at room temperature and pressure about 10^5 Pa, typical for low troposphere, the continuum regime theory can adequately (with overestimation not more 4 % in mass fluxes) predict the condensation rates of spherical water droplets of about > 1 μm in size. The value of this size limit, changing due to changes in ambient pressure and temperature, increases as the height increases, reaching about 5-7 μm in ***Cs*** and more than 10-15 μm stratospheric clouds. This means that quite a big part of the atmospheric aerosol particles grow in either transitional or free molecular regimes, in which the continuum regime equations are not valid. However, accurate theoretical description of the particle growth processes is important, as models used for simulating the cloud formation are sensitive to growth rates, which strongly influence the formation and properties of clouds.

The purpose of this study is to modify the Fuchs theory to predict the growth of non-spherical atmospheric aerosol particles with taking all growth regimes into account. The motivation to develop this more generalised form of the Fuchs theory derives from numerous observations and laboratory experiments, which indicated the overestimation of the growth rates (*e.g.* Pruppacher and Klett (1997)), given by existing condensation models.

As has been pointed out above, quite a big part of non-spherical atmospheric aerosol particles can grow or evaporate in the transitional and free molecular regimes. For the spherical droplets a transitional regime theory was presented for the first time by Fuchs (1964), who matched the free molecular and continuum mass fluxes at the boundary of the Knudsen layer and derived the correction factor, accounting for all growth regimes, to the continuum regime mass flux, in the following form:

$$I = \beta_M I_c, \tag{1}$$

where

$$\beta_M = \left[\frac{0.75\alpha\left(1 + Kn\left(\frac{\Delta}{\lambda}\right)\right)}{0.75\alpha + Kn + \left(\frac{\Delta}{\lambda}\right)Kn^2} \right], \tag{2}$$

where α is condensation coefficient, I is actual mass flux, I_c is continuum mass flux, β_M is transitional correction factor, Kn is the Knudsen number and Δ is the length of the Knudsen layer. As may be seen, equation (1) rightly predicts mass fluxes in both free molecular and continuum regimes with arbitrarily chosen value of Δ. Dahneke (1979) suggested $\Delta = \lambda$, where λ is a means free path of the vapour molecules and reached quite good agreement with results of Fuchs (1964) and Sutugin (1970). Note that only theories by Fuchs (1964), Fuchs and Sutugin (1970) and Dahneke (1979) are consistent for full range of the Knudsen numbers and condensation coefficients. Theories proposed by Sitarski and Novakovski (1979) and Loyalka (1983) give incorrect results near the free molecular regime. Application of the boundary conditions with temperature and (or) concentration jumps also gives incorrect solution in the free molecular regime.

To start the derivation consider a convex motionless particle of arbitrary shape surrounded by a Knudsen layer, which has the same shape and space orientation as particle shape and orientation. We also specify Δ as an average normal distance between the particle surface and the boundary of the Knudsen layer, assuming Δ uniform for whole range of the particle shapes. After that one matches the free molecular and continuum mass fluxes at the boundary of the Knudsen layer.

Free molecular mass flux is given (Shchukin *et al.* (1979)) by

$$I_{fm} = \alpha S\left(\frac{\mu_v}{2\pi R_g}\right)^{\frac{1}{2}}\left[\left(\frac{p_{vS}}{T_s^{\frac{1}{2}}}\right)-\left(\frac{p_{v1}}{T_1^{\frac{1}{2}}}\right)\right], \tag{3}$$

where T_s is particle surface temperature, T_1 is a temperature at the boundary of the Knudsen layer, S is particle surface area, p_{vS} is vapour pressure at the particle surface and p_{v1} is condensable vapour pressure at the boundary of the Knudsen layer. Equations for continuum mass flux to the particle surface and to the boundary of the Knudsen layer are given (*e.g.* Nadykto *et al.* (2000)) by

$$I_{cb} = C_{t-c}F_b\left(\frac{D\mu_v}{R_gT_\infty}\right)\left[p_{v1} - p_{v\infty}\right], \tag{4}$$

where D is diffusion coefficient, T_∞ is ambient gas temperature, $F_b = 4\pi C_b$ is shape factor (see *e.g.* Pruppacher and Klett (1997)), C_b is capacitance of a body, which bounds the Knudsen layer, $C_{t\text{-}c}$ is correction factor, accounting for the temperature and composition dependence of the transport coefficients. Correction factor $C_{t\text{-}c}$ can be regarded as unity at the atmospheric conditions because the atmospheric aerosols are weakly non-isothermal ($|(T_S - T_\infty)/T_\infty| << 1$) dilute gas mixtures (*e.g.* Fuchs (1959), Seinfeld and

Pandis (1998)). Matching the free molecular and continuum mass fluxes at the boundary of the Knudsen layer, one obtains, after some algebra, the following equation

$$I = F_b(\Delta)\left[\frac{\psi(\Delta)}{1+\psi(\Delta)}\right]\left(\frac{D\mu_v}{R_g T_\infty}\right)\left[p_{vs} - p_{v\infty}\right], \tag{5}$$

where

$$\psi(\Delta) = \left[\frac{\alpha S}{F_b(\Delta)}\right]\left(\frac{\bar{c}_A}{4D}\right), \tag{6}$$

where $\bar{c}_A$ is mean speed of the vapour molecules. Equation (6) can be rewritten as

$$I = \beta_M^{(n)} I_c, \tag{7}$$

where

$$\beta_M^{(n)} = \frac{F_b(\Delta)}{F_p}\left[\frac{\psi(\Delta)}{1+\psi(\Delta)}\right], \tag{8}$$

is the transitional correction factor for a particle of arbitrary shape. Notice that in case of spherical particle of radius R, $F_p = 4\pi R$, $F_b(\Delta) = 4\pi(R+\Delta)$ and Eq. (7) directly reduces to Eq.(1), derived by Fuchs (1964).
Employing $\Delta = \lambda$ and making use of Eq. (7), one obtains

$$I = \beta_M^{(n)} I_c, \tag{9}$$

where

$$\beta_M^{(n)} = \frac{F_b(\lambda)}{F_p}\left[\frac{\psi(\lambda)}{1+\psi(\lambda)}\right], \tag{10}$$

Now one has, instead two equations describing just the free molecular and continuum regimes, a single equation, predicting the growth rates of a non-spherical particle in all growth regimes. Although Eq.(10) gives accurate limits in both continuum and free molecular regimes, one has to perform an additional testing to confirm the validity of the derived theory in the transitional regime. Since no data on and theoretical predictions of the growth/evaporation rates of non-spherical particles in the transitional regime have been published, spherical particles are considered. Comparison of the transition corrections given by different theories (this study, Fuchs and Sutugin (1970), and Dahneke (1983)) as functions of the Knudsen numbers, given in Nadykto (2001) and Nadykto *et al.* (2001), shows that the theory presented in this paper agrees well with theory by Fuchs and

Sutugin (1970). The theory by Dahneke (1983) gives a bit smaller values of the correction compared to both Fuchs and Sutugin (1970) and this study mainly due to special definition of the mean free path $\lambda = 2D / \bar{c}_a$.

To derive the equation for the change in particle mass, one has to multiply the equation (9) by ventilation coefficient (see Pruppacher and Klett (1998)) and apply the mass conservation law to obtain

$$\frac{dM}{dt} = \bar{f}_v \beta_M^{(n)} I_c , \tag{11}$$

where $\bar{f}_v$ is ventilation coefficient and M is particle mass.

3.CONDENSATIONAL GROWTH OF A MULTICOMPONENT NON-SPHERICAL PARTICLE

This section presents a model of the growth rates of a multicomponent non-spherical particle. The model is based on analytical expressions for the growth rates of a non-spherical particle in the non-isothermal binary gas mixture, obtained in previous section. An analytical expression for the particle surface temperature is derived (see also Nadykto (2001)) by eliminating of the particle surface temperature from the equation for the energy conservation (Mattila *et al.* (1997) and Lehtinen *et al.* (1998)). Transition correction, derived in the last section, is applied to describe the growth of a non-spherical particle of arbitrary size.

When a snow crystal evaporates or grows in humid air, there are two dominant mechanisms (see *e.g.* Mason (1971), Hobbs (1974)) that govern the evaporation or growth rate: diffusion of water vapour to/away from to the droplet surface and different surface effects, with influence on the efficiency of mass accommodation. Both surface physics of ice crystals and crystal morphologies are very complicated. At the moment there is no theory, which could account for a full description of such phenomena. In this chapter the surface effects are not considered in detail.

To accurately describe the growth processes in multicomponent systems, one has to solve a set of coupled differential equations, describing heat and mass transfer processes. The problem is very complicated mathematically and thus, there does not exist a general solution to this problem. However, growth processes in atmospheric clouds typically occur in dilute gas mixtures, in which the vapour –vapour interaction and coupling between mass fluxes of the species can be neglected. At such conditions, the diffusion of the species i in a multicomponent gas mixture can be approximated by the binary diffusion of the species being considered in a carrier gas (*e.g.* Mattila *et al.* (1997)).

Assuming temperature gradients in the aerodisperse system low enough to regard the thermal conductivity of the gas mixture and all binary diffusion coefficients as constants, one obtains, (after the multiplication by transitional correction factor) the expression for mass flux of species i in a multicomponent gas mixture

$$I_i = F\frac{\beta_{M,i} D_{im} M_i (p_{is}(T_S) - p_{iv\infty}(T_\infty))}{R_g T_\infty}, \tag{12}$$

where p_i is a partial vapour pressure of species i, D_{im} is the binary diffusion coefficient , characterizing the diffusion of the species i in a carrier gas. The vapour pressure at the droplet surface can be approximated using Clausius-Clapeyron equation

$$\frac{dp_i^{(sat)}}{dT} = \frac{p_i^{(sat)} L_i M_i}{R_g T^2}, \tag{13}$$

Assuming temperature dependence of latent heat to be linear

$$L_i(T) = L_{io} + L_{i1} T, \tag{14}$$

where L_{io} and L_{i1} are constants, one obtains

$$\frac{p_i^{(sat)}(T_s)}{p_i^{(sat)}(T_\infty)} = exp\left[\frac{L_{io} M_i}{R_g T_\infty}\frac{(T_S - T_\infty)/T_\infty}{1 + (T_S - T_\infty)/T_\infty} + \frac{L_{i1} M_i}{R_g} ln\left[1 + \frac{(T_S - T_\infty)}{T_\infty}\right]\right]. \tag{15}$$

Expanding the right hand side to a Taylor series and keeping the first three terms, one obtains

$$p_{iS}(T_s) = S_{i,s} p_i^{(sat)}(T_s) \approx S_{i,s} p_i^{(sat)}(T_\infty)$$

$$\left[1 + \frac{L_i(T_\infty) M_i}{R_g T_\infty}\frac{(T_S - T_\infty)}{T_\infty} + \frac{1}{2}\left[\left(\frac{L_i(T_\infty) M_i}{R_g T_\infty}\right)^2 - 2\frac{L_i(T_\infty) M_i}{R_g T_\infty} + \frac{L_{i1} M_i}{R_g}\right]\left(\frac{T_S - T_\infty}{T_\infty}\right)^2\right]. \tag{16}$$

Inserting the right hand side of the equation (16) into (12), one obtains the following equation

$$I_i = -F\frac{D_{im} M_i p_i^{(sat)}(T_\infty)}{R_g T_\infty}$$

$$\left\{S_{i,\infty} - S_{i,s}\left[1 + \frac{L_i(T_\infty) M_i}{R_g T_\infty}\frac{(T_S - T_\infty)}{T_\infty} + \frac{1}{2}\left[\left(\frac{L_i(T_\infty) M_i}{R_g T_\infty}\right)^2 - 2\frac{L_i(T_\infty) M_i}{R_g T_\infty} + \frac{L_{i1} M_i}{R_g}\right]\left(\frac{T_S - T_\infty}{T_\infty}\right)^2\right]\right\}. \tag{17}$$

Equation (17) can be rewritten as

$$I_i = a_0 + a_1 x + a_2 x^2, \tag{18}$$

where

$$x = \frac{\left(T_S - T_\infty\right)}{T_\infty}, \tag{19}$$

$$a_0 = -F\frac{\beta_{M,i}D_{im}M_i p_i^{(sat)}(T_\infty)}{R_g T_\infty}\left(S_{i,\infty} - S_{i,s}\right), \tag{20}$$

$$a_1 = F\frac{\beta_{M,i}D_{im}M_i^{\,2}p_i^{(sat)}(T_\infty)L_i(T_\infty)}{R_g^{\,2}T_\infty^{\,2}}S_{i,s}, \tag{21}$$

$$a_2 = \frac{1}{2}F\frac{\beta_{M,i}D_{im}M_i p_i^{(sat)}(T_\infty)}{R_g T_\infty}S_{i,s}\left[\left(\frac{L_i(T_\infty)M_i}{R_g T_\infty}\right)^2 - 2\frac{L_i(T_\infty)M_i}{R_g T_\infty} + \frac{L_{i1}M_i}{R_g}\right], \tag{22}$$

If no internal heat release and radiant heat exchange occur in the aerodisperse system, the particle surface temperature can be found from the following equation

$$\int_{T_\infty}^{T_S}\kappa dT + \sum_{i=1}^{n}L_i(T_S)\frac{I_i}{F} = 0. \tag{23}$$

Assuming κ to be constant and taking (18)-(22) into account, one obtains

$$A_2 x^2 + \left(A_1 + A_{11}\right)x + A_0 = 0 \tag{24}$$

$$A_0 = \sum_{i=1}^{n}\frac{L_i(T_\infty)a_0}{F}, \tag{25}$$

$$A_1 = \sum_{i=1}^{n}\frac{L_i(T_\infty)a_1}{F}, \tag{26}$$

$$A_2 = \sum_{i=1}^{n}\frac{L_i(T_\infty)a_2}{F}, \tag{27}$$

$$A_{11} = \kappa T_\infty, \tag{28}$$

The solution of equation (24) is given by

$$x = \frac{-(A_1 + A_{11}) + \sqrt{(A_1 + A_{11})^2 - 4A_0 A_2}}{2A_2}. \tag{29}$$

Equation (29) can be inserted directly to equation (12) to obtain completely analytical expression for the mass fluxes of different condensing species. The vapour pressure of species i at the particle surface can be expressed as

$$p_{iS} = p_i^{(sat)}(T_s) f_i \chi_{iS}, \tag{30}$$

where f_i is activity coefficient for species i and

$$\chi_{iS} = \frac{n_{iS}}{\sum\limits_{n=1}^{n} n_{iS}}, \tag{31}$$

where n_{iS} is amount of moles of species i in the particle.

The presented model is derived in closed completely analytical form with explicit accounting for all growth regimes, internal heat release and radiation and , thus, it can be used for a wide class of problems related to atmospheric physics such as formation and properties of clouds, fogs and smogs. Expressions obtained here can be utilized directly a part of the solution routines, including **GDE** (Seinfrld (1986)) solution routine. No time-consuming solution routines are needed and the solution thus becomes much faster.

4.MOBILITY OF IONS: REVISED KELVIN-THOMSON EQUATION AND MILLIKAN-FUCHS CORRELATION

Different mobility-size relationships for nanoparticles and clusters has been considered by numerous authors (*e.g.* Hopke *et al.* (1992), Ramamurthi and Hopke (1989), Tammett (1995), Makela *et al.* (1996)). As usual, the mobility diameter (the size derived from the ion mobility data with some mobility -size relationship) compares with the Kelvin -Thomson diameter. The Kelvin-Thomson diameter derives from Kelvin-Thomson equation for ion induced nucleation

$$ln S = \frac{M}{\kappa T \rho}\left(\frac{4\gamma}{D_p^*} - \left[1 - \frac{1}{\varepsilon_r}\right]\frac{(ie_o)^2}{2\pi^2 \varepsilon_o D_p^{*4}}\right), \tag{32}$$

where S is saturation ratio, M is a mass of the vapour molecules, ρ is a liquid density, ε_0 is vacuum permittivity, ε_r is relative permittivity of a liquid, γ is a surface tension, e_o is elementary charge, i is a number of elementary charges in a cluster, T is a temperature and D_p^* is equilibrium particle diameter. When S<1, only one real solution of equation (32) exists. This is called Kelvin-Thomson diameter (see also Makela *et al.* (1996)) in this paper.

However, the Kelvin - Thomson equation includes incomplete description of effect of the particle charge on the equilibrium vapour pressure. More accurate theory was derived by Korshunov (1980), who accounted for a polarisation of the vapour molecules in the vicinity of the charged particle and derived a revised Kelvin -Thomson equation in the following form

$$ln S = \frac{M}{\kappa T\rho}\left(\frac{4\gamma}{D_p^*} - \left[1 - \frac{1}{\varepsilon_r}\right]\frac{(ie_o)^2}{2\pi^2\varepsilon_o D_p^{*4}}\right) - \frac{1}{2\pi^2\kappa T\varepsilon_0^2}\left[\alpha + \frac{l^2}{3\kappa T}\right]\frac{(ie_o)^2}{D_p^{*4}}, \quad (33)$$

where l is dipole moment of the vapour molecule. For the definition of α see well - known Lorentz formula (*e.g.* Tamm (1976)). Additional effect of the particle charge can be (Korshunov (1980)) much higher than the Thomson effect. Therefore,. this effect should be taken into account. However, existing comparisons (see *e.g.* Makela *et al.* (1996)) of the mobility -equivalent size with theoretical particle size are performed neglecting the effect obtained by Korshunov (1980). This can be a reason for the existing disagreement between mobility-equavalent size and theoretical particle size, indicated in *e.g.* Makela *et al.* (1996).

A treatment of the experimental data by Makela *et al.* (1996) shows that the theoretical particle size, derived from the equation (33), agrees with the mobility-equivalent diameter within less than 3.5 % for ethanol, isopropanol and n-butanol and within less than 15 % for acetone, when the Millikan-Fuchs (*e.g.* Makela *et al.*(1996)) mobility-size relationship is used. This means the Millikan-Fuchs relationship can provide quite good correlation between ion mobility and ion size when the particle charge effect is taken into account properly.

5.ACKNOWLEDGMENTS

Support of this work by the Academy of Finland is gratefully acknowledged. The author is also indebted to Prof. A.Laaksonen, Prof. E.R.Shchukin and Dr. J.Makela for fruitful discussions.

REFERENCES

1. Dahneke, B.(1983) Simple kinetic theory of Brownian diffusion in vapors and aerosols, in *Theory of Dispersed Multiphase Flow,* (Edited by R.E.Meyer), Academic Press, New York, 97-133
2. Fuchs, N.A. (1959), Evaporation and Growth of Droplets in a Gaseous Media, Pergamon Press, New-York
3. Fuchs, N.A. and Sutugin, A.G.(1970) *Highly Dispersed Aerosols*, Ann Arbor Science Publishers., Ann Arbor.
4. Fuchs , N.A. (1964). Mechanics of Aerosols, Pergamon Press, New-York.

5. Hirschfelder J.O., Curtiss C.F. and Bird, R.B. (1954). Molecular Theory of Gases and Liquids, Wiley, New York.
6. Hobbs, P.V. (1974), Ice Physics, Clarendon Press, Oxford
7. Hopke, P.K. *et al.* (1992). *Health Physics*, **63**, 560
8. Korshunov , V.K. (1980).O ravnovesii zaryzenoi kapli s parom, *Izv.AN USSR*, Fizika atmofery i okeana, **16** (1), 92-94
9. Lehtinen K., Kulmala M., Vesala T. and Jokiniemi J. (1998). Analytical method to calculate condensation rates of a multicomponent droplet, *J. Aerosol Science*, **29**, 1041-1052.
10. Lie, W. and Davis, E.J. (1995) Aerosol evaporation in the transitional regime, *Aerosol Sci. Technol.*, **25**, 11-21
11. Loyalka, S.K.(1983) Modeling of condensation on aerosols. *Prog. Nucl. Energy*, 12, 1-8
12. Makela J.M. *et al.* (1996). Comparison of mobility equivalent diameter with Kelvin-Thomson diameter using ion mobility data. J. Chem. Phys., 105(4), 1562-1571
13. Mason B.J.(1971). *The Physics of Clouds*, 2 ed. Clarendon Press, 671
14. Mattila T., Kulmala M. and Vesala T. (1997). On the condensational growth of a multicomponent droplets. *J. Aerosol Science*, **28**, 553-564.
15. Nadykto, A.B. (2001). Ph.D. thesis, University of Kuopio, Kuopio, Finland.
16. Nadykto, A.B. and E.R.Shchukin (1999). Vaporization and growth of aerosol particles, given internal heat release and radiant heat exchange. In *Mathematical Models of Non-Linear Excitations, Dynamics, Transfer and Control in Condensed Systems and Other Media* (Edited by Uvarova L.A. *et al.*), Kluwer Academic/Plenum Publishers, New York
17. Nadykto, A.B., Shchukin E.R., M.Kulmala and A.Laaksonen. A generalised reformulation of Fuchs condensation theory to predict a transitional condensational growth of ice particles, (to be published).
18. Pruppacher , H.R. and Klett, J.D. (1978) Microphysics of Clouds and Precipitation, Kluwer Academic/Plenum Publishers, Dordgerht/ Boston/London.
19. Pruppacher H.R. and Klett, J.D. (1997) Microphysics of Clouds and Precipitation, Second edition, Kluwer Academic Publishers, Dordgerht/ Boston/London.
20. Ramamurthi, M. and Hopke, P.K. (1989). *Health Phys.*, **56**, 189
21. Seinfeld , J.H. (1986), Atmospheric Chemistry and Physics of Air Pollution, Wiley, New York.
22. Seinfeld, J.H. and Pandis, S.N. (1998), Atmospheric Chemistry and Physics, Wiley, New York.
23. Shchukin E.R.. and Nadykto A.B (1999). Diffusive vaporization and growth of assermbly of N large particles. In *Mathematical Models of Non-Linear Excitations, Dynamics, Transfer and Control in Condensed Systems and Other Media* (Edited by Uvarova L.A. *et al.*), Kluwer Academic/Plenum Publishers, New York.
24. Shchukin E.R., Yalamov Y.I. and Bahtilov V.I. (1979). in *Fizika aerodispersnih system*, **4** (1), MRPI Publisher, Moscow, 130-143.
25. Sitarsky, M. and B.Nowakovski (1979). Condensation rate of trace vapor on Knudsen aerosols from solution of the Bolzmann equation, *J. Colloid Interface Sci.*, **72**, 113-122
26. Tamm, I.E. (1976). Osnovi Teorii Electrichestva, Nauka, Moskva.
27. Tammett, H. (1995). J.Aerosol Science, **25**, 459
28. Williams, M.M.R. and Loyalka, S.K. (1991). Aerosol Science: Theory and Practice, Pergamon, New York.

SOLUTION OF SOME NON-LINEAR PROBLEMS IN THE THEORY OF HEATING, VAPORIZATION AND BURNING OF SOLID PARTICLES AND DROPS

Eugene R. Shchukin *

1.ABSTRACT

There are many different applications where it is interesting to work out methods for the solution of non-linear systems of equations concerning of transport theory and the theory of heating, vaporization or burning of solid particles and drops. In this connection our paper presents solutions of some non-linear systems of equations and relevant boundary-value problems of transport theory.

2.MATHEMATICAL APPROACH TO THE SOLUTION OF SOME NON-LINEAR SYSTEM OF EQUATIONS

Consider first solution for the two boundary- value problems of transport theory . The problems were solved by method developed by the author (Shchukin, 1979 1995). In specific cases these problems describe the quasi steady state processes of evaporation (growth) (Shchukin, 1979, 1995; Shchukin et al. 1979, 1999, 2000; 1996) and heterogeneous burning (Shchukin 1982, 1988; Shchukin et al.1996) that occur at high temperature difference by both-single and array of particles with the arbitrary surface shapes S_j.

The first boundary- value problem has the following form:

* Eugene R. Shchukin Institute for High Temperature Russian Academy of Science, Moscow, Russia, 127412

$$\operatorname{div}\left(\sum_{a=1}^{F} f_{k,a}^{(c)}\nabla c_a + f_k^{(T)}\nabla T_e\right) = 0, \operatorname{div}\left(\sum_{a=1}^{F} \varphi_a^{(c)}\nabla c_a + \varphi^{(T)}\nabla T_e\right) = 0, \quad (2.1)$$

$$\operatorname{div}\left(\sum_{m=1}^{W} h_{b,m}^{(j)}\nabla \gamma_m^{(j)} + h_b^{(T,j)}\nabla T_i^{(j)}\right) = 0, \quad (2.2)$$

$$\mathrm{c}_a\big|_{S_j} = \mathrm{c}_{aS}^{(j)}\big|_{S_j}, \mathrm{T}_e\big|_{S_j} = \mathrm{T}_{i0}, \mathrm{T}_i^{(j)}\big|_{S_j} = \mathrm{T}_{i0}, \gamma_m^{(j)}\big|_{r_j} = \gamma_{m0}^{(j)}, \gamma_m^{(j)} \neq \infty, \ \mathrm{T}_i^{(j)} \neq \infty, \quad (2.3)$$

$$\sum_{m=1}^{W}\left(\Theta_{p,m}^{(j)}\nabla_{\perp}\gamma_m^{(j)} + \Theta_p^{(T,j)}\nabla_{\perp}T_i^{(j)}\right)\Big|_{S_j} = 0, \mathrm{c}_a\big|_{S^{(L)}} = \mathrm{c}_a^{(L)}, \ \mathrm{T}_e\big|_{S^{(L)}} = \mathrm{T}_e^{(L)}, \quad (2.4)$$

where indices a=1,…,F; k=1,…,F; b=1,…,W+1; m=1,…, W; p=1,…,W; j=1,…,N; function c_a and T_e depend on spatial coordinates in all points of the space out wards the closed domains bounded by the arbitrary surface S_j and $S^{(L)}$. The closed surface S_j are in a volume $V^{(L)}$ bounded by an arbitrary surface $S^{(L)}$. Coefficients $f_{k,a}^{(c)}$, $f_k^{(T)}$, $\varphi_a^{(c)}$, $\varphi^{(T)}$ are functions dependent on $c_1,\dots,c_F$ and T_e ; $h_{b,m}^{(j)}$, $h_b^{(T,j)}$, $\Theta_{p,m}^{(j)}$, $\Theta_p^{(T,j)}$ depend on $\gamma_1^{(j)},\dots,\gamma_W^{(j)}$ and $T_i^{(j)}$; $c_{aS}^{(j)} = c_{aS}(T_{i0}, \gamma_1^{(j)}, \dots, \gamma_W^{(j)})$; T_{i0}=const is known value ; $c_a^{(L)}$=const and $T_e^{(L)}$=const are known value of c_a and T_e at each point of the surface $S^{(L)}$; $\gamma_{m0}^{(j)}$ is known value of $\gamma_m^{(j)}$ at some point r_j inside S_j. Functions $T_i^{(j)}$ and $\gamma_m^{(j)}$ are defined in all points inside the closed domains restricted of surface S_j.

The boundary – value problem (2.1) – (2.4) can be solved analytically under the condition that $\gamma_m^{(j)}\big|_{r_j} = \gamma_{m0} = \mathrm{const}$. Then inside of the surfaces S_j $\gamma_m^{(j)} = \gamma_{m0}$, $T_i^{(j)} = T_{i0}$ and functions c_a depend only on T_e. In this from the equations (2.1) to equations (2.5) we arrive at

$$\operatorname{div}(\Omega_k \nabla T_e) = 0, \qquad \operatorname{div}(\Omega^{(T)}\nabla T_e) = 0, \quad (2.5)$$

where:

$$\Omega_k = \sum_{a=1}^{F} f_{k,a}^{(c)}\frac{dc_a}{dT_e} + f_k^{(T)}, \qquad \Omega^{(T)} = \sum_{a=1}^{F} \varphi_a^{(c)}\frac{dc_p}{dT_e} + \varphi^{(T)}. \quad (2.6)$$

Let us simplify the first equation from (2.5) representing it in the following form:

$$\operatorname{div}\left(\left(\Omega_k/\Omega^{(T)}\right)\left(\Omega^{(T)}\nabla T_e\right)\right) = 0. \quad (2.7)$$

Taking into account the second equation from (2.5) equations (2.7) lead to:

$$(\nabla T_e)^2 \frac{d}{dT_e}\left(\frac{\Omega_k}{\Omega^{(T)}}\right)=0, \text{ or } \frac{d}{dT_e}\left(\frac{\Omega_k}{\Omega^{(T)}}\right)=0. \quad (2.8)$$

The ratio $\Omega_k/\Omega^{(T)}$ depends only on T_e. Performing integration (2.8) with respect to T_e we get more simple equations (2.10):

$$\sum_{a=1}^{F} b_{k,a}\frac{dc_a}{dT_e}=d_k, \quad (2.10)$$

where B_k are constants of integration;

$$b_{k,a}=\left(f_{k,a}^{(c)}-\varphi_a^{(c)}B_k\right),\ d_k=\left(\varphi^{(T)}B_k-f_k^{(T)}\right).$$

Solving the system equations (2.10) for $\frac{dc_a}{dT_e}$ we obtain:

$$\frac{dc_a}{dT_e}=E_a, E_a=\frac{D_a}{D}, \quad (2.11)$$

$$D_a=\begin{vmatrix} b_{1,1} & \dots & d_1 & \dots & b_{1,F} \\ b_{2,1} & \dots & d_2 & \dots & b_{2,F} \\ \dots & \dots & \dots & \dots & \dots \\ b_{F,1} & \dots & d_F & \dots & b_{F,F} \end{vmatrix}, D=\begin{vmatrix} b_{1,1} & \dots & b_{1,F} \\ b_{2,1} & \dots & b_{2,F} \\ \dots & \dots & \dots \\ b_{F,1} & \dots & b_{F,F} \end{vmatrix},$$

where E_a are functions dependent on $c_1,\dots,c_F$, T_e and $B_1,\dots,B_F$.

With the aid of equation (2.11) and the second equation from (2.5) the boundary-value problem (2.1) – (2.4) is reduced to the following two more simple problems (2.12) and (2.13):

$$\frac{dc_a}{dT_e}=E_a, c_a\Big|_{T_e=T_{i0}}=c_{aS}\left(T_{i0},\gamma_{10},\dots,\gamma_{W0}\right), c_a\Big|_{T_e=T_e^{(L)}}=c_a^{(L)}, a=1,\dots,F; \quad (2.12)$$

$$\operatorname{div}\left(\Omega^{(T)}\nabla T_e\right)=0, T_e\Big|_{S_j}=T_{i0}, \ T_e\Big|_{S^{(L)}}=T_e^{(L)}, \quad (2.13)$$

The solution of the boundary - value problem (2.12) is performed independently on the problem (2.13). Performing either numerical or analytical solution of (2.12) one should find a dependence c_a on T_e and value of constants $B_{1\le k\le F}$ simultaneously. It should be noted that the boundary problem (2.12) is independent on the spatial coordinates.

Knowing the dependence c_a on T_e one can perform the integration of (2.13). In order to do this let us represent (2.13) in the form

$$\operatorname{div}\operatorname{grad}\psi_1^{(T)}(T_e)=0, \quad \psi_1^{(T)}(T_e)\big|_{S_j}=\psi_1^{(T)}(T_{i0}), \quad \psi_1^{(T)}\big|_{S^{(L)}}=0, \qquad (2.14)$$

$$\psi_1^{(T)}(T_e)=\int_{T_e^{(L)}}^{T_e}\Omega^{(T)}dT_e. \qquad (2.15)$$

It follows from (2.14) that

$$\psi_1^{(T)}(T_e)=\psi_1^{(T)}(T_{i0})U(x_f), \quad \psi_1^{(T)}(T_{i0})=\psi_1^{(T)}(T_e)\big|_{T_e=T_{i0}}, \qquad (2.16)$$

where x_f are coordinates of space points. Dependence $U(x_f)$ on x_f may be found by the solution of the Laplace equation under the boundary condition of the first kind

$$\Delta U=0, \qquad U\big|_{S_j}=1, \qquad U\big|_{S^{(L)}}=0. \qquad (2.17)$$

In case of infinite volume $V^{(L)}$ ($V^{(L)}=\infty$) the problem (2.17) for the function $U(x_f)$ is reduced to

$$\Delta U=0, \qquad U\big|_{S_j}=1, \qquad U\big|_{\infty}=0. \qquad (2.18)$$

If $F \geq 2$ the non-linear system of equations (2.12) can be integrated in quadratures when the function E_a can be represented in the following two forms:

$$E_a=\Delta^*\left\{\sum_{i=1}^{F}\Delta_{ai}c_i+\Delta_a\right\}, \qquad (2.19)$$

$$E_a=A_a^{(1)}c_a^{1+\alpha_a}-A_a^{(2)}c_a, \qquad (2.20)$$

where $\Delta^*=\Delta^*(T_e)$ is a function dependent on T_e and B_k; Δ_{ai} and Δ_a are constants dependent on B_k; $A_a^{(1)}$ and $A_a^{(2)}$ depend on T_e and B_k ; α_a are constant coefficients. In the first case after introducing new independent variable $\theta(T_e)=\int_{T_e^{(L)}}^{T_e}\Delta^* dT_e$ .then the non-linear boundary-value problem (2.12) is reduced to the following linear problem

$$\frac{dc_a}{d\theta}=\sum_{i=1}^{F}\Delta_{ai}c_a+\Delta_a, \quad c_a\big|_{\theta=\theta^{(i)}}=c_{as}^*, c_a\big|_{\theta=0}=c_a^{(L)}, \qquad (2.21)$$

where $\theta^{(i)}=\theta(T_{io}); c_{as}^*=c_{as}(T_{io},\gamma_{10},...,\gamma_{w0})$. Then the dependence c_a on T_e and value of constants B_k are determined by integration of system of linear equations (2.21) by the Eiler method (Matveev, 1976). In the second case (2.20) solutions of equations (2.12) are:

$$c_a = \left(\left(c_a^{(L)}\right)^{-\alpha_a} - \alpha_a \psi_1^{(a)}(T_e)\right)^{-1/\alpha_a} \exp\left(-\psi_2^{(a)}(T_e)\right), \tag{2.22}$$

where:

$$\psi_1^{(a)} = \int_{T_e^{(L)}}^{T_e} A_a^{(1)} \exp\left(-\alpha_a \psi_1^{(a)}(T_e)\right) dT_e, \quad \psi_2^{(a)}(T_e) = \int_{T_e^{(L)}}^{T_e} A_a^{(2)} dT_e. \tag{2.23}$$

When F=1(Shchukin,1995) then the problem (2.12) and the coefficient $\Omega^{(T)}$ are:

$$\frac{dc_1}{dT_e} = E_1, \qquad c_1\Big|_{T_e=T_{io}} = c_{1S}\left(T_{i0}, \gamma_{10}, \ldots, \gamma_{W0}\right), \; c_1\Big|_{T_e=T_e^{(L)}} = c_1^{(L)} \tag{2.24}$$

$$E_1 = -\frac{\left(f_1^{(T)} - B_1 \varphi^{(T)}\right)}{\left(f_{1,1}^{(c)} - B_1 \varphi_1^{(c)}\right)}, \quad \Omega^{(T)} = \frac{\left(f_{1,1}^{(c)} \varphi^{(T)} - \varphi_1^{(c)} f_1^{(T)}\right)}{\left(f_{1,1}^{(c)} - B_1 \varphi_1^{(c)}\right)}. \tag{2.25}$$

Non-linear equation (2.24) can be integrated in quadratures when the function E_1 (2.25) can be represented in the form (2.20) or the following two forms (Shchukin,1995)

$$E_1 = E_1^{(T)} E_1^{(c)}, \quad E_1^{(c)} = E_1^{(c)}(c_1); \quad E_1 = A_4^{(1)} c_2 - A_3^{(1)} c_2^{1+\nu_1}, \tag{2.26}$$

where $E_1^{(T)}$, $A_3^{(1)}$, $A_4^{(1)}$ are coefficients dependent on T_e and B_1; $c_2 = 1 - c_1$; ν_1 is a constant. Substitute expressions (2.26) into equation (2.24) we obtain

$$\int_{c_1^{(L)}}^{c_1} \frac{1}{E_1^{(c)}} dc_1 = \int_{T_e^{(L)}}^{T_e} E_1^{(T)} dT_e, \; c_1 = 1 - \left[\left(c_2^{(L)}\right)^{-\nu_1} - \nu_1 \Psi_3^{(1)}(T_e)\right]^{-1/\nu_1} \exp\left(-\Psi_4^{(1)}(T_e)\right), \tag{2.27}$$

$$\Psi_3^{(1)}(T_e) = \int_{T_e^{(L)}}^{T_e} A_3^{(1)} \exp\left(-\nu_1 \Psi_4^{(1)}(T_e)\right) dT_e, \; \Psi_4^{(1)}(T_e) = \int_{T_e^{(L)}}^{T_e} A_4^{(1)} dT_e, \; c_2^{(L)} = 1 - c_1^{(L)}.$$

If the function c_1 is equal to (2.22), (2.27) then the value of the integral coefficient B_1 may be found from the solution of the algebraic equation: $c_1\Big|_{T_e=T_{io}} = c_{1S}\left(T_{i0}, \gamma_{10}, \ldots, \gamma_{W0}\right)$.

If the distributions c_a and T_e are described by the boundary-value problems (2.12)-(2.13) then the value integral fluxes $Q_J^{(j)}$ of the vector $\vec{J} = -\left(\sum_{a=1}^{F} b_a^{(c)} \nabla c_a + b_a^{(T)} \nabla T_e\right)$ can be found by the formulas

$$Q_J^{(j)} = \left(\sum_{a=1}^{F} b_a^{(c)} \frac{dc_a}{dT_e} + b^{(T)} \right) \frac{1}{\Omega^{(T)}} \Psi_1^{(T)}(T_{i0}) Q_U^{(j)}, \quad Q_U^{(j)} = -\oint_{S_j} \nabla U d\vec{S}_j \,. \quad (2.28)$$

where $d\vec{S}_j$ is the differential vector element of the j-th surface whose direction coincides with that of the outer normal . It follows from the expression for $Q_u^{(j)}$ (2.28) that the value of $Q_u^{(j)}$ is always positive.

If the boundary-value problem (2.1)-(2.4) is solved with taking into account additional boundary condition then distributions c_a, T_e and values T_{io} and b_a may be found simultaneously. Additional condition may be integral one, for example,

$$Q_i = \sum_{j=1}^{N} \sum_{p=1}^{P} Q_p^{(j)}, Q_p^{(j)} = \oint_{S_j} \vec{B}_p^{(j)} d\vec{S}_j, \vec{B}_p^{(j)} = -\left(\sum_{a=1}^{F} b_{p,a}^{(j)} \nabla c_a + b_T^{(j)} \nabla T_e \right), \quad (2.29)$$

where Q_i is known value;

$$b_{p,a}^{(j)} = b_{p,a}^{(j)}(T_e, T_i^{(j)}; c_1,\dots,c_F; \gamma_1,\dots,\gamma_W), b_T^{(j)} = b_T^{(j)}(T_e, T_i^{(j)}; c_1,\dots,c_F; \gamma_1,\dots,\gamma_W).$$

If $\gamma_m^{(j)}\Big|_{r_j} = \gamma_{m0}$, then the boundary-value problem (2.1)-(2.4), (2.29) is reduced to the more cimple boundary-value problem including the boundary-value problems (2.12),(2.13) and additional condition (2.30)

$$Q_i = \left\{ \sum_{j=1}^{N} Q_U^{(j)} \sum_{p=1}^{P} \left(\sum_{a=1}^{F} b_{p,a}^{(j)} E_a + b_T^{(j)} \right) \right\} \frac{1}{\Omega^{(T)}} \Psi_1(T_{io}) \Big|_{T_e = T_{io}} \,. \quad (2.30)$$

In applications it is interesting to consider solution of the non-linear boundary-value problems

$$\mathrm{div}\left(\sum_{k=1}^{F} f_{k,a}^{(c)} \nabla c_a + f_k^{(T)} \nabla T_e \right) = 0, \mathrm{div}\left(\sum_{a=1}^{F} \varphi_a^{(c)} \nabla c_a + \varphi^{(T)} \nabla T_e \right) = 0, \quad (2.31)$$

$$\mathrm{div}\left(\sum_{m=1}^{W} h_{b,m}^{(j)} \nabla \gamma_m^{(j)} + h_b^{(T,j)} \nabla T_i^{(j)} \right) = 0, \quad (2.32)$$

$$c_a\Big|_{S_j} = c_{aS}^{(j)}\Big|_{S_j}, T_e\Big|_{S_j} = T_{i0}^{(j)}\Big|_{S_j}, \sum_{m=1}^{W} \left(\Theta_{p,m}^{(j)} \nabla_\perp \gamma_m^{(j)} + \Theta_p^{(T,j)} \nabla_\perp T_i^{(j)} \right)\Big|_{S_j} = 0, \quad (2.33)$$

$$\left(\sum_{a=1}^{F} v_a^{(c,j)} \nabla_\perp c_a + v^{(T,j)} \nabla_\perp T_e \right)\Bigg|_{S_j} = \left(\sum_{m=1}^{W} \omega_m^{(\gamma,j)} \nabla_\perp \gamma_m^{(j)} + \omega^{(T,j)} \nabla_\perp T_i^{(j)} \right)\Bigg|_{S_j}, \quad (2.34)$$

$$\gamma_m^{(j)}\Big|_{r_j} = \gamma_{m0}^{(j)}, c_a\Big|_{S^{(L)}} = c_a^{(L)}, T_e\Big|_{S^{(L)}} = T_e^{(L)}, \gamma_m^{(j)} \neq \infty, T_i^{(j)} \neq \infty \,. \quad (2.35)$$

where indices k=1,...,F; a=1,...,F; b=1,...,W+1; m=1,..., W; p=1,...,W; j=1,...,N; functions c_a and T_e depend on spatial coordinates in all points of the space autwards the closed domains bounded by the arbitrary surface S_j and $S^{(L)}$; coefficients $f^{(c)}_{k,a}$, $f^{(T)}_{k}$, $\varphi^{(c)}_a$, $\varphi^{(T)}$ are functions dependent on $c_1,\ldots,c_F$ and T_e ; $h^{(j)}_{b,m}$, $h^{(T,j)}_b$, $\omega^{(\gamma,j)}_m$, $\omega^{(T,j)}$, $\Theta^{(j)}_{p,m}$, $\Theta^{(j)}_{,p,T}$ depend on $\gamma^{(j)}_1,\ldots,\gamma^{(j)}_W$ and $T^{(j)}_i$; coefficients $v^{(c,j)}_a = v^{(c)}_a(T_e, T^{(j)}_i, c_1,\ldots, c_F, \gamma^{(j)}_1,\ldots, \gamma^{(j)}_W)$, $v^{(T,j)} = v^{(T)}(T_e, T^{(j)}_i, c_1,\ldots, c_F, \gamma^{(j)}_1,\ldots, \gamma^{(j)}_W)$, $c^{(j)}_{aS} = c_{aS}(T^{(j)}_i, \gamma^{(j)}_1, \ldots, \gamma^{(j)}_m)$; $\gamma^{(j)}_{m0}$ is known value of $\gamma^{(j)}_m$ in some point r_j insid S_j ; $c^{(L)}_a = const, T^{(L)}_e = const$. Function $T^{(j)}_i$ and $\gamma^{(j)}_m$ are defined in all points inside closed domain bounded by the surface S_j.

The boundary – value problem (2.31) – (2.35) may be solved analytically under the condition that $\gamma^{(j)}_m\Big|_{r_j} = \gamma_{m0} = \mathrm{const}$. Then $\gamma^{(j)}_m = \gamma_{m,o}$, $T^{(j)}_i = T_{i0}$ and the functions c_a depend only on T_e. In this case the boundary-value problem (2.31) - (2.35) similarly to the problem (2.1) – (2.4) is reduced to the solution of two problems

$$\frac{dc_a}{dT_e} = E_a \,, c_a\Big|_{T_e=T_{i0}} = c_{aS} \,, \left.\left(\sum_{a=1}^{F} v^{(c)}_a \frac{dc_a}{dT_e} + v^{(T)}\right)\right|_{T_e=T_{i0}} = 0 \,, c_a\Big|_{T_e=T^{(L)}_e} = c^{(L)}_a \,, \quad (2.36)$$

$$\operatorname{div}\operatorname{grad}\Psi^{(T)}_1(T_e) = 0 \,, \quad \Psi^{(T)}_1(T_e)\Big|_{S_j} = \Psi^{(T)}_1(T_{i0}) \,, \Psi^{(T)}_1(T_e)\Big|_{\infty} = 0 \,, \quad (2.37)$$

where $c_{aS} = c_{aS}(T_{i0}, \gamma_{10},\ldots,\gamma_{W0})$; E_a and $\Psi^{(T)}_1$ are similar to (2.11) and (2.15). The solution of the boundary-value problem (2.36) is performed independently on the problem (2.37).

Performing either numerical or analytical solution of (2.36) one should find a dependence c_a on T_e and value of T_{i0} and constant B_k simultaneously. It follows from (2.36) that value of T_{i0} and B_k is independent on form of the surfaces S_j, $S^{(L)}$ and distances between the surfaces S_j.

The boundary-value problem (2.37) is similar to the problem (2.14). At known value of T_{i0} and B_k the distribution of T_e is described by the formula (2.16). In the explicit form, the equation (2.36) can be integrated when the function E_a can be represented in the forms (2.19), (2.20), (2.26). Then values of T_{i0} and $B_k$can be obtained in course of the solution of the following system of algebraic equation:

$$c_a\Big|_{T_e=T_{i0}} = c_{aS}(T_{i0}, \gamma_{10},\ldots,\gamma_{W0}), \quad \left(\sum_{a=1}^{F} v^{(c)}_a E_a + v^{(T)}\right)\Big|_{T_e=T_{i0}} = 0 \,. \quad (2.38)$$

In specific cases with the help of the boundary-value problem (2.31)-(2.35) one can estimate the non-linear quasi-steady state processes of the free diffusive and convective (Stefan) evaporation (growth) (Shchukin,1979,1982,1985,1995,2000; Shchukin et.al.

1982,1985,1996,1999) and heterogeneous burning (Shchukin, 1982,1988; Shchukin et.al. 1996) that occur at large temperature difference by both single and array of large drops or particles with the arbitrary surface shapes S_j. Then in equations (2.31)-(2.32) c_a are relative concentrations of gaseous components $\gamma_m^{(j)}$ are densities of substances inside the drops, T_e and $T_i^{(j)}$ are temperatures of gas and drops or particles, c_{as} are relative concentrations of saturated vapours or values of relative concentrations of molecules of oxidizers. Free evaporation and burning of drops and particles take place when evaporation and burning occur without internal heat release and radiant heat exchange. It follows from (2.36) that when free diffusive or convective evaporation (growth) of drops is described by boundary-value problem (2.31)-(2.35) and $c_{as} = c_{as}(T_i^{(j)}, \gamma_{10}, \ldots, \gamma_{W0})$ is the function depending only on the surface temperature $T_i^{(j)}$ (for example, in cases of the drops of pure liquid and solutions with γ_{mo} non-dependent on time) then steady-state evaporation or growth occurs at the constant temperature $T_i^{(j)} = T_{io}$=const non-dependent on time, the forms of drops and distances between the surface S_j. Free heterogeneous burning of particles with $c_{as}^{(j)} = c_{as}(T_i^{(j)})$ occur at $T_i^{(j)} = T_{io}$ non-dependent on time as in the case of drops.

3.HEATING OF AN ARRAY OF LARGE INTERACTING PARTICLES

Consider now quasi-steady state heating in gaseous medium of N immobile highly conducting solid particles of arbitrary surface shape S_j . Particles are in a volume $V^{(L)}$, boundary by the arbitrary surface $S^{(L)}$. The gravitational convection effect on the heat exchange with the environment is minor. In this case the temperature distribution T_e in the vicinity of N particles is described by following boundary-value problem:

$$\operatorname{div} \kappa_e \nabla T_e = 0, T_e\Big|_{S_{1\le j\le N}} = T_i^{(j)}, T_e\Big|_{S^{(L)}} = T_e^{(L)} , \tag{3.1}$$

where κ_e is coefficient of thermal conductivity of a gas media; $T_i^{(j)}$ =const and $T_e^{(L)} = \text{const}$ are the j-th particle surface temperature and known temperature of the surface $S^{(L)}$; indices j=1,...,N. Introduce a new unknown function $\psi_2^{(T)}(T_e) = \int_{T_e^{(L)}}^{T_e} \kappa_e dT_e$. Then the above boundary-value problem (3.1) is reduced to

$$\Delta\psi_2^{(T)}(T_e) = 0, \psi_2^{(T)}(T_e)\Big|_{S_j} = \psi_2^{(T)}(T_i^{(j)}), \psi_2^{(T)}(T_e)\Big|_{S^{(L)}} = 0 , \tag{3.2}$$

Represent the function $\psi_2^{(T)}(T_e)$ in the form

$$\psi_2^{(T)}(T_e) = \sum_{k=1}^{N} \psi_2^{(T)}(T_i^{(k)}) U^{(k)} , \tag{3.3}$$

where $U^{(k)}=U^{(k)}(x_f)$ are functions dependent on the spatial coordinates x_f ; k=1,...,N. Then the problem (3.2) is reduced to

$$\sum_{k=1}^{N}\psi_2^{(T)}(T_i^{(k)})\Delta U^{(k)}=0,\sum_{k=1}^{N}\psi_2^{(T)}(T_i^{(k)})\Delta U^{(k)}\Big|_{S_j}=\psi_1^{(T)}(T_i^{(j)}),$$
$$\sum_{k=1}^{N}\psi_2^{(T)}(T_i^{(k)})U^{(k)}\Big|_{S^{(L)}}=0. \tag{3.4}$$

For the functions $U^{(k)}(x_f)$ it follows from (3.4) :

$$\Delta U^{(k)}=0, U^{(k)}\Big|_{S_j}=\delta_{kj,}U^{(k)}\Big|_{S^{(L)}}=0 \quad , \tag{3.5}$$

where δ_{kj} is the Kroneker delta. To compute particle heating it is necessary to know expressions for the total heat fluxes $Q_T^{(j)}$ (withdrawn from the particle surface due to molecular heat condition) are equal to

$$Q_T^{(j)}=-\oint_{S_j}\kappa_e\nabla T_e d\vec{S}_j=\sum_{K=1}^{N}\psi(T_i^{(k)})Q_U^{(j,k)}, Q_U^{(j,k)}=-\oint_{S_j}\nabla U^{(k)}d\vec{S}_j \quad , \tag{3.6}$$

where $d\vec{S}_j$ is differential vector element of the j-th surface which direction coincides with that of the outer normal. Knowing the expression for $Q_U^{(j,k)}$ the time dependence of $T_i^{(j)}$ is found by integration of the set of equations of conservation of energy

$$M^{(j)}\alpha_P^{(j)}\frac{dT_i^{(j)}}{dt}=Q_W^{(j)}-Q_T^{(j)}-Q_L^{(j)}, 1\leq j\leq N \quad , \tag{3.7}$$

where $\alpha_p^{(j)}$ is j-th particle material specific heat capacities ; $Q_W^{(j)}$ is the total power of the heat sources of j-th particle; $Q_L^{(j)}$ is heat flux withdrawn from the surface due to radiant heat withdrawal.

In case of two spherical particles and $V^{(L)}\rightarrow\infty$ the non-lineal value-boundary problem (3.1) has been solved (Shchukin 1979).

4.ON ANALYTICAL SOLUTIONS OF THE LAPLASE EQUATION FOR THE FUNCTIONS $U^{(k)}$

The distribution of functions c_a and T_e depend on functions $U(x_f)$ (2.16) and $U^{(k)}(x_f)$ (3.3). These functions are completely characterized by geometrical properties of the considered systems. The function $U(x_f)$ (2.16) and the coefficients $Q_U^{(j)}$ (2.28) are equal to

$$U(x_f)=\sum_{k=1}^{N}U^{(k)}(x_f),\quad Q_U^{(j)}=\sum_{k=1}^{N}Q_U^{(j,k)},\tag{4.1}$$

where $Q_U^{(j,k)}$ are equal to (3.8). Functions $U^{(k)}(x_f)$ are determined from solution of the boundary value problems (3.5) Computation of the function $U^{(k)}(x_f)$ is an independent problem of applied mathematics which can be solved numerically in the case of an arbitrary system of the surfaces S_j. In some special cases the problems (3.5) may be solved analytically, for example, when the volume $V^{(L)}$ contains of one spherical, spheroidal or ellipsoidal surface S_1. If the centers of these single surfaces S_1 coincide with the center of the domain $V^{(L)}$ and surfaces S_1 and $S^{(L)}$ are sofocal then the solution $U^{(1)}(x_f)$ of the problem (3.5) are equal to

$$U^{(1)}(x_f)=\left(F-F^{(L)}\right)/\left(F^*-F^{(L)}\right).\tag{4.2}$$

In case of spherical and spheroidal surfaces the functions F are represented as follows:

$$F(r)=R_1/r,\quad F^*=1,\quad F^{(L)}=R_1/R^{(L)};$$

$$F(\xi)=\ln(1+\operatorname{ch}\xi)/|1-\operatorname{ch}\xi|,\ F^*=F(\xi)\big|_{\xi=\xi_1},\quad F^{(L)}=F(\xi)\big|_{\xi=\xi^{(L)}},a_1>b_1;$$

$$F(\xi)=\operatorname{arcctg}(\operatorname{sh}\xi),F^*=F(\xi)\big|_{\xi=\xi_1},F^{(L)}=F(\xi)\big|_{\xi=\xi^{(L)}},a_1<b_1;$$

where R_1 and $R^{(L)}$ are the radii of surfaces of S_1 and $S^{(L)}$, r is a radial coordinate in the spherical coordinate system whose origin coincides with the center of surfaces S_1 and $S^{(L)}$; a_1 and b_1 are semiaxes of prolate ($a_1 > b_1$) and oblate ($a_1 < b_1$) spheroidal surfaces S_1 ; the semiaxes a_1 is parallel to the axis of symmetry of spheroidal surface, $\xi_1=(1/2)\ln|(a_1+b_1)/(a_1-b_1)|$; $c_1=\sqrt{|a_1^2-b_1^2|}$; ξ is spheroidal coordinate; $a^{(L)}=c_1\operatorname{ch}\xi^{(L)}$, $b^{(L)}=c_1\operatorname{sh}\xi^{(L)}$ and $a^{(L)}=c_1\operatorname{sh}\xi^{(L)}$, $b^{(L)}=c_1\operatorname{ch}\xi^{(L)}$ are lengths of semiaxes of corressponding sofocal spheroidal surfaces $S^{(L)}$ (Happel and Branner, 1965).

When the center of a spherical particle do not concede with the center of the spherical domain $V^{(L)}$, then the function $U^{(1)}(x_f)$ is

$$U^{(1)}=\sqrt{2(\operatorname{ch}\xi-\cos\eta)}\sum_{n=0}^{\infty}\left\{\frac{\operatorname{sh}\left[\left(\xi-\xi^{(L)}\right)\left(n+\frac{1}{2}\right)\right]}{\operatorname{sh}\left[\left(\xi_1-\xi^{(L)}\right)\left(n+\frac{1}{2}\right)\right]}\right\}\exp\left(-\left(n+\frac{1}{2}\right)|\xi_1|\right)P_n,$$

where $P_n = P_n(\cos\eta)$; ξ, η are bispherical coordinates (Tichonov, Samarskii, 1971); h is the distance between the centers of the surfaces S_1 and $S^{(L)}$; $\operatorname{ch}\xi_1 = \left(h^2 + R_1^2 - R^{(L)2}\right)/2hR_1$, $\operatorname{ch}\xi^{(L)} = \left(h^2 + R^{(L)2} - R_1^2\right)/2hR^{(L)}$; $\xi_1 > 0$, $\xi^{(L)} > 0$; R_1 and $R^{(L)}$ are radii of surfaces S_1 and $S^{(L)}$.

In case of two spherical surfaces (S_1 and S_2) and $V^{(L)} \to \infty$ the solutions $U^{(j)}(\xi,\eta)$ of the problem (3.5) is also estimated analytically (Shchukin, 1979):

$$U^{(j)} = \sqrt{2(\operatorname{ch}\xi - \cos\eta)}\,\varphi^{(j)}(\xi,\eta), \tag{4.3}$$

where ξ, η are bispherical coordinates; j = 1, 2; R_1 and R_2 are radii of surfaces; h is the distance between the centers of the surfaces;

$$\varphi^{(j)}(\xi,\eta) = \sum_{n=0}^{\infty}\left(\left(\operatorname{sh}\left(\left(n+\frac{1}{2}\right)\left(|\xi_{2^{2-j}}| + (-1)^{j+1}\xi\right)\right)\exp\left(-\left(n+\frac{1}{2}\right)|\xi_j|\right)\right)\Big/G_n\right)P_n$$

$$G_n = \operatorname{sh}\left((n+1/2)(\xi_1 + |\xi_2|)\right), \operatorname{ch}\xi_j = \left(h^2 + (-1)^j\left(R_2^2 - R_1^2\right)\right)/2hR_j,\ \xi_1 > 0,$$
$$\xi_2 < 0, P_n = P_n(\cos\eta) \tag{4.4}$$

The coefficients $Q_U^{(j,k)}$ (3.6) are positive at j=k and negative at j≠k. Their values depend on the form of surface S_j and $S^{(L)}$ and the distances between surfaces of a considered system. After substituting of the functions (4.2) into the expression for $Q_U^{(1,1)}$ we obtain the following formulas for $Q_U^{(1,1)}$

$$Q_U^{(1,1)} = 4\pi R_1/\left(1 - F^{(L)}\right); \qquad Q_U^{(1,1)} = 4\pi c_1 I\left(\xi_1, \xi^{(L)}\right),$$

$$I\left(\xi_1,\xi^{(L)}\right)\Big|_{a_1>b_1} = 2/\left(F^* - F^{(L)}\right),\ I\left(\xi_1,\xi^{(L)}\right)\Big|_{a_1<b_1} = 1/\left(F^* - F^{(L)}\right); \tag{4.5}$$

$$Q_U^{(1,1)} = 4\pi a\sum_{n=0}^{\infty}\left[\exp\left(-\left(n+\frac{1}{2}\right)\left(\xi_1 + \xi^{(L)}\right)\right)\right]\Big/\operatorname{sh}\left[\left(n+\frac{1}{2}\right)\left(\xi_1 - \xi^{(L)}\right)\right],$$

where $a = \left[h^4 + R_1^4 + R^{(L)4} - 2\left(h^2\left(R_1^2 + R^{(L)2}\right) + R_1^2R^{(L)2}\right)\right]^{1/2}/2h$.

In case of two spherical surfaces (j = 1, 2) in infinite volume $V^{(L)}$ the value of $Q_U^{(j,k)}$ is determined by the analytical relations (Shchukin 1979, 1995,1999).

In case of N ≥ 3 surfaces the expression for the functions $U^{(k)}(x_f)$ can be found by the expansion into the series on irreducible tensors (Traytak,1990,1992).

5.CONCLUSIONS

Paper presents the solution of problems which are of interest for the theory of heat and mass transfer. In particles it was shown that the solution of non-linear problem on heating of finite array of solid particles may be reduced to the solution of the corresponding linear problem. Obtained results by solving of non-linear systems of equations allows to determine the conditions under which the steady-state evaporation and heterogeneous burning of drops and particles at constant surfaces temperature give possibility.

REFERENCES

1. Happel,J., Brenner,G.,1965, Low Reynolds number hydrodynamics. Prentice. Hall.
2. Matveev N.M., 1976, Differential equations. Higher institutes Publishing House, Minsk.
3. Shchukin,E.R.,1979, Theory of the evaporation of interacting large particles, in collection of articles: Physics of aerodispertion systems and physical kinetics, instalment No. 4, part.1, N. Krupsraya MOPI, deposition in VINITI No. 3828-79, Moscow, pp. 176-215
4. Shchukin,E.R., 1995, Quasi stationary evaporation and growth of drops of pure substances at considerable drops of temperature in their vicinity. Institute for High Temperature Russian Academy of Science, Manuskript Deposition in VINITI No.412-95, Moscow, 87p.
5. Shchukin,E.R., Shulimanova, Z.L., 1996, On vaporization and growth of finite assemble of immovable large interacting drops. In Collection of articles: Chemistry and chemical technology of unorganic substances, D.Mendeleev Russian Chem. Tech. University, v. 171, Moscow, pp.141-143.
6. Shchukin,E.R., 2000, Quasi-stationary diffusive vaporization and growth of finite assembly of large drops of pure liqwuid, in: The Theses of Reports of The International Conference Dedicated of The Memory of The Professor of Sutugin A.G., Moscow, pp. 82-83.
7. Shchukin,E.R., 1982, On burning of two particles disposed on arbitrary distance against each other, in: Thesis of papers II All-Union Meeting on spreading of Laser radiation in dispersed media, v.2, Obninsk, pp. 149-152.
8. Shchukin,E.R., Baxtilov,V.T.,1979, Heating and vaporization of spheroidal particles warmed by means of internal heat sources, in: collection of articles: Physics of aerodispertion systems and physical kinetics, instalment No. 4, part 1, N. Krupskaya MOPI, deposition in VINITI No.3828-79,Moscow, pp. 144-165.
9. Shchukin,E.R., Shulimanova, Z.L.,1996, Influence of non-linear characteristic on heat and mass exchange with gaseous medium of chemical active immovable particles. Shadrinsk state pedagogical institute, Manuscript deposition in VINITI No. 30-B96, Shadrinsk, 118p.
10. Shchukin,E.R., Nadykto,A.B., 1999, Diffusive vaporization and growth of assembly of N-large particles. In Mathematical Models of Non-Linear Excitations, Transfer, Dynamics and Control in Condensed Systems and Other Media. Ed. by L.A.Yvarova et.al, Kluwer Academic/Plenum Publishers. New York , pp. 355-368.
11. Tixonov, A.N., Samarskii A.A., 1972, Equations of mathematical physics. Nauka, Moscow.
12. Traytak,S.D., 1990, Theory of recondensation of N drops. Theor. Osnovi Khim. Tekh. v.24, p.473. (English transition (1991), Theor.Found.Chem.Eng., v..24, p.320).
13. Traytak,S.D., 1992, The diffusive interaction in diffusion-limited reactions: the steady-state case. Chem. Phys. Letters,v.197, p.247.

ON THE IRREDUCIBLE TENSORS METHOD IN THE THEORY OF DIFFUSIVE INTERACTION BETWEEN PARTICLES

Sergey D. Traytak*

1. INTRODUCTION

In the papers by Elperin and Krasovitov [1–3] the authors claim that they have suggested a new method called "modified method of irreducible multipoles" in order to solve the quasi steady state heat and mass transfer equations for systems with many interacting burning particles. "A method of solution of the Laplace equation in a region exterior to N arbitrarily located spheres of different radii is suggested. The method is based on the expansion of the solution into irreducible multipoles." (p. 79) [1]. "The present study extends the modified method of irreducible multipoles expansion, suggested by Elperin and Krasovitov (1994) to combustion of random char/carbon particles of different radii." (p. 167) [2]. "The method of expansion into irreducible multipoles which was developed in our previous works (Elperin and Krasovitov, 1994a, 1994b) is applicable to more realistic problems and is particularly suitable for dense random clusters of droplets (particles)." (p. 288) [3]. It is important to note here that mathematically the irreducible tensors approach gives the possibility of finding the solution of linear boundary-value problems (BVP) for the Laplace equation in a three-dimensional multi-connected domain.

The goal of this paper is twofold: (a) to indicate the numerous mistakes which have been made in the above papers by Elperin and Krasovitov and (b) to show that they reproduced (almost without any changes) the approach based on Cartesian irreducible tensors suggested earlier in our papers [4,5]. For the sake of definiteness we shall consider here mostly the Ref. 2 (hereafter we refer to it as I) using the notation of I.

* Department of Mathematical Analysis of Moscow Pedagogical University, 10a Radio St., Moscow 107005, Russia; E-mail: sergtr@dataforce.net

2. SOLUTION OF THE LAPLACE EQUATION IN A MULTI-CONNECTED DOMAIN

First of all we note that it is not clear from the text of I why the method is called "method of irreducible multipoles"? One can easily see that, apart from the title of the method, there is no mention irreducible multipoles at all in the text of I. Moreover the authors used expansions into irreducible tensors but not into irreducible multipoles.

On p. 166 of I we read: "2) method of images - suitable for symmetrical arrays of less than 20 monosized particles (drops)". However in the review [6] one can read: "(2) Method of images - suitable for arrays of less than 20 drops". Indeed it is well-known that this method may be used for particles of different size and any location of them [7]. There is only one limitation of the method of images: " a large number of sources if many drops are considered" [6].

Consider the "method of irreducible multipoles". For definiteness we shall treat here the quasi steady state heat transfer in a system of N burning spherical particles with radii R_i (it is clear that appropriate concentrations fields may be expressed through the known gas mixture temperature dependence). The heat equation as respect to the gas mixture temperature T_g reads

$$\nabla\left(k_g \nabla T_g\right)=0 .$$

Assuming that the thermal conductivity of burning particles is much lager than the gas mixture thermal conductivity k_g, we supplement this equation with the boundary condition (BC) on i-th particle surface and condition at infinity, respectively

$$T_g\Big|_{S_i}=T_s^{(i)}=const\,,\quad T_g\Big|_{r_i\to\infty}\to T_\infty\,,$$

where r_i is the radial coordinate connected with i-th particle; $T_s^{(i)}$ and T_∞ are the temperature of i-th particle and temperature T_g at infinity. Note that $T_s^{(i)}$ should be found with the help of the energy conservation condition written on the i-th particle surface[a]

Using the well-known substitution $k_g \nabla T_g=\nabla U$ the above BVP may be reduced to the following

$$\nabla^2 U=0\,, \tag{1}$$

$$U\big|_{S_i}=U_s^{(i)}\,, \tag{2}$$

$$U\big|_{r_i\to\infty}\to 0\,, \tag{3}$$

where $U_s^{(i)}=\int\limits_{T_\infty}^{T_s^{(i)}} k_g dT_g$. Thus it is necessary to solve the Dirichlet problem for the

---

[a] In order to save space hereafter we do not dwell on details concerning physical aspects of the problem.

Laplace equation (Eqs. 1-3) in a three-dimensional unbounded multi-connected domain outside the particles.

Due to the linearity of the posed BVP its solution may be presented as a superposition

$$U = U_i + \delta U_i, \ \delta U_i = \sum_{j(\neq i)=1}^{N} U_j, \tag{4}$$

where U_i contains irregular solid spherical harmonics which vanish at infinity. It is well-known that in spherical coordinates connected with the center of i-th sphere U_i is

$$U_i = \sum_{n=0}^{\infty}\sum_{m=0}^{n}\left[\alpha^1_{nm} Y^e_{nm}(\theta_i,\varphi_i) + \beta^1_{nm} Y^o_{nm}(\theta_i,\varphi_i)\right] r_i^{-(n+1)}. \tag{5}$$

Here α^1_{nm} and β^1_{nm} are some constants and $Y^e_{nm} = \cos(m\varphi_i) P^m_n(\cos\theta_i)$, $Y^o_{nm} = \sin(m\varphi_i) P^m_n(\cos\theta_i)$, where P^m_n are associated Legendre polynomials.

On p. 170 of I the authors claim that "in this case the boundary conditions cannot be satisfied exactly at the surface of i-th particle, because the solution in the neighbourhood of the i-th particle has singularities at the centers of other $N-1$ particles". First of all this is an elementary ignorant assertion, because the solution in a neighbourhood of the i-th particle cannot be considered at the center of any finite j-th particle by definition. It is also worth noting that Eq. (4) is written in the whole three-dimensional space outside particles without introducing any coordinate systems. So at this stage of the treatment there is no meaning to mention about the solution in a neighbourhood of the i-th particle. Moreover, and more importantly, the above explanation is incorrect. To show that it is enough to consider the specific case $N = 2$. In this case "the solution in a neighbourhood of the first particle" ($\delta U_1 = U_2$) has a singularity at the second particle center, but in spite of this it is well-known that the problem may be solved exactly with the help of bispherical coordinates[8]. This is because the complex structure of the domain boundary for $N > 2$ makes it impossible to introduce a global coordinate system in order to solve BVP (Eqs. 1-3).

Hereafter we use the Cartesian coordinates (x^i_1, x^i_2, x^i_3) with origin O_i at i-th particle center and in local coordinates connected with i-th particle one can write

$$U_i = \sum_{n=0}^{\infty} A^i_{\gamma_1\cdots\gamma_n} x_i^{-(2n+1)} \overline{x^i_{\gamma_1}\cdots x^i_{\gamma_n}}, \tag{6}$$

$$\delta U_i = \sum_{n=0}^{\infty} B^i_{\gamma_1\cdots\gamma_n} \overline{x^i_{\gamma_1}\cdots x^i_{\gamma_n}}, \tag{7}$$

where $A^i_{\gamma_1\cdots\gamma_n}$ and $B^i_{\gamma_1\cdots\gamma_n}$ are unknown tensor coefficients to be determined from the BC (Eq. 2) and $\overline{x^i_{\gamma_1}\cdots x^i_{\gamma_n}}$ is the irreducible tensor of rank n defined by[9]

$$\overline{x^i_{\gamma_1} \dots x^i_{\gamma_n}} = \frac{(-1)^n}{(2n-1)!!} x_i^{2n+1} \frac{\partial^n}{\partial x^i_{\gamma_1} \dots \partial x^i_{\gamma_n}} \left(\frac{1}{x_i} \right), \tag{8}$$

where $(2n-1)!! = 1 \cdot 3 \cdot \dots \cdot (2n-1)$; $x^i_{\gamma_\nu} = r^i_{\gamma_\nu} / R_i$, $\gamma_\nu = \overline{1,3}$, $\nu = \overline{1,n}$.

Then on p. 170, concerning the tensor $\overline{x^i_{\gamma_1} \dots x^i_{\gamma_n}}$, the authors of I write: "A set of these components provides a linear combination of the spherical harmonics Y_{lm}:

$$Y^{(o)}_{lm} = \cos(m\varphi) P^m_l(\cos\varphi), \; Y^{(e)}_{lm} = \sin(m\varphi) P^m_l(\cos\varphi), \tag{9}$$

where P^m_l - associated Legendre polynomials of the first kind." Ignoring the comments the statement about Legendre polynomials "of the first kind" we note that the given assertion is incorrect for the tensor (Eq. 8). Really it is evident that in the appropriate spherical coordinates function δU_i may be expressed as follows:

$$\delta U_i = \sum_{n=0}^{\infty} \sum_{m=0}^{n} \left[\alpha^2_{nm} Y^e_{nm}(\theta_i, \varphi_i) + \beta^2_{nm} Y^o_{nm}(\theta_i, \varphi_i) \right] r^n_i,$$

where α^2_{nm} and β^2_{nm} are some constants. Whence the tensor $\overline{x^i_{\gamma_1} \dots x^i_{\gamma_n}}$ is a linear combination of the regular solid spherical harmonics $x^n_i Y_{nm}$, but not the surface ones Y_{nm}.

The use of the BC (Eq. 2) yields

$$A^i_{\gamma_1 \dots \gamma_n} = U^{(i)}_s \delta_{0n} - B^i_{\gamma_1 \dots \gamma_n}, \; n = \overline{0, \infty}. \tag{10}$$

Taking Eq. (10) into account one can obtain

$$U = U^{(0)}_i + \sum_{n=0}^{\infty} B^i_{\gamma_1 \dots \gamma_n} \left(1 - x_i^{-(2n+1)} \right) \overline{x^i_{\gamma_1} \dots x^i_{\gamma_n}}. \tag{11}$$

On p. 171 of I after this expression we read: "the first term $U^{(0)}_i = U^{(i)}_s / x_i$ in expansion (Eq. 11) coincides with the solution in the point sources approximation." This assertion is wrong since the first term $U^{(0)}_i$ is an unperturbed solution for i-th isolated particle and it does not include any influence from other particles ($B^i_{\gamma_1 \dots \gamma_n} = 0$ means that $\delta U_i = 0$). However the point source approximation (PSA) includes this influence as it considers the other $N-1$ particles as point sources, i.e. the PSA for U is

$$U_{PSA} = U^{(0)}_i + B^i_0 \left(1 - x_i^{-1} \right).$$

It is emphasized that to find the PSA we should also limit ourselves by terms with B^i_0 in the corresponding set of linear equations (see Eq. (14) below).

Assume that Cartesian coordinates of j-th and i-th particles are connected by translations, i.e. the directions of the coordinate axes are the same. Using these Cartesian coordinates it is a simple matter to show that in a neighbourhood of origin O_i (not O_j as in Refs. 1-3) the following expansion holds true (see Discussion for details)

$$x_j^{-(2n+1)}\overline{x^j_{\gamma_1}\dots x^j_{\gamma_n}} = \sum_{k=0}^{\infty}\omega_{kn}\varepsilon_j^{n+1}\varepsilon_i^k\Omega_{\gamma_1\dots\gamma_n\mu_1\dots\mu_k}\left(\mathbf{L}_{ij}\right)\overline{x^i_{\mu_1}\dots x^i_{\mu_k}}. \tag{12}$$

Here

$$\omega_{kn} = (-1)^n\left[2(n+k)-1\right]!!/k!(2n-1)!!, \tag{13}$$

$$\Omega_{\gamma_1\dots\gamma_n\mu_1\dots\mu_k}\left(\mathbf{L}_{ij}\right) = L_{ij}^{-(n+k)}\overline{L^{ij}_{\gamma_1}\dots L^{ij}_{\gamma_n}L^{ij}_{\mu_1}\dots L^{ij}_{\mu_k}},\ \varepsilon_{i(j)} = R_{i(j)}/L_{ij},$$

where $L^{ij}_{\gamma_\nu}$ are the Cartesian coordinates of the vector $\mathbf{L}_{ij} = \mathbf{r}_0^{(j)} - \mathbf{r}_0^{(i)}$.

It is important that in our work[5] the expression (Eq. 13) was written as $(-1)^k\,\omega_{kn}$ (this misprint has been corrected in Ref. 10 and paper I and this error is reproduced in the related papers of the same authors[1,3]. Hence, all of the calculations depending on dipoles and higher multipoles (i.e. on ω_{kn} if $n>0$) performed in Refs. 1-3 are incorrect (see Discussion for details).

Expansion (Eq. 12) is a form of the well-known addition theorem[11] for solid spherical harmonics written in terms of Cartesian irreducible tensors. It is clear that one can find the solution of the posed BVP using the addition theorem for the relevant spherical coordinates, however the use of Cartesian irreducible tensors leads to this theorem in the simplest way.

Using the expansion (Eq. 12) one gets for $B^i_{\gamma_1\dots\gamma_n}$ the set of linear equations of the form

$$B^i_{\gamma_1\dots\gamma_n} = B^{i(0)}_{\gamma_1\dots\gamma_n} - \sum_{j(\neq i)=1}^{N}\sum_{k=0}^{\infty}B^j_{\mu_1\dots\mu_k}\omega_{nk}\varepsilon_j^{k+1}\varepsilon_i^n\Omega_{\gamma_1\dots\gamma_n\mu_1\dots\mu_k}\left(\mathbf{L}_{ij}\right), \tag{14}$$

where

$$B^{i(0)}_{\gamma_1\dots\gamma_n} = \sum_{j(\neq i)=1}^{N}U_s^{(j)}\omega_{n0}\varepsilon_j\varepsilon_i^n\Omega_{\gamma_1\dots\gamma_n}\left(\mathbf{L}_{ij}\right),\ n=\overline{0,\infty}.$$

The total heat flux is determined by the surface integral

$$Q_T^{(i)} = \int_{S_i}k_g\nabla T_g d\mathbf{s}. \tag{15}$$

Performing integration over the unit sphere $x_i = 1$ the authors of I found

$$Q_T^{(i)} = 4\pi R_i\left[U_s^{(i)} - \sum_{n=0}^{\infty}\frac{(-1)^n}{(2n-1)!!}(2n+1)B^i_{\gamma_1\dots\gamma_n}\nabla^i_{\gamma_1}\dots\nabla^i_{\gamma_n}\left(\frac{1}{x_i}\right)\Bigg|_{x_i=1}\right]. \tag{16}$$

According to I this formula (Eq. (38) on p. 173) describes the heat flux when radiation effect is important and after factoring out $U_s^{(i)}$ it describes the case when effect of radiation is negligible (see formula (Eq. 39) of I). However the point is that formula (Eq. 16) +and other related formulae ((Eq. 39) and (Eq. 41) of I) are wrong for a very simple reason. Due to the orthonormal property of irreducible tensors after integration over the unit sphere only one term with B_0^i survives. This is evident from the fact that for the tensor $\overline{x_{\gamma_1}^i \dots x_{\gamma_n}^i}$ on the unit sphere "set of components provides a linear combination of the spherical harmonics Y_{nm}". Thus for the total flux into i -th particle the integral (Eq. 15) leads to

$$Q_T^{(i)} = 4\pi R_i \left(U_s^{(i)} - B_0^i \right). \tag{17}$$

It is interesting that even using wrong formula the authors of I obtained good agreement with experimental results (see Fig. 6 on p. 177). Furthermore in presenting the numerical results, the authors did not indicate the accuracy of their calculation for the solution of the system (Eq. 14) nor the system of differential equations describing the dependence of the burning particles radii on time. This is important, because for dense clusters containing many particles (e.g., $N = 100$ as in I) a known problem concerning convergence of the truncated system corresponding to the infinite system (Eq. 14) naturally arises [12]. Particularly one should investigate the validity of the iteration method for the solution of (Eq. 14). It may be shown that the solution of the set of Eqs. (14) is equivalent to the relevant solution of the problem by the known reflection method [13].

It is worth noting that with the help of addition theorem for spherical harmonics (Eq. 12) one can solve the BVP (Eqs. 1-3) using only expression (Eq. 6) without introducing the expansion (Eq. 7). This derivation leads to the simple formula for the total flux straightforwardly. Indeed, substitution of the function $U = \sum_{i=1}^{N} U_i$ into the BC (Eq. 2) with the aid of the addition theorem (Eq. 12) leads to the set of Eqs.

$$A_{\gamma_1 \dots \gamma_n}^i = U_s^{(j)} \delta_{0n} - \sum_{j(\neq i)=1}^{N} \sum_{k=0}^{\infty} A_{\mu_1 \dots \mu_k}^j \omega_{nk} \varepsilon_j^{k+1} \varepsilon_i^n \Omega_{\gamma_1 \dots \gamma_n \mu_1 \dots \mu_k} \left(\mathbf{L}_{ij} \right)$$

and the corresponding formula for the total flux is simplified to $Q_T^{(i)} = 4\pi R_i A_0^i$.

Nevertheless derivations given in Refs. 1-3 completely coincide with those done before [5].

In addition note that on p. 169 of I the authors write "A detailed description of the modified method of irreducible multipoles expansion and its application to the evaporation and combustion of random cluster of droplets can be found in Elperin and Krasovitov, (1994)"(i.e. in Ref. 1). Unfortunately, three misprints in indices for the contractions of relevant tensors were made in our formulae (Eq. 11) of Ref. 5 which were carefully reproduced in the relevant formulae (Eq. 25) and (Eq. 26) of Ref. 1. Moreover Eqs. for $B_{\gamma_1 \dots \gamma_n}^{i(0)}$ and $Q_{\gamma_1 \dots \gamma_n}^{i(0)}$ after Eqs. (37), (38) on p. 292 of Ref. 3 correspond to the similar Eqs. after Eqs. (25) of Ref. 5. One can see there that factor $\omega_{k0}\varepsilon_j\varepsilon_i^k$ in Ref. 3 contains index k as in Ref. 5, but not n as should be in the notation of Ref. 3. In addition similar to Ref. 5 the notation $i=\overline{1,N}$ [b] is used on pp. 290, 291 after Eq. (24) and in Table 3 on p. 292, but after Eq. (27) on p. 291 the different notation $k = 1,\ldots, N$ is utilized.

3. DISCUSSION

The main text of this paper was also submitted to the Journal of Combustion Science and Technology as a Comment and recently we received the Reply to it[14]. So we shall discuss questions pointed out in this Reply.

"The declared aim of the Comment is to demonstrate that our paper I 1) reproduces the earlier published papers by the author of the Comment[5,10] and 2) contains numerous errors. The goal of this Reply is to show that both claims are maliciously false." Note that we cited our paper [10] after Eq. (13) only concerning correction in the sign of ω_{kn}. It is also worth noting here that our claim concerns only the irreducible tensors approach "suggested by Elperin and Krasovitov (1994)" (see Introduction).

"1) In our papers[1-3] we used the formulation of the method of expansion into irreducible multipoles for the analysis of the interactive evaporation of clusters of droplets as it was presented in Ref. 15 and in a monograph[16] submitted in 1990[c]. Therefore in study we referred to Ref. 15 for the description of the method and do dot reproduced it from Refs. 5, 10 as claimed in the Comment." Indeed there are several references to the thesis[15] in Refs. 1, 2 namely: on p. 79, pp. 91-92 of Ref. 1 and on p. 166 of Ref. 2. However it is necessary to stress that all of them refereed to Ref. 15 concerning only the case of two particles (droplets) interaction.

"Note in passing that contrary to the ardent claims of the author of Comment we were not aware about his papers[5,10] which for some reason were published in the literature not related to this subject. However, it is of interest to note also that Eqs. (6)-(8), (10)-(14) in the Comment coincide exactly with those in Refs. 15, 16 (see, e.g. Ref. 15, Chapter 4, Sec. 4.3, Evaporation of a Finite System of N Drops under Arbitrary Temperature Differences). Eq. (12) in the Comment is identical to that derived in Ref. 24 which is cited in Refs. 15, 16[d] and which was not referenced by the author of the Comment. Curiously, Eq. (10) in the Comment contains the same misprint as the same equation in the Refs. 15,

---

[b] Note that this notation is of seldom use in literature on physics.

[c] Actually this "monograph" is the manuscript of the thesis (Ref. 15). Scientific council of Moscow Pedagogical Institute recommended it for the depositing in VINITI on 27 December, 1990. This means that a manuscript without reviewing was submitted to the institution called VINITI (USSR Institute of Scientific Information) which publishes abstract in a special book of abstracts. If the requested, a copy of the manuscript of interest is sent by mail. So it is not available even in libraries and, strictly speaking, this is not even a preprint.

[d] Simple inspection of Refs. 15, 16 show that this assertion is not in accordance with the truth.

16, namely, its RHS is the difference of tensors of different ranks. In our subsequent studies [1–3] this misprint was corrected." One can see that the above formulas are the same as in our paper [5] submitted in 1987 and, therefore, it is really interesting why they "coincide exactly with those in" Refs. 15, 16?

"There are a number of studies employing the method of multipole expansion which were published before. Some of the studies using the method of expansion into irreducible multipoles in the form similar to that used in our studies are listed below in the References list in a chronological order (see References 9, 17-25)." One can see that Refs. 17-21 are simply taken from the References list of Ref. 9 and they are very remotely connected with the subject at issue. Such kind of references may be easily taken, e.g., from the known book [26]. It is worth noting that Ref. 22 investigates the hydrodynamic interactions with the help of the method of induced forces which was applied to the problem of diffusive interaction between drops in Ref. 27.

"Note in passing that in the studies [5,10] these investigations (including Refs. 15, 16, 24, 25) are not mentioned at all." References 24, 25 deal with the investigation of the hydrodynamic interaction between particles in Stokes flow and simple comparison with Ref. 5, 10 shows that these works were performed independently. On the other hand it is obvious that all derivations of Refs. 1-3, 15, 16 concerning "the method of irreducible multipoles" reproduce the relevant derivations given in Ref. 5 including even misprints. As regards Refs. 15, 16 it is absolutely incredible that paper [5] submitted in 1987 and published in 1990 could contain any references to a manuscript deposited in 1991 and corresponding thesis of the same year.

"...in Ref. 5 it is written that "major computational difficulties arise when this method is employed for direct study of heat and mass transfer in real dispersed systems." In our studies (see Refs. 1-3) we resolved the arising computational difficulties and obtained detailed numerical solutions to the original nonlinear heat and mass transfer problems (e.g., evaporation of clusters of droplets under large temperature and concentration differences) which to the best of our knowledge were not solved before." First of all we should note that given quotations from our paper misrepresented our main idea. Indeed, one can read below: "However, there is as a rule no need to have detailed information on the heat and mass fluxes at the surface of each particle in order to describe systems containing large number of particles. It is sufficient for this purpose to know the system statistical characteristics. In this sense, the Cartesian representation method solves a problem analogous to the N-body problem in classical mechanics. The results of the present study can therefore be utilized to construct a rigorous statistical theory of heat and mass transfer in dispersed systems and to verify the conclusions derived from it, by numerical solution of a problem with a sufficiently large number of particles." [5] It is evident from this that, saying about computational difficulties, we mean that similar to the case of statistical mechanics "detailed numerical solutions" are unfeasible for real systems of interacting particles. Nevertheless there is no doubt that correct numerical solution of the problem for large number of particles is of great scientific interest. However we underline once again that our aim is a consideration of "the modified method of irreducible multipoles expansion, suggested by Elperin and Krasovitov (1994)" [2].

"2) Let us now consider the second claim about the correctness of our results. We will skip all the undergraduate stuff in the Comment and the thoughtful analysis of the eigenfunctions of Laplace operator." We agree with the authors of the Reply that almost all mistakes made in Refs. 1-3 are of the undergraduate level.

"Consider first the substitution which allows to reduce the BVP for heat conduction to the solution of Dirichlet problem for Laplaće equation (Eqs. (1-3) in the Comment). This substitution is absolutely standard and appears in many text books on heat and mass transfer (see, e.g., Ref. 28, Part 2, Chapter 3). However, the latter substitution cannot be applied for the solution of simultaneous nonlinear system of equations of heat conduction and diffusion which was considered in our investigations". One can see that exactly this substitution was applied in Refs. 1-2 and we just followed Ref. 2 describing the method under consideration for study of N burning particles with large temperature differences.

"The total heat flux is indeed determined by the surface integral (see Eq. (15) in the Comment). Integration yields:

$$Q_T^{(i)} = 4\pi R_i \int_{T_\infty}^{T_S} k_g dT_g \left(1 - \tilde{B}_0^{(i)}\right)." \tag{18}$$

Below in spite of our explanation the authors give their own explanation of this well-known fact: "The reason that all the other terms vanish is that spherically symmetric boundary condition at the particle's surface in the heat conduction equation:

$$\int_{T_\infty}^{T_s^{(i)}} k_g dT_g = const ." \tag{19}$$

It is evident that one should skip this undergraduate stuff referring the authors of the Reply to any standard textbook on mathematical physics.

"The correct equation (1) was used in our calculations, and it is written explicitly in our studies (see Eq. (44) in Ref. 1 and Eqs. 40 in Ref. 3)." To show that this assertion is in contradiction with those given in the authors papers, let us read, e.g., Results and Discussion sections of papers [1,2]. On p. 91 of Ref. 1 we read: "The results of calculations of the correction factor by expression (48)

$$\eta_i = 1 - \sum_{n=0}^{\infty} \frac{(-1)^n}{(2n-1)!!}(2n+1) B^i_{\gamma_1 \ldots \gamma_n} \nabla^i_{\gamma_1} \ldots \nabla^i_{\gamma_n} \left(\frac{1}{x_i}\right)\Bigg|_{x_i=1}$$

for the various symmetric monosized arrays are shown in Fig. 2." On p. 174 of Ref. 2 we read: "The results of calculation of the local correction factor $\bar{\eta}_i$ by expression (41)

$$\eta_i = 1 - \sum_{n=0}^{\infty} \frac{(-1)^n}{(2n-1)!!}(2n+1) W^i_{\gamma_1 \ldots \gamma_n} \nabla^i_{\gamma_1} \ldots \nabla^i_{\gamma_n} \left(\frac{1}{x_i}\right)\Bigg|_{x_i=1}$$

for the monosized coal/char spherical clouds, containing $N = 50$ and 100 particles, uniformly distributed in space as a function of relative distance from the center of a cloud are shown in Figure 3."

"The sign in the formula for the coefficients ω_{kn} in our study[1] is correct." In order to show that this statement is wrong let us give here the derivation of Eq. (12) and, therefore, ω_{kn}. It is clear from the definition of the irreducible tensors that

$$r_j^{-(2n+1)}\overline{r_{\gamma_1}^j \dots r_{\gamma_n}^j} = \frac{(-1)^n}{(2n-1)!!}\frac{\partial^n}{\partial r_{\gamma_1}^j \dots \partial r_{\gamma_n}^j}\left(\frac{1}{r_j}\right). \tag{20}$$

Consider the local coordinates $\mathbf{r}_i$ and $\mathbf{r}_j$ which obey the evident connection $\mathbf{r}_j = \mathbf{r}_i - \mathbf{L}_{ij}$. Taking into account that for Cartesian coordinates of j-th and i-th spheres the directions of the coordinate axes are the same, we have the following relationships

$$\frac{\partial}{\partial r_{\gamma_\nu}^j} = \frac{\partial}{\partial r_{\gamma_\nu}^i} = -\frac{\partial}{\partial L_{\gamma_\nu}^{ij}}.$$

With the help of these relationships it follows from Eq. (20) that

$$r_j^{-(2n+1)}\overline{r_{\gamma_1}^j \dots r_{\gamma_n}^j} = \frac{1}{(2n-1)!!}\frac{\partial^n}{\partial L_{\gamma_1}^{ij} \dots \partial L_{\gamma_n}^{ij}}\left(\frac{1}{r_j}\right).$$

Using here the well-known expansion[9]

$$\frac{1}{r_j} = \frac{1}{\left|\mathbf{L}_{ij} - \mathbf{r}_i\right|} = \sum_{k=0}^{\infty}\frac{(-1)^k}{k!}\frac{\partial^k}{\partial L_{\gamma_1}^{ij} \dots \partial L_{\gamma_k}^{ij}}\left(\frac{1}{L_{ij}}\right)\overline{x_{\gamma_1}^i \dots x_{\gamma_k}^i}$$

one can readily obtain the Eq. (12) and, therefore, the desired coefficients ω_{kn}.

Below we read: "Anyway, since we employed an expansion of the third order of accuracy with respect to the small parameter ε_{ij} (for details see Ref. 1) our system of the equations for the coefficients $B_{\gamma_1 \dots \gamma_n}^i$ contains only terms with $n = 0$." One can see that this statement means that authors considered "the third order of accuracy with respect to the small parameter ε_{ij}" for the PSA (see above), i.e. approximation for the point source approximation! Taking this into account, it is very strange to read, e.g., on p. 179 in the conclusions of Ref. 2: "The main advantage of the developed method is that it allows to determine the effect of particle interaction in large random clusters of particles with different sizes which is beyond the scope of methods suggested in the previous studies."

"However, it should be noted that expressions for the coefficients $B_{\gamma_1 \dots \gamma_n}^i$ (for notations see Refs. 1-3 and Ref. 9) in the Comment (Eqs. 12-13) are wrong: indices in the RHS of Eq. 12 near x must be $\mu_1 \dots \mu_n$ and not $\gamma_1 \dots \gamma_n$; the power near L_{ij} in Eqs. (13) must be $-(n+k+1)$ otherwise the RHS is dimensionless while the RHS has a dimension of the length." One can see that Eqs. (12), (13) do not contain coefficients $B_{\gamma_1 \dots \gamma_n}^i$ and some elementary algebra shows that the last assertion is incorrect.

Finally let us dwell on the cited Chapter 4 of the thesis[15]. There is a misprint in our paper[5] in Eq. (10): instead $B_{\gamma_1 \dots \gamma_n}^j$ should be $B_{\mu_1 \dots \mu_l}^j$. This misprint was reproduced in

Eq. (4.3.19) of the thesis [15] and moreover the index n was used there instead of l.

It is evident that the method of irreducible tensors is applicable only to the linear BVP (see Eq. (4)). The use of the expansion (Eq. 4.4.8) of thesis [15]

$$C_{1s}\left(T_g^{(i)}\right)=C_{1s}^{(i)}\left(T_{gs}^{(i)}\right)+\left.\frac{\partial C_{1s}}{\partial T_g^{(i)}}\right|_{T_g^{(i)}=T_{gs}^{(i)}}\left(T_g^{(i)}-T_{gs}^{(i)}\right)+\ldots$$

(where $C_{1s}\left(T_g^{(i)}\right)$ is the relative concentration of saturated vapor at the temperature $T_g^{(i)}$ and the $T_{gs}^{(i)}$ is the temperature of the drop surface) in the boundary condition (Eq. 4.4.5) at the drops surface (i.e. when $T_g^{(i)}=T_{gs}^{(i)}$) leads to the non-linear function $C_{1s}^{(i)}\left(T_{gs}^{(i)}\right)$. So the corresponding BVP cannot be solved with the help of the method at issue. Nevertheless this BVP "was solved" and expressions similar to the expressions of our paper [5] were obtained.

4. SUMMARY

In summary, it should be noted that although the derivations in Refs. 1-3, 15, 16 essentially coincide with those presented in the Ref. 5, there is, in addition to the numerous mistakes mentioned above, two crucial mistakes. (a) The expression for ω_{nk} Eq. (13) is incorrect. (b) The local flux was expressed in the form of a series of Cartesian irreducible tensors, but after integration over a unit sphere the resulting series still contains all multipoles despite the fact that irreducible tensors represent an orthonormal set of functions on the unit sphere. Therefore the main formulae of Refs. 1-3 are in error and numerical results, particularly the good agreement with experimental data, are questionable. In addition all results of Refs. 15, 16 concerning the case of evaporation of the moderate size drops are wrong due to nonlinear boundary condition for the concentration used there.

REFERENCES

1. T. Elperin, and B. Krasovitov, Analysis of evaporation and combustion of random clusters of droplets or particles by the modified method of irreducible multipoles expansion, *Atomization and Sprays*, **4**, 79-97 (1994).
2. T. Elperin, and B. Krasovitov, Combustion of cylindrical and spherical random clusters of char/carbon particles, *Combust. Sci. and Tech.*, **102**, 165-180 (1994).
3. T. Elperin, and B. Krasovitov, Evaporation and growth of multicomponent droplets in random dense clusters, *Trans. ASME*, **119**, 288-297 (1997).

4. S.D. Traytak, Theory of quasi steady state recondensation of N drops, *Book of abstracts of the XV conference on actual problems on physics of aerosol systems*, Odessa **1**, 30 (1989) (in Russian).
5. S.D. Traytak, Theory of recondensation of N drops, *Theor. Osnovy. Khim. Tekh.*, **24**, 473-482 (1990); (English translation *Theor. Found. Chem. Eng.*, **24**, 320-328 (1991)).
6. K. Annamalai and W. Ryan, Interactive processes in gasification and combustion. Part I: Liquid drop arrays and clouds, *Prog. Energy Combust. Sci.*, **18**, 221-295 (1992).
7. J.H. Jeans, *The mathematical theory of electricity and magnetism* (Cambridge, 1925).
8. P.M. Morse and H. Feshbach, *Methods of theoretical physics*. (McGraw-Hill, New York, 1953).
9. S. Hess and W. Kohler, *Formeln zur Tensor-Rechnung*. (Palm and Enke, Erlangen, 1980).
10. S.D. Traytak, The diffusive interaction in diffusion-limited reactions: the steady-state case, *Chem. Phys. Letters*, **197**, 247-254 (1992).
11. E.W. Hobson, *The theory of spherical and ellipsoidal harmonics*. (Chelsea Publ. Comp. New York, 1965)
12. B.Z. Vulikh, *Introduction to functional analysis for scientists and technologists*. (Pergamon Press, 1963).
13. S.D. Traytak, Solution of some problems of the potential theory in multiconnected domains (in preparation).
14. T. Elperin and B. Krasovitov, Reply to Comment "On the irreducible tensors method in the theory of interaction between burning particles" by S. Traitak (Submitted to *Combust. Sci. and Tech.*, 2001).
15. B. Krasovitov, *Evaporation and condensation of large size and moderate size droplets in gaseous media with arbitrary temperature gradients*, Ph. D. Thesis, Moscow (in Russian).
16. E.R. Schukin, B. Krasovitov and Yu.I. Yalamov, *Evaporation and condensation of large size and moderate size droplets in gaseous media with arbitrary temperature gradients*, Deposited in VINITI No. 3706-B91, Moscow, 1991 (in Russian).
17. S. Hess, Birefringence caused by the diffusion of macromolecules or colloidal particles, *Physica* **74A**, 277 (1974).
18. P. Brunn, The effect of Brownian motion for a suspension of spheres, Rheol. Acta **15**, 104-119 (1976).
19. P. Brunn, The behavior of a sphere in non-homogeneous flow of a viscoelastic fluid, Rheol. Acta **15**, 589-611 (1976).
20. W.E. Kohler, Current induced Kerr effect in weakly ionized gases in the presence of a magnetic field, Physica **86A**, 159-168 (1977).
21. I. Pardowitz and S. Hess, On the theory of irreducible processes in molecular liquids and liquid crystals, Nonequilibrium phenomena associated with the second and fourth rank alignment tensors, Physica **100A**, 540-562 (1980).
22. P. Mazur and W. Van Saarlos, Many-sphere hydrodynamic interactions and mobilities in a suspension, Physica **115A**, 21-57 (1982).
23. Yu.N. Kulbitsky and V.V. Struminsky, General solution of the problem of motion N dispersed particles using Stokes Approximation, Dept. Of Mechanics of Inhomogeneous Media, Preprint No. 17 (1988) (in Russian).
24. V.A. Belozertsev and A.V. Terzian, Calculation of the average velocity of gravitational settling of cluster of droplets with hydrodynamic interactions in nonisothermal fluid, Deposited in VINITI No. 7132-B88, 90-105, Moscow (1988) (in Russian).
25. M.N. Gaydukov and V.A. Belozertsev, Gravitational setting of cluster of droplets with hydrodynamic interactions in nonisothermal fluid, Deposited in VINITI No. 7132-B88, 106-120, Moscow (1988)(in Russian).
26. B.L. Silver, *Irreducible tensor methods* (Academic press New-York, 1976).
27. S.D. Traytak, Heat and mass exchange in a spatially ordered array of drops, Teplofiz. Vys. Temp., **27**, 969-975 (1989).
28. S.S. Kutateladze and V.M. Borishansky, *Handbook on heat transfer* (Dayton, Ohio, 1963).

EVAPORATION AND GROWTH OF SINGLE DROPS AND FINITE ARRAY OF INTERACTING DROPS OF PURE LIQUIDS AND HYGROSCOPIC SOLUTIONS

Eugene R. Shchukin *

1.INTRODUCTION

The paper is devoted to theoretical description of the diffusive and convective evaporation and growth of large and moderately large drops and particles spherical and non-spherical in shape at small and large relative temperature differences in a drops-medium system. Solution of the non-linear transport equation system makes allowance for the arbitrary dependence of the molecular transport factors and thermal diffusion on gaseous medium temperature and gaseous component concentrations.

Theoretical description of quasi –steady state diffusive ($c_1 << 1$) evaporation of large and moderately large spherical drop of pure liquid at large temperature differences in her vicinity has been given by Shchukin (Shchukin et al.,979,1980). It was obtained the expressions for the heat and molecules fluxes and distributions of gas temperature T and relative concentration of drop vapors c_1

$$Q_T = 4\pi R \psi_T(T_p), Q_1 = \{[c_{1S}(T_i) - c_{1\infty}]/[1+\varphi_1(T_p)]\}Q_T/\psi_1(T_p),$$

$$\psi_T(T) = \frac{r}{R}\psi_T(T_p), c_1 - c_{1\infty} = \frac{Q_1}{Q_T}\psi_1(T),$$

$$\psi_T(T) = \int_{T_\infty}^{T}\kappa dT, \psi_1(T) = \int_{T_\infty}^{T}\frac{\kappa}{nD_{12}}dT,$$

where R is the drop radius ; r is the radial coordinate; κ and D_{12} are heat conductivity and diffusion factors ; $\kappa_p = \kappa\big|_{T=T_p}, D_p = D_{12}\big|_{T=T_p}, n_p = n\big|_{r=R}$; T_i is the tem-

* Eugene R. Shchukin Institute for High Temperature Russian Academy of Science, Moscow, Russia, 127412

perature of the drop surface; $T_p = T\big|_{r=R}; c_{1S}(T_i) = n_{1S}(T_i)/n\big|_{r=R}, n_{1S}(T_i)$ is the saturated vapor concentration at temperature T_i ; K_T and K_c are the coefficients of the temperature and concentration jumps respectively;

$\varphi_1(T_p) = (K_C\lambda_p / n_p RD_p)\psi_T(T_p)/\psi_1(T_p)$.

Values of the temperatures T_i and T_p are found by solution of algebraic system

$$T_p - T_i = -(K_T\lambda_p / \kappa_p R)\psi_T(T_p), Q_i = Q_T + Lm_1 Q_1.$$

where Q_i is total power of the heat sources; L specific vaporization heat at temperature T_i; m_1 is the mass molecules of vapor.

For the first time the analytical solution of problem of diffusive evaporation of two large spherical interacting drops of pure liquid at small and large temperature differences in their vicinity has been obtained by Shchukin (1979). At small temperature differences the following expressions for the distribution of gas temperature T_e, relative concentration c_1 and the fluxes of heat $I_T^{(j)}$ and molecules $I_1^{(j)}$ were found (Shchukin,1979)

$$T_e = T_{e\infty} + \sum_{k=1}^{2}(T_i^{(k)} - T_{e\infty})U^{(k)}(\xi,\eta),\ c_1 = c_{1\infty} + \sum_{k=1}^{2}(c_i^{(k)} - c_{1\infty})U^{(k)}(\xi,\eta),$$

$$I_T^{(j)} = \kappa_e \sum_{k=1}^{2}(T_i^{(k)} - T_{e\infty})Q_U^{(j,k)}, I_1^{(j)} = nD_{12}\sum_{k=1}^{2}(c_i^{(k)} - c_{1\infty})Q_U^{(j,k)},$$

where $T_i^{(j)}$=const is the temperature of the j-th drop surface (j=1,2); $c_i^{(j)}$ is relative concentration of saturated vapor at temperatures $T_i^{(j)}$. The analytical expressions for the functions $U^{(j)}$ and the coefficients $Q_u^{(j,k)}$ obtained by Shchukin, (1979) are

$$U^{(j)} = \sqrt{2(\mathrm{ch}\,\xi - \cos\eta)}\varphi^{(j)}(\xi,\eta),$$

$$Q_U^{(1,1)} = 4\pi a\psi_1,\quad Q_U^{(1,2)} = Q_U^{(2,1)} = -4\pi a\psi_3,\ Q_U^{(2,2)} = 4\pi a\psi_2,$$

$$\varphi^{(j)}(\xi,\eta) = \sum_{n=0}^{\infty}\left(\left(\mathrm{sh}\left(\left(n+\frac{1}{2}\right)(|\xi_{2^{2-j}}| + (-1)^{j+1}\xi)\right)\exp\left(-(n+\frac{1}{2})|\xi_j|\right)\right)/G_n\right)P_n$$

$$\psi_{j=1,2} = \sum_{n=0}^{\infty}\left(\exp(-1)^j(n+1/2)(\xi_1 - |\xi_2|)\right)/G_n, P_n = P_n(\cos\eta),$$

$$\psi_3 = \sum_{n=0}^{\infty}\left(\exp(-(n+1/2)(\xi_1 + |\xi_2|))\right)/G_n,$$

$$G_n = \mathrm{sh}((n+1/2)(\xi_1 + |\xi_2|)), \mathrm{ch}\,\xi_j = \left(h^2 + (-1)^j(R_2^2 - R_1^2)\right)/2hR_j,\ \xi_1 > 0,$$

$$\xi_2 < 0;\ a = \left(h^4 + R_1^4 + R_2^4 - 2\left(h^2\left(R_1^2 + R_2^2\right) + R_1^2 R_2^2\right)\right)^{1/2} / 2h,$$

where ξ, η are bispherical coordinates; $j = 1, 2$; R_1 and R_2 are radii of surfaces; h is the distance between the centers of the surfaces.

. Values of temperatures $T_i^{(j)}$ are found by solution of algebraic equations

$$Q_1 = I_T^{(1)} + Lm_1 I_1^{(1)},\ Q_2 = I_T^{(2)} + Lm_1 I_1^{(2)},$$

where Q_1 and Q_2 are the total power of heat sources of drops.

At large temperature differences the solution of the problem of evaporation of two interacting drops was obtained by method developed by Shchukin in 1978 year (Shchukin, 1979). In the above paper (Shchukin, 1979) particularly it was shown that in absence no internal heat release and radiant heat exchange of drops with gaseous medium:

1. In steady state interacting drops either evaporate or grow simultaneously at identical temperature of the surface ($T_i^{(j)} = T_i$).
2. The steady-state evaporating or growth temperature is independent on distances between drops.

The expressions for the distribution of c_1, T_e and for the fluxes of heat and molecules of vapor are

$$I_T^{(j)} = \psi_T(T_i)\sum_{k=1}^{2} Q_U^{(j,k)},\ I_1^{(j)} = [c_{1S}(T_i) - c_{1\infty}] I_T^{(j)} / \psi_1(T_i),$$

$$\psi_T(T_e) = \int_{T_{e\infty}}^{T_e} \kappa_e dT_e,\ \psi_1(T_e) = \int_{T_{e\infty}}^{T_e} \left(\frac{\kappa_e}{nD_{12}}\right) dT_e,$$

$$\psi_T(T_e) = \psi_T(T_i)\sum_{k=1}^{2} U^{(k)}(\xi,\eta),\ c_1 = c_{1\infty} + [c_{1S}(T_i) - c_{1\infty}]\frac{\psi_1(T_e)}{\psi_1(T_i)},$$

where T_i is the temperature of drops. The value of T_i is found from solution to algebraic equation

$$1 + Lm_1[c_{1S}(T_i) - c_{1\infty}]/\psi_1(T_i) = 0.$$

The results of the paper by Shchukin (1979) were included in articles (Shchukin et al., 1980,1985).

The quasi-steady state problem of the convective (Stefan) evaporation and growth of large and moderately large spherical drop at large temperature differences in hear vicinity was considered in articles (Shchukin et al., 1979,1982). It was given there the analytical solution of the following problem

$$\frac{d}{dr} r^2 I_T = 0,\ \frac{d}{dr} r^2 I_1 = 0,\ (T - T_i) = K_T^T \frac{dT}{dr} + K_T^c \frac{dc_1}{dr}\Big|_{r=R},\ T\big|_{r\to\infty} = T_\infty,$$

$$c_1 - c_{1S}(T_i) = K_c^c \frac{dc_1}{dr} + K_T^c \frac{dT}{dr}, Q_i = Lm_1 Q_1 + Q_T, c_1\big|_{r\to\infty} = c_{1\infty},$$

where $c_1=n_1/n$, $c_2=n_2/n$, $n=(n_1+n_2)$; $\rho = m_1 n_1 + m_2 n_2$; n_1, n_2 and m_1, m_2 are concentrations and mass of molecules; $D_{12}=D_0D_1(T)D_2(c_1)$, $\kappa = \kappa_0 \kappa_1(T)\kappa_2(c_1)$;D_1 (T), $\kappa_1(T)$ and $D_2(c_1)$, $\kappa_2(c_1)$ dependent on T and c_1 ; γ_p is thermal capacity; $K_T{}^T$, $K_c{}^T$, $K_c{}^c$, $K_T{}^c$ are the gas-kinetic factors of the temperature and concentration jumps; $I_T = -\kappa \frac{dT}{dr} + \gamma_p kTI_1, I_1 = -(n^2/n_2)D_{12}$ are heat and first kind molecules flux density.

More general results of the theory of evaporation obtained by Shchukin were used in monographs by Shchukin (1995), Shchukin et al.,(1990,1991,1997) and articles Shchukin et al., (1993,1996,1998,1999,2001). The results of monograph by(Shchukin ,1995) were used in articles by Shchukin et al., (1996,1998,1999,2001).

2.PECULIARITIES OF FREE QUASI-STADY STATE PROCESS OF EVAPORATION OR GROWTH

First consider a finite array of N immobile large interacting drops (or evaporating solid particles). The drops have arbitrary surface shape S_j. They may contain hygroscopic salts (for example, sodium chloride) and small particles. The drops are in a volume $V^{(L)}$ bounded by an arbitrary surface $S^{(L)}$.The first component molecules of drops occur phase transition on the drops surfaces. Quasi-stationary evaporation and growth of drops occurs in multi-component gaseous medium with large temperature differences in a drops-gaseous medium system. The f-th component molecules of the gas with $f \geq 2$ occur no phase transition on the drops surfaces. Therefore, one can estimate the parameters of evaporation if one assumes that the molecules flux densities of these components (f=2,...,F)are equal to zero in a vicinity of the drops (or particles). We consider of steady-state drops evaporation and growth. Evaporation and growth occurs without internal heat release and radiation heat exchange (free regime of drops evaporation). Then the distribution of the relative concentrations c_f of gaseous components, the densities $\rho_b^{(j)}$ of drops components (b=1,...,W), the gas and drops temperature T_e, $T_i^{(j)}$ are described by system of non-linear equations (2.1),(2.2) under the boundary conditions (2.3) – (2.5):

$$\mathrm{div}\left[(nD_{1f}/c_f)\left(\nabla c_f + K_{T,f}\frac{1}{T_e}\nabla T_e\right)\right] = 0 \;,\; f=2,\ldots,F, \tag{2.1}$$

$$\mathrm{div}\vec{q}_T = 0, \qquad \mathrm{div}\vec{q}_b^{(j)} = 0, \qquad \mathrm{div}\vec{q}_T^{(j)} = 0, \tag{2.2}$$

$$\sum_{f=2}^{F} c_f \Big|_{S_j} = [1 - c_1^{(j,S)}]\Big|_{S_j}, T_e\Big|_{S_j} = T_i^{(j)}\Big|_{S_j}, \rho_b^{(j)}\Big|_{r_j} = \rho_{b0}^{(j)}, \vec{q}_{b\perp}^{(j)}\Big|_{S_j} = 0, \tag{2.3}$$

$$\left[\left(\vec{q}_{T\perp}^{(e)} - m_1 h_1 \vec{q}_{1\perp}\right) + L_1^{(j)} m_1 \vec{q}_{1\perp}\right]_{S_j} = \left[\vec{q}_{T\perp}^{(j)}\right]_{S_j}, \quad \rho_b^{(j)} \neq \infty, T_i^{(j)} \neq 0, \tag{2.4}$$

$$c_f\Big|_{S^{(L)}} = c_f^{(L)}, \qquad T_e\Big|_{S^{(L)}} = T_e^{(L)}, \tag{2.5}$$

where $\vec{q}_1$, and $\vec{q}_T{}^{(e)}$, $\vec{q}_T^{(j)}$ are the first kind molecules and heat flux densities (Hirschfelder et al.,1954;Williams,1971); $\vec{q}_b^{(j)}$ is the mass flux density of b-th component of j-th drop; $c_f = n_f / n, c_1 = 1 - \sum_{f=2}^{F} c_f, n = \sum_{f=1}^{F} n_f$; m_f, n_f are mass and concentration for f-th molecules of the gas; $G_1 = m_1 h_1 + e_1$, $e_1 = V_1 \left(K_{T,1}/c_1\right)$; h_1 is specific enthalpy of vapor molecules;V_1 is the coefficient, including the Dufour effect; $c_1^{(j,S)} = c_{1S}\left(T_i^{(j)}, \gamma_1^{(j)}, \ldots, \gamma_W^{(j)}\right)$ is relative concentrations of the saturated vapors dependent on $T_i^{(j)}, \rho_1^{(j)}, \ldots, \rho_W^{(j)}$; $L_1^{(j)} = L_1(T_i^{(j)}, \rho_1^{(j)}, \ldots, \rho_W^{(j)})$ are the specific evaporation heat; F is the number of gaseous components; W is the number of substanses drops; $\rho_1^{(j)}$ is density for evaporated component of j-th drop; $\rho_{b0}^{(j)}$ is known meaning of $\rho_b^{(j)}$ at some point r_j inside j-th drops; $h_b^{(j)}$ is the specific enthalpy of molecules or small particles inside j-th drop; j=1,…,N; $c_f^{(L)} = \text{const}$, $T_e^{(L)} = \text{const}$ are known values of c_f and T_e at the surface $S^{(L)}$; D_{af}, $D_{bf}^{(j)}$ and κ_e, $\kappa_i^{(j)}$ are binary diffusion coefficients and heat conductivity factors; subscript "$\perp$" denotes projections of vectors normal to the surface S_j; the coefficients $K_{T,f}$ and the flux densities of $\vec{q}_1, \vec{q}_T, \vec{q}_T^{(j)}, \vec{q}_b^{(j)}$ are equal to:

$$K_{T,f} = c_f K_{T,f}^*, K_{T,f}^* = c_f \sum_{r=1}^{F} \left(\frac{c_r}{D_{fr}}\right)\left(\frac{D_{T,f}}{n_f m_f} - \frac{D_{T,r}}{n_r m_r}\right); \tag{2.6}$$

$$\vec{q}_1 = -\left[n\Big/\sum_{f=2}^{F}\left(c_f/D_{1f}\right)\right]\left(\nabla c_1 + K_{T,1}\frac{1}{T_e}\nabla T_e\right) = 0,$$

$$\vec{q}_1 = \left[nD_{1f}/c_f\right]\left(\nabla c_f + K_{T,f}\frac{1}{T_e}\nabla T_e\right), \; f\geq 2; \; \vec{q}_T^{(e)} = G_1\vec{q}_1 - \kappa_e \nabla T_e,$$

$$\vec{q}_T^{(j)} = \sum_{f=1}^{W} h_b^{(j)}\vec{q}_b^{(j)} - \kappa_i^{(j)}\nabla T_i^{(j)}, \vec{q}_b^{(j)} = -\sum_{p=1}^{W} \psi_{bp}^{(j)}\left(\nabla\rho_p^{(j)} + K_p^{(j)}\nabla T_i^{(j)}/T_i^{(j)}\right),$$

In the general case the coefficients D_{af}, κ_e, $K_{T,f}$ depend on T_e, c_f; ; $D_{bf}^{(j)}$, $\kappa_i^{(j)}$, $K_p^{(j)}$ depend on $T_i^{(j)}, \rho_b^{(j)}$; functions $\psi_{bp}^{(j)}$ depend on $T_i^{(j)}, \rho_b^{(j)}$, $D_{bf}^{(j)}$; $h_l^{(e)}$, V_l and $h_f^{(j)}$ depend on T_e and $T_i^{(j)}$ respectively. Because of small pressure differences in particles vicinity, the molecule concentration n can be found by formula: $n = p/kT_e$, where p is gaseous pressure; k is Boltzman constant.

The boundary – value problem (2.1) – (2.5) can be solved with involvement of analytical formulas under the condition that $\rho_b^{(j)}\big|_{r_j} = \rho_{b0}, \rho_{b0}$ .is known value. Then inside of drops $\rho_b^{(j)} = \rho_{b0} = \text{const}$, $T_i^{(j)} = T_{i0} = \text{const}$ and the relative concentrations c_f depend only on T_e. In this case with help of the method developed by Shchukin (1979,1995,2001) boundary-value problem (2.1) - (2.5) can be reduced to the solution of two problems including the problem (2.7)-(2.8) and the problem (2.9)

$$\frac{dc_f}{dT_e} = \varphi_f, \tag{2.7}$$

$$(1-\sum_{f=2}^{F} c_f)\Big|_{T_e=T_{i0}} = c_{1S}, [1+B_1(L_1^{(S)} - h_1^{(S)})m_1 = 0, c_f\Big|_{T_e=T_e^{(L)}} = c_f^{(L)}, \tag{2.8}$$

$$\operatorname{div}\nabla\Psi_1^{(T)}(T_e) = 0, \Psi_1^{(T)}(T_e)\Big|_{S_j} = \Psi_1(T_{i0}), \qquad \Psi_1^{(T)}(T_e)\Big|_{S^{(L)}} = 0, \tag{2.9}$$

where: $c_{1S} = c_{1S}(T_{i0}, \rho_{10}, ..., \rho_{W0})$; $L_1^{(S)} = L_1(T_{i0}, \rho_{10}, ..., \rho_{W0})$, $h_1^{(S)} = h_1\big|_{T_e=T_{i0}}$, B_1 is a constant of integration;

$$\varphi_f = -[(c_f a / nD_{1f})B_1 \quad (K_{Tf}/T_e)], \tag{2.10}$$

$$\Psi_1^{(T)}(T_e) = \int_{T_e^{(L)}}^{T_e} a\,dT, \quad a = \kappa_e/(1-B_1G_1). \tag{2.11}$$

The solution of the boundary-value problem (2.7)-(2.8) may be performed independently on the problem (2.9).

Performing either numerical or analytical solution (2.7)-(2.8), one can find a dependence c_f on T_e and value of T_{i0} and constant B_1 simultaneously. If follows from (2.7)-(2.8) that when the dependencies of c_f on T_e are described by the boundary-value problem

(2.7)-(2.8) and $c_{1S}=c_{1S}(T_{i0},\rho_{10},\ldots,\rho_{W0})$ is the functions depending only on the surface temperature T_{i0} (for example, in cases of the drops of pure liquid and solutions with ρ_{bo} non-dependent on time) then steady-state evaporation (or growth) occurs at the constant temperature $T_i^{(j)}=T_{io}$=const non-dependent on time, the forms of drops and distances between the drops surfaces S_j. The boundary- value problem (2.9) is similar to the problem (2.14) in paper by Shchukin(2001a). At known value of T_{i0} and B_1 the distribution of T_e is described by formula

$$\Psi_1^{(T)}(T_e)=\Psi_1^{(T)}(T_{i0})U(x_f), \tag{2.12}$$

where x_f are coordinates of space point. Dependence $U(x_f)$ on x_f may be found by the solution of the Laplace equation under the boundary condition of the first kind (Shchukin, 1979, 1995, 2001, Shchukin et al., 1985,1991,1998,1999)

$$\Delta U=0, \qquad U\big|_{S_j}=1, \qquad U\big|_{S^{(L)}}=0. \tag{2.13}$$

Knowing the distribution of c_1 and T_e the value of integral fluxes of molecules $Q_1^{(j)}$ and heat $Q_T^{(j,1)}$ and $Q_T^{(j,2)}$ can be found by the formulas

$$Q_1^{(j)}=\oint_{S_j}\vec{q}_1 d\vec{S}_j, \qquad Q_T^{(j,1)}=\oint_{S_j}\vec{q}_T d\vec{S}_j,\quad Q_T^{(j,2)}=-\oint_{S_j}\kappa_e\nabla T_e d\vec{S}_j, \tag{2.14}$$

where $Q_T^{(j,1)}$ is the total heat flux; $Q_T^{(j,2)}$ is the heat flux due to molecular heat conduction. Taking into account the expressions for $\vec{q}_1$ and $\vec{q}_T^{(e)}$ (2.6) the formulas (2.14) lead to

$$Q_1^{(j)}=B_1\psi_1^{(T)}(T_{io})Q_U^{(j)}, \qquad Q_T^{(j,1)}=\psi_1^{(T)}(T_{io})Q_U^{(j)},$$

$$Q_T^{(j,2)}=(1-B_1G_1^{(S)})\psi_1^{(T)}(T_{io})Q_U^{(j)},\quad Q_U^{(j)}=-\oint_{S_j}\nabla U d\vec{S}_j, \tag{2.15}$$

where $G_1^{(S)}=G_1\big|_{T_e=T_{io}}$; $d\vec{S}_j$ is the differential vector element of the j-th surface whose direction coincides with that of the outer normal. The expressions for $Q_U^{(j)}$ were adduced in articles by Shchukin (1979, 1995, 2001a,2001b) and Shchukin et al.,(1985, 1998, 1999). From the expression for $Q_U^{(j)}$ (2.15) it follows that the value of $Q_U^{(j)}$ is always positive. Therefore in case of $\rho_m^{(j)}=\rho_{m0}$ simultaneously either evaporation or growth of all interacting drops of an array occur.

Non-linear equation (2.7) can be integrated in quadrates when $K_{T,f}=K_{T,f}^{(1)}c_f+K_{T,f}^{(2)}c_f^{1+\nu_f}$ and simultaneously the coefficients κ_e, D_{1f}, G_1 depend only on $T_e(\nu_f=\text{const})$. Then

$$\varphi_f = A_f^{(1)} c_f^{1+\nu_f} - A_f^{(2)} c_f, A_f^{(1)} = -K_{T,f}^{(2)} \frac{1}{T_e}, A_f^{(2)} = \left[(a/nD_{1f}) B_1 + K_{T,f}^{(1)} \frac{1}{T_e} \right],$$

$$c_f = \left(c_f^{(L)-\nu_f} - \nu_f \psi_f^{(1)}(T_e) \right)^{-1/\nu_f} \exp\left(-\psi_f^{(2)}(T_e)\right), \tag{2.16}$$

$$\psi_f^{(1)}(T_e) = \int_{T_e^{(L)}}^{T_e} A_f^{(1)} \exp\left(-\nu_f \psi_f^{(2)}(T_e)\right) dT_e, \qquad \psi_f^{(2)}(T_e) = \int_{T_e^{(L)}}^{T_e} A_f^{(2)} dT_e.$$

Knowing the expression for c_f (2.16) the values of the temperature T_{i0} and the coefficient B_1 can be obtained by solving of a system of algebraical equations :

$$\left(1 - \sum_{f=2}^{F} c_f \right) \Bigg|_{T_e = T_{i0}} = c_{1S}, \quad \left[1 + B_1 \left(L_1^{(S)} - h_1^{(S)} \right) m_1 \right] = 0 \quad . \tag{2.17}$$

In case of F=2 the boundary-value problem (2.7)-(2.8) is reduced to the problem

$$\frac{dc_1}{dT_e} = \varphi_1(c_1, T_e), c_1 \Big|_{T_e = T_{io}} = c_{1S}, [1 + B_1 (L_1^{(S)} - h_1^{(S)}) m_1] = 0, c_1 \Big|_{T_e = T_e^{(L)}} = c_1^{(L)},$$

$$\varphi_1(c_1,T_e)=[(c_2 a/nD_{12}) - K_{T,1}/T_e]. \tag{2.18}$$

Non-linear equation (2.18) can be integrated in quadratures when the function $\varphi_1(c_1,T_e)$ can be represented in the following three forms (Shchukin,1995;Shchukin et al.,1999)

$$\varphi_1 = \varphi_1^{(T)} \varphi_1^{(c)}, \varphi_1^{(c)} = \varphi_1^{(c)}(c_1); \varphi_1 = A_1 c_1^{1+\alpha_1} - A_2 c_1, \varphi_1 = A_4 c_2 - A_3 c_2^{1+\nu_1}, \tag{2.19}$$

where $\varphi_1^{(T)}$, A_1, A_2, A_3, A_4 are coefficients dependent on T_e and B; $c_2 = 1 - c_1$; α_1, ν_1 are constant coefficients. Substituting is expressions (2.19) into equation (2.18)we obtain:

$$\int_{c_1^{(L)}}^{c_1} \frac{dc_1}{\varphi_1^{(c)}} = \int_{T_e^{(L)}}^{T_e} \varphi_1^{(T)} dT_e, \tag{2.20}$$

$$c_1 = \left(c_1^{(L)^{-\alpha_1}} - \alpha_1 \psi_1(T_e) \right)^{-1/\alpha_1} \exp\left(-\psi_2(T_e)\right),$$

$$c_1 = 1 - \left(c_2^{(L)^{-\nu_1}} - \nu_1 \psi_3(T_e) \right)^{-1/\nu_1} \exp\left(-\psi_4(T_e)\right), \tag{2.21}$$

$$\psi_1(T_e)=\int_{T_e^{(L)}}^{T_e} A_1\exp(-\alpha_1\psi_2(T_e))dT_e,\qquad \psi_2(T_e)=\int_{T_e^{(L)}}^{T_e} A_2 dT_e, \tag{2.22}$$

$$\psi_3(T_e)=\int_{T_e^{(L)}}^{T_e} A_3\exp(-\nu_1\psi_4(T_e))dT_e,\qquad \psi_4(T_e)=\int_{T_e^{(L)}}^{T_e} A_4 dT_e. \tag{2.23}$$

If vaporization and growth occurs in gaseous medium with $c_1 \ll 1$, then $\varphi_1 = A_1 - A_2 c_1$,

$$\kappa_e = \kappa_e^{(0)} + \kappa_e^{(1)}(c_1 - c_1^{(L)}),\qquad K_{T,1} = K_{T,1}^{*} c_1,\ e_1 = V_1 K_{T,1}^{*}, \tag{2.24}$$

$$A_1 = \frac{aB_1}{nD_{12}\kappa_e}[\kappa_e^{(0)} - \kappa_e^{(1)} c_1^{(L)}], A_2 = (K_T^{(1)}\frac{1}{T_e} + [\kappa_e^{(0)} - \kappa_e^{(1)}]\frac{aB_1}{nD_{12}\kappa_e}),$$

where $\kappa_e^{(0)}$, $\kappa_e^{(1)}$, $K_{T,1}^{*}$ depend on T_e

With help of expressions (2.24) one can estimate the time and velocity of evaporation of drops and in multi-component gas. Then in (2.24) the coefficient $D_{12} = 1/\sum_{f=2}^{F}(c_f^{(L)}/D_{1f})$. In case of considerable differences of vapors concentrations the function φ_1 (2.18) can be represented in the forms (2.19), for example, if ratio (κ_e/D_{12}), $K_{T,1}$ and e_1 may be estimated by the following formulas (Shchukin,1995;Shchukin et.al.,1982,1999)

$$\kappa_e/D_{12} = (\kappa_e^{(L)}/D_{12}^{(L)})f^{(T)}f^{(c)},\ K_{T,1}=0,\ e_1=0, f^{(T)}=f^{(T)}(T_e), f^{(c)}=f^{(c)}(c_1); \tag{2.25}$$

$$\kappa_e/D_{12} = (\kappa_e^{(L)}/D_{12}^{(L)})(\Delta_1 - \Delta_2 c_2^{\nu_1}), K_{T,1} = (1-c_2^{\nu_1})c_2\Delta_3, e_1 = \bar{e}_1, \nu_1 > 0, \Delta_{t=1,2,3} = \Delta_t(T_e);$$

where functions $f^{(T)}(T_e)$, $\Delta_t(T_e)$ and $f^{(c)}(c_1)$ depend on T_e and c_1 respectively ;

$$\kappa_e^{(L)} = \kappa_e\Big|_{T_e=T_e^{(L)}, c_1=c_1^{(L)}}, D_{12}^{(L)} = D_{12}\Big|_{T_e=T_e^{(L)}, c_1=c_1^{(L)}};$$

$$\bar{e}_1 = e_1\Big|_{c_1=\bar{c}_1}, \bar{c}_1 = (c_1^{(L)} + c_{1S})/2).$$

The evaporation or growth rate of the j-th of drops is determined by the equation

$$\frac{dM^{(j)}}{dt} = -m_1 B_1 \psi_1^{(T)}(T_{io}) Q_U^{(j)}. \qquad (2.26)$$

where $M^{(j)}$ -mass of the j-th particle, t- time.

Numerical or analytical integration (2.26) with the help of the above formulas may be performed if during the drops evaporation the condition holds $\gamma_{b0}^{(j)} = \gamma_{b0}$ (where γ_{bo} may be dependent on time as well as T_{io}). Moreover if the drops volumes $V^{(j)}$ and surface shape are related with one-to-one correspondence, then one can perform integration (2.26) to determine the drops a evaporation and growth time (Shchkin,1995;Shchukin et al.,1999).

In this case provided drops evaporate at the constant temperature T_{io} equation system (2.25) allows for integration in quadratures when each volume $V^{(j)}$, density $\rho^{(j)}$ of drops and integral factor $Q_U^{(j)}$ are a functions of a single independent variable Δ common to all drops. This condition is met, for example, at vaporization or growth of separate drops of spherical , spheroidal surface shape and N large drops identical in shape and equal in value ($V^{(j)} = V, Q_U^{(j)} = Q_U$), symmetrically located at vertices of regular polygons and polyhedrons (Shchukin,1995; Shchukin et al.,1999). In such systems integrate (2.26) to obtain expression (2.27) allowing to find time variation of the drop sizes

$$E(\Delta,\Delta_0) = -m_1 B_1 \psi_1^{(T)}(T_{io})t \ , \ E(\Delta,\Delta_0) = \int_{\Delta_0}^{\Delta} \frac{1}{Q_U}\left(\frac{dM}{d\Delta}\right) d\Delta \, , \qquad (2.27)$$

where $M = \rho_i V; \rho_i$ is drop material density Δ and Δ_0 are values of the variable Δ at current, t , and initial, t=0, time respectively.

3.EVAPORATION OF DROPS HEATED BY THE INTERNAL HEAT SOURCES

Consider now the vaporization of drops heated by the internal heat sources. The heat conductivity factor of the drops $\kappa_i^{(j)}$ is considerably higher then that of gaseous medium κ_e . If $\kappa_i^{(j)} >> \kappa_e$, then the temperature distribution across the drop surface can be considered uniform. In this case the problem on evaporation of N large drops can be solved with involvement of analytical formulas at the conditions that $\rho_{bo}^{(j)} = \rho_{bo}$ and the distance between the drops is simultaneously much shorter than the dimensions of the smallest drop. In this case, when making the estimations, the drops surface temperature values be assumed equal (i.e. $T_i^{(j)} = T_{i0}$). Here the distributions of relative concentration c_f and temperature T_e in the drops vicinity are described with the boundary problem

$$\frac{dc_f}{dT_e} = \varphi_f \, , \quad \left(1 - \sum_{f=2}^{F} c_f\right)\Bigg|_{T_e = T_{i0}} = c_{1S}\left(T_{i0}, \gamma_{10}, \ldots, \gamma_{W0}\right), c_f\Big|_{S^{(L)}} = c_f^{(L)} \ , \qquad (2.28)$$

where φ_f are equal to (2.10). Considering T_{i0} of the known value then the dependence of c_f on T_e and value of B_1 are found at integration of the value problem (2.28) with the functions φ_f equaled to (210). Knowing the dependence c_f on T_e the distribution of T_e can be found by formula (2.12).The dependence of temperature T_{io} and drop dimensions on time t is determined by integration of a system composed of the following N+1 differential equations of conservation of energy and mass (2.28)

$$\sum_{j=1}^{N} Q_W^{(j)} = \sum_{j=1}^{N} \left\{ \alpha_p^{(j)} M^{(j)} \frac{dT_{io}}{dt} + (L_1^{(S)} m_1 + e_1^{(S)}) Q_1^{(j)} + (Q_T^{(j,1)} - G_1^{(S)} Q_1^{(j)}) + Q_L^{(j)} \right\},$$

$$\frac{dM^{(j)}}{dt} = -m_1 Q_1^{(j)}, \qquad (2.29)$$

where $M^{(j)}$ is mass of the j-th drop; $\alpha_p^{(j)}$ is drop specific heat capacity; $e_1^{(S)} = e_1 \big|_{T_e = T_{io}}$; $Q_W^{(j)}$ is total power of heat sources of j-th drop; $Q_1^{(j)}$ and $Q_T^{(j,1)}$ are fluxes equaled to (2.15); $Q_L^{(j)}$ is heat flux withdrawal from the surface of the j-th drop due to the heat exchange radiation (Shchukin,1995, 2001b;Shchukin et al.,1999).At steady-state drops evaporation temperatures $T_i^{(j)}$ are found by solving of a system of equations with $dT_{i0}/dt=0$.

To compute drops evaporation and growth rate it is convenient to represent constant B_1 in the following form (Shchukin, 1995, 2001): $B_1 = 1/(B_1^* + G_1^{(S)})$, where B_1^* is unknown constant.Then from the expressions (36) it is follow that the effect of the transversal convective heat transfer on the distribution of molecule temperature and concentrations c_f in the vicinity of drops can be neglected if following condition is met (Shchukin, 1995) $|B^*| >> |G_1^{(S)} - G_1|$.At free drops evaporation the convective heat transfer may be not taken into account when (Shchukin, 1995):

$$m_1 L_1^{(S)} >> |G_1 - h_1^{(S)} m_1|$$

4.CONCLUSIONS

Paper presents the theory of quasi-steady evaporation and growth of single drops and finite array of interacting drops of pure liquids and hygroscopic solutions at considerable differences of gaseous component concentrations and temperature. With help of the obtained results one can find the rate and the time of drops evaporation and growth. The obtained results allows to determine the conditions under witch the free steady-state evaporation or growth of drops at constant surfaces temperature give possibility.

REFERENCES

1. Chapmen, S., Cowling,T.G.,1952, The Mathematical Theory of Non-Uniform Gases.Cambridge University Press, Cambridge.

2. Hirschfelder,J.O., Curtiss,C.F., Bird,R.B.,1954, Molecular Theory of Gases and Liquids. John Wiley and Sons, New York.
3. Shchukin,E.R., Kutukov, Yu.I, Yalamov,Yu.I., 1977, On diffusive vaporization of drops in the electromagnetic field at considerable drops of temperature. Izv. Acad. Nauk SSSR. Ser. Thermophys. High Temp., v.15, No.2, pp. 434-436.
4. Shchukin,E.R., Yvarova,L.A. ,1979, Quasi-stationary vaporization of large and moderately large drops of refractory matter under the influence of internal heat sources. In collection of articles: Physics of aerodispertion systems and physical kinetics, installment, No. 4, part 1, N. Krupskay MOPI, deposition in VINITI No. 3828-79, Moscow, pp. 216-229
5. Shchukin,E.R.,1979, Theory of the evaporation of interacting large particles. In collection of articles: Physics of aerodispertion systems and physical kinetics, installment No. 4, part.1, N. Krupskaya MOPI, deposition in VINITI No. 3828-79, Moscow, pp. 176-215.
6. Shchukin,E.R., Barinova, M.F., Yvarova,L.A.,1980, On evaporation in diffusive regime of two interacting drops warmed by means of internal heat sources, Inzener.Fiz. Zh., v.39,No.1, pp.148.
7. Shchukin,E.R., Sanasaryan,A.S., Yalamov, Yu.I., 1982, On quasi-stationary vaporization of drops at arbitrary temperature and concentrations differences. Zh. Tekh. Fiz, v, 52, pp. 581-582.
8. Shchukin,E.R., Barinova,M.F., Yalamov, Yu. I., 1985, Theory of vaporization of two drops located at arbitrary distances from each other. DAN SSSR, Fizika, v. 284,pp. 341-344.
9. Shchukin,E.R., Shulimanova,Z.L., 1990, The influence of thermo-diffusion on velocity of vaporization of drops. Shadrinsk state teachers training institute, Typescript deposition in VINITI No. 6423-B90, Shadrinsk, 31 p.
10. Shchukin,E.R., 1995, Quasi stationary evaporation and growth of drops of pure substances at considerable drops of temperature in their vicinity. Institute for High Temperature Russian Academy of Science, Manuskript deposition in VINITI No. 412-95,Moscow, 87p.
11. Shchukin,E.R., Shulimanova, Z.L., 1996, On vaporization and growth of finite assemble of immovable large interacting drops. In Collection of articles: Chemistry and chemical technology of unorganic substances, D.Mendeleev Russian Chem. Tech. University. , Moscow ,v.. 171, pp.141-143.
12. Shchukin,E.R., Yumashev,A.V., Shulimanova,Z.L.,1997, Influence of unlinear properties of gaseous medium on evaporation and growth of large and moderately large drops. Institute for High Temperature Russian Academy of Science, Typescript deposition in VINITI No. 3083-B97, Moscow, 111p.
13. Shchukin,E.R., Shulimanova,Z.L., 1993, On peculiarities of vaporization of spheroidal drops at considerable drops of temperature. Shadrinsk state training institute, Typescript deposition in VINITI No. 1813-B93. , Shadrinsk , 20p.
14. Shchukin,E.R., Nadykto,A.B., 1998, Kinetics of large particle vaporization processes, given internal heat release. VANT. Th. and apply phys.,v.1, pp. 39-42.
15. Shchukin,E.R., 2000, Diffusive vaporization and growth of large drop of solution. The Theses of Reports of The International Conference Dedicated of the Memory of The Professor of Sutugin A.G., Moscow, pp. 76-77.
16. Shchukin,E.R., Nadykto,A.B.,1999, Vaporization and growth of large and moderately large particles at considerable differences of gaseous component concentrations. In Mathematical Models of Non-Linear Excitations, Transfer, Dynamics and Control in Condensed Systems and Other Media. Ed. by L.A.Yvarova et al., Kluwer Academic/Plenum Publishers, New York, pp. 339-355.
17. Shchukin,E.R., Nadykto,A.B., 1999, Diffusive vaporization and growth of assembly of N-large particles. In Mathematical Models of Non-Linear Excitations, Transfer, Dynamics and Control in Condensed Systems and Other Media. Ed. by L.A.Yvarova et al., Kluwer Academic/Plenum Publishers, New York, pp. 355-368.
18. Shchukin,E.R., 2001, Solution of some non-linear problems in the theory of heating, vaporization and burning of solid particles and drops. In Mathematical Modeling: Problems. Methods. Applications Ed. by L.A.Uvarova et al.., Kluwer Academic/Plenum Publishers, New York (to be published).
19. Shchukin,E.R., 2001, To method of solution of some non-linear systems of mathematical physics equations and the quasi-stationary theory of vaporization and growth of single drop and finite assembly of drops of pure liquid, Russian Proceeding of the 1V Intern. Conference on Math. Modeling, Publishing House STANKIN, Moscow,v.1, pp.159-179.
20. Shchukin,E.R., 2001, On simplification and solution of some non-linear systems of equations of transport theory of multi-component media and their application, Russian Proceeding of the 1V Intern. Conference on Math. Modeling, Publishing House STANKIN, Moscow,v.1.
21. 6Williams F.A.,1971, Combustion Theory. Nauka, Moscow.

INDEX